Compressive Sensing in Healthcare

Advances in Ubiquitous Sensing Applications for Healthcare

Compressive Sensing in Healthcare

Volume Eleven

Volume Editors

Mahdi Khosravy
Osaka University, Osaka, Japan

Nilanjan Dey
Techno International New Town, Kolkata, India

Carlos A. Duque
Federal University of Juiz de Fora, Juiz de Fora, MI, Brazil

Series Editors

Nilanjan Dey

Amira S. Ashour

Simon James Fong

Academic Press is an imprint of Elsevier
125 London Wall, London EC2Y 5AS, United Kingdom
525 B Street, Suite 1650, San Diego, CA 92101, United States
50 Hampshire Street, 5th Floor, Cambridge, MA 02139, United States
The Boulevard, Langford Lane, Kidlington, Oxford OX5 1GB, United Kingdom

Library of Congress Cataloging-in-Publication Data
A catalog record for this book is available from the Library of Congress

British Library Cataloguing-in-Publication Data
A catalogue record for this book is available from the British Library

ISBN: 978-0-12-821247-9

For information on all Academic Press publications
visit our website at https://www.elsevier.com/books-and-journals

Publisher: Mara Conner
Acquisitions Editor: Fiona Geraghty
Editorial Project Manager: John Leonard
Production Project Manager: Nirmala Arumugam
Designer: Matthew Limbert

Typeset by VTeX

Contents

Contents

List of contributors

S. Aasha Nandhini
Department of ECE, SSN College of Engineering, Kalavakkam, Chennai, India

Faramarz Asharif
School of Earth, Energy and Environmental Engineering, Kitami Institute of Technology, Hokkaido, Japan
Kitami Institute of Technology, Kitami, Japan

Noboru Babaguchi
Media Integrated Communication Laboratory, Graduate School of Engineering, Osaka University, Suita, Osaka, Japan
Computer Science and Engineering, Oakland University, Rochester, MI, United States

Ayan Banerjee
Impact Lab, Arizona State University, Tempe, AZ, United States

Thales W. Cabral
Department of Electrical Engineering, Federal University of Juiz de Fora, Juiz de Fora, Brazil

Thales Wulfert Cabral
Department of Electrical Engineering, Federal University of Juiz de Fora, Juiz de Fora, Brazil

Mir Sayed Shah Danish
Strategic Research Projects Center, University of the Ryukyus, Senbaru, Okinawa, Japan

Sumit Datta
Department of Electronics and Communication Engineering, Tezpur University, Tezpur, Assam, India

Luciano Manhaes de Andrade Filho
Department of Electrical Engineering, Federal University of Juiz de Fora, Juiz de Fora, Brazil

Bhabesh Deka
Department of Electronics and Communication Engineering, Tezpur University, Tezpur, Assam, India

List of contributors

Leonardo de Mello Honório
Department of Electrical Engineering, Federal University of Juiz de Fora, Juiz de Fora, Brazil

Mateus M. de Oliveira
Department of Electrical Engineering, Federal University of Juiz de Fora, Juiz de Fora, Brazil

Felipe M. Dias
Department of Electrical Engineering, Federal University of Juiz de Fora, Juiz de Fora, Brazil

Carlos A. Duque
Department of Electrical Engineering, Federal University of Juiz de Fora, Juiz de Fora, Brazil

Denise Fonseca Resende
Department of Electrical Engineering, Federal University of Juiz de Fora, Juiz de Fora, Brazil

Neeraj Gupta
School of Engineering and Computer Science, Oakland University, Rochester, MI, United States
Computer Science and Engineering, Oakland University, Rochester, MI, United States

Sandeep K.S. Gupta
Impact Lab, Arizona State University, Tempe, AZ, United States

K. Keerthana
Department of ECE, SSN College of Engineering, Kalavakkam, Chennai, India

Mahdi Khosravy
Media Integrated Communication Laboratory, Graduate School of Engineering, Osaka University, Suita, Osaka, Japan
Graduate School of Engineering, Osaka University, Osaka, Japan

Sushant Kumar
Department of Electronics and Communication Engineering, Tezpur University, Tezpur, Assam, India

Marcelo A.A. Lima
Department of Electrical Engineering, Federal University of Juiz de Fora, Juiz de Fora, Brazil

Leandro R. Manso Silva
Department of Electrical Engineering, Federal University of Juiz de Fora, Juiz de Fora, Brazil

Katia Melo
Department of Electrical Engineering, Federal University of Juiz de Fora, Juiz de Fora, Brazil

Felipe Meneguitti Dias
Department of Electrical Engineering, Federal University of Juiz de Fora, Juiz de Fora, Brazil

Henrique L.M. Monteiro
Department of Electrical Engineering, Federal University of Juiz de Fora, Juiz de Fora, Brazil

Henrique Luis Moreira Monteiro
Department of Electrical Engineering, Federal University of Juiz de Fora, Juiz de Fora, Brazil

Rayen Naji
Medical School, Federal University of Juiz de Fora, Juiz de Fora, Brazil

Kazuaki Nakamura
Media Integrated Communication Laboratory, Graduate School of Engineering, Osaka University, Suita, Osaka, Japan

Naoko Nitta
Media Integrated Communication Laboratory, Graduate School of Engineering, Osaka University, Suita, Osaka, Japan
Graduate School of Engineering, Osaka University, Osaka, Japan

Nilesh Patel
Department of Electrical Engineering, Federal University of Juiz de Fora, Juiz de Fora, Brazil
Computer Science and Engineering, Oakland University, Rochester, MI, United States

S. Radha
Department of ECE, SSN College of Engineering, Kalavakkam, Chennai, India

Daniel Ramalho
Department of Electrical Engineering, Federal University of Juiz de Fora, Juiz de Fora, Brazil

List of contributors

Nassim Ravanshad

Faculty of Electrical and Biomedical Engineering, Sadjad University of Technology, Mashhad, Iran

Hamidreza Rezaee-Dehsorkh

Faculty of Electrical and Biomedical Engineering, Sadjad University of Technology, Mashhad, Iran

CHAPTER

1 Compressive sensing theoretical foundations in a nutshell

Mahdi Khosravy, Naoko Nitta, Kazuaki Nakamura, Noboru Babaguchi
Media Integrated Communication Laboratory, Graduate School of Engineering, Osaka University, Suita, Osaka, Japan

1.1 Introduction

Nowadays, in the era of the Internet revolution, digital signals and images have become a substantial part of human life because of their applications. They are in use in a wide range of daily life activities from entertainment to advanced medical diagnosis. Without any doubt, digital data in the form of a signal, an image or any other form, is essential in today's life, and nobody can deny its vital role. Radios, audio players and recorders, television systems, cell phones, modems, etc. all show a comprehensive use of digital signals as the main carrier of information. All these devices use the signals, individually or collaboratively, especially after the revolution of Internet. The revolution of Internet connected the computer systems of people all over the world, leading to a global data network which again is based on signals. This leads to the comprehensive presence of signals, and therefore a huge amount of circulation of information.

In order to give you an idea of this global circulation of data, we invite you to pay attention to the following recent statistics. The clock of the current world population [1] at the very moment of writing this chapter shows over 7.676 billions human being on the earth. According to Global Digital 2019 reports [2], 56% of this population is urbanized. 5.112 billion people possess an individual mobile set where 67% actively use it. The number of Internet users is 4.388 billion, where 57% of them penetrate the network. There are 3.484 billion social media users where 45% of them are active, and finally there are 3.256 billion of social media users on smart phones with 42% active penetration. Besides all these statistics, it is very interesting that these numbers show a considerable yearly growth. In the year 2018 the number of mobile set users has increased by two percent, which means up to 100 million, the number of Internet users grew by nine percent (366 million), the total number of social media users grew by nine percent (288 million), and the number of social media users on mobiles has grown by ten percent. Fig. 1.1 gives an info-graphic presentation of the above-mentioned statistics, and Table 1.1 summarizes the same data.

Compressive Sensing in Healthcare. https://doi.org/10.1016/B978-0-12-821247-9.00006-8

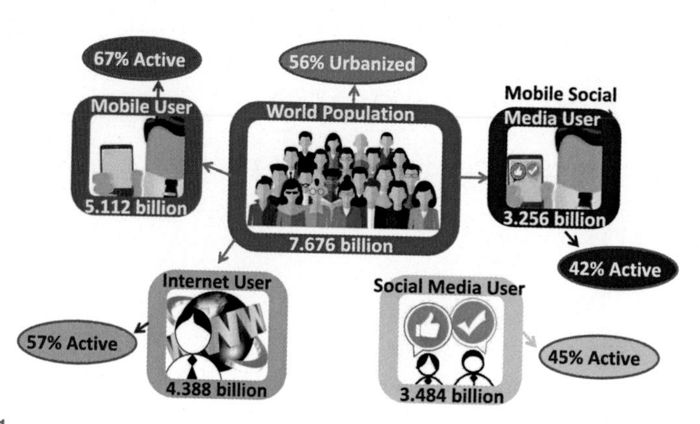

FIGURE 1.1

Infography of statistics of global Internet use.

Table 1.1 Statistics of global increasing rate of Internet use.

	Population	Penetration percentage	Growth rate (2018)
People of the world	7,676,000,000	56% urbanized	1.1%
Mobile users	5,112,000,000	67% active	2%
Internet users	4,388,000,000	57%	9%
Social media users	3,484,000,000	45% active	9%
Social media users on mobile	3,256,000,000	42% active	>10%

Observing the above-mentioned big number of social media and Internet users, as well as considering their growth rate, let us pay attention to the huge amount of information generated in the shape of signals by this enormous number of users. Keep in mind that this is just one example of a globally extended application, Internet in connection with social networks. If we add the novel revolutionary increasing use of Internet of Things (IoT) to Internet and social media, then we will observe an exponential growth of the above data sources, and therefore an enormous increase in signals carrying the data. As every day millions people come online for the first time in their life, billions of devices will also be connected to the IoT network as a first connection. In short, we face a huge volume of data and a long length of signal records.

A classic solution of the problem of managing the enormous volume of the data has been data compression, wherein firstly the data is collected as its original big size; then, for the sake of storage in memory or transmission, its format changes to the compressed format of lower size. This classic data compression solution means acquisition of data in high volume bulky signals, and then applying compression.

Compressive sensing is a novel approach which senses the signal from the very first stage in a compressed format using a sub-Nyquist sampling rate, and when the data of the compressed sensed signal is needed, it reconstructs the signal. In the modern era of information technology, and increasing use of digital signals and images as the carriers of information, compressive sensing is a very attractive topic which helps a lot to avoid a huge volume of data and the resulting costs of memory, processing and time complexities. Although compressive sensing is quite a new topic, it has gained a wide attention of researchers in recent years.

This chapter aims to give a simple review of the foundations of compressive sensing theory and implementations. It is particularly meant for non-senior researchers who want to start work on compressive sensing.

1.2 Digital signal acquisition

As mentioned earlier, we are in the era of Internet revolution where data in the form of signals are increasingly circulating amongst the users. Sensing data as a signal plays a critical role in the technical form the information is given. The theoretical base of digital sensing systems dates back to the Nyquist–Shannon sampling theorem for sensing the continuous-time band-limited signals [3,4]. The Nyquist–Shannon sampling theorem shows that a signal can be completely reconstructed by samples uniformly sensed at a rate which is at least twice higher than the highest frequency content of the signal. This sampling rate is called the Nyquist rate.

Although the Nyquist–Shannon sampling theorem assisted us a lot as regards sensing the analogue signal and transferring them to the digital domain, resulting in a digital revolution, in some cases the signals made based on this theory are bulky with too many samples. Apart from huge data amounts in the current era of Internet, for some devices, such a bulky signal is impossible to manage in view of different aspects: storage, processing, transition, projecting, etc.

A classical approach to managing such a situation of limitations for managing a signal with an enormous number of samples is looking for a compressed representation of the signal. The most popular compression method is transform coding [5–7], which looks for a basis wherein the signal is compressible or in other words '*sparse*'. In such a basis, a signal of length n can be represented by k non-zero coefficients where $k << n$. As we have the signal in its sparse basis, impressively, the whole n sample length of the signal can be represented by having just k measure amplitudes and locations.

Compressive sensing as it came from the pioneering work of Candes, Romberg, and Tao and of Donoho [8–13] directly samples/senses the signal with much fewer samples than required by Nyquist–Shannon theorem of sampling when there exists a basis for a sparse representation of the signal. This brings about the capability of reducing the sampling operations at the sensory level to a much more moderate sensing operation. Note that compressive sensing is possible for signals which are of a sparse nature, which means that there is a sparse representation for them in some basis. As a matter of fact most of the signals containing information from natural sources have the advantage of sparseness because the information has an inherited sparsity by nature. This is the main idea behind the compressive sensing, and its challenges are: (*i*) how to compressively measure a signal, (*ii*) how to extend it to practical cases, and (*iii*) how to reconstruct the signal from the compressed measurements. Compressive sensing can be applied in a variety of application fields, like power line communications [14], power quality analysis [15,16], public transportation systems [17], power system planning [18,19], location-based services [20], smart environments [21], text data processing [22], texture analysis [23], signal processing [24,25], blind source separation [26–31], medical image processing [32–34], image enhancement [35,36], image adaptation [37], human motion analysis [38], human–robot interaction [39], data mining [40,41], electrocardiogram processing [42–48], acoustic OFDM [49], sentiment mining [50], software intensive systems [51], telecommunications [52–57], and agriculture machinery [58,59].

1.3 Vectorial representation of signal

As the compressive sensing works with signals in a vectorial representation, here we give a brief introduction to the vectorial representation of signals. As most of the systems under study in different ranges of engineering are considered as linear time invariants, the signals produced by these natural or artificial systems are considered to be of a linear nature. A signal can be represented as a 'vector' in a 'vector space'.

Definition (Signal vector)

A signal of length n is represented as an n-length vector in a n-dimensional vectorial space $\mathbb{R}^n$. We write

$$\boldsymbol{x} = \begin{bmatrix} x_1 \\ x_2 \\ \dots \\ x_n \end{bmatrix} \in \mathbb{R}^n. \tag{1.1}$$

A useful measure of the signal vector is its l_p norm where p is a real value from 1 to infinity; $p \in [1, \infty]$.

Definition (l_p norm of a signal vector)

$$||\boldsymbol{x}||_p := \begin{cases} \sqrt[p]{\sum_{i=1}^{n} |x_i|^p}, & p \in [1, \infty), \\ \max\limits_{i=1,2,\dots,n} |x_i|, & p = \infty. \end{cases} \tag{1.2}$$

Indeed the l_p norm is a measure of the vector magnitude. It can be used as the distance between two vectors too. Here, we present a brief review of the main l_p norms, l_1, l_2, and l_∞, and give you an intuition of these measures.

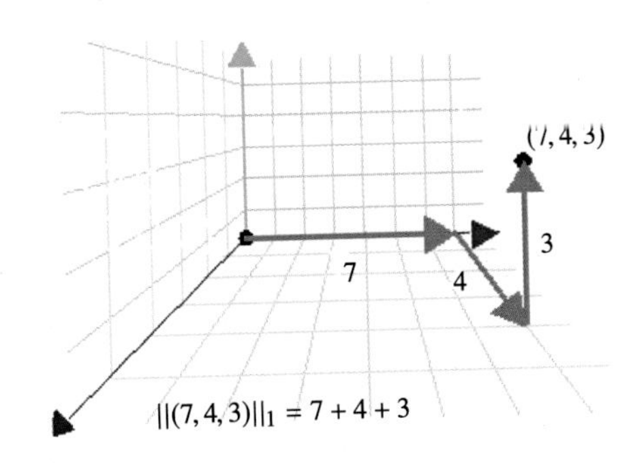

FIGURE 1.2

Example of l_1 norm.

l_1 norm

The measure of the l_1 norm is called '*Manhattan distance*' or '*Taxicab norm*'. As in Definition 1.2, substituting 1 for p, the l_1 norm is the summation of the vector/distance elements magnitudes:

$$\|\boldsymbol{x}\|_1 := |x_1| + |x_2| + \cdots + |x_n|. \tag{1.3}$$

Indeed, it is the distance obtained by just moving in the direction of the current space dimensions. It is like a person walking in a Manhattan quarter where he can just walk in the directions provided by the streets. That is the reason why it is called Manhattan distance. Fig. 1.2 shows an example of the l_1 norm.

l_2 norm

The l_2 norm is for the shortest distance indicated by a vector. It is called a *Euclidean* norm too. As in Definition 1.2, substituting 2 for p, the l_2 norm is the square root of the summation of vector/distance squared element magnitudes:

$$\|\boldsymbol{x}\|_2 := \sqrt{x_1^2 + x_2^2 + \cdots + x_n^2}. \tag{1.4}$$

It can also be expressed by the inner product in the vector space $\mathbb{R}^n$ as $\|\boldsymbol{x}\|_2 = \sqrt{\langle \boldsymbol{x}, \boldsymbol{x} \rangle}$ (Fig. 1.3).

l_∞ norm

As observed in Definition 1.2, the l_∞ norm gives the largest absolute value magnitude in the vector. As a matter of fact the element with the largest absolute value is the only one which has impact on the l_∞ norm distance. Fig. 1.4 shows an example of a l_∞ norm.

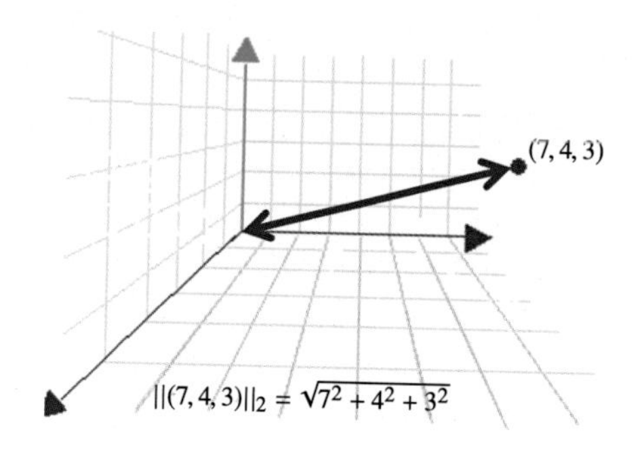

FIGURE 1.3

Example of l_2 norm.

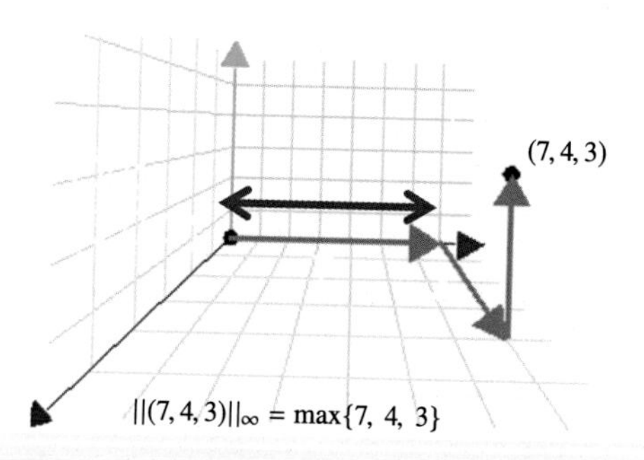

FIGURE 1.4

Example of l_∞ norm.

Spheres made by different l_p norms as distance criterion

To give you an intuition of using different l_p norms as the distance criterion in a vectorial space, we invite you to pay attention to the following intuitive example. As is well known in Cartesian geometry, a sphere is the locus of points with equal distance from a central point, the center of the sphere. The distance in Cartesian geometry is defined by the l_2 norm. Let us see how the sphere will be when different norms are used as the distance criteria. Fig. 1.5 depicts the spheres made by l_p norms of $p = \frac{1}{3}, \frac{1}{2}, 1, 2, 4,$ and ∞. Interestingly, while all the shapes are symmetric, they are very different from the sphere we know except for $p = 2$. Even a cube is a sphere when the distance criterion is l_∞.

Definition (l_0 norm of a signal vector)

The l_0 norm of a signal vector is defined as follows:

$$\|\boldsymbol{x}\|_0 := |supp(\boldsymbol{x})|. \tag{1.5}$$

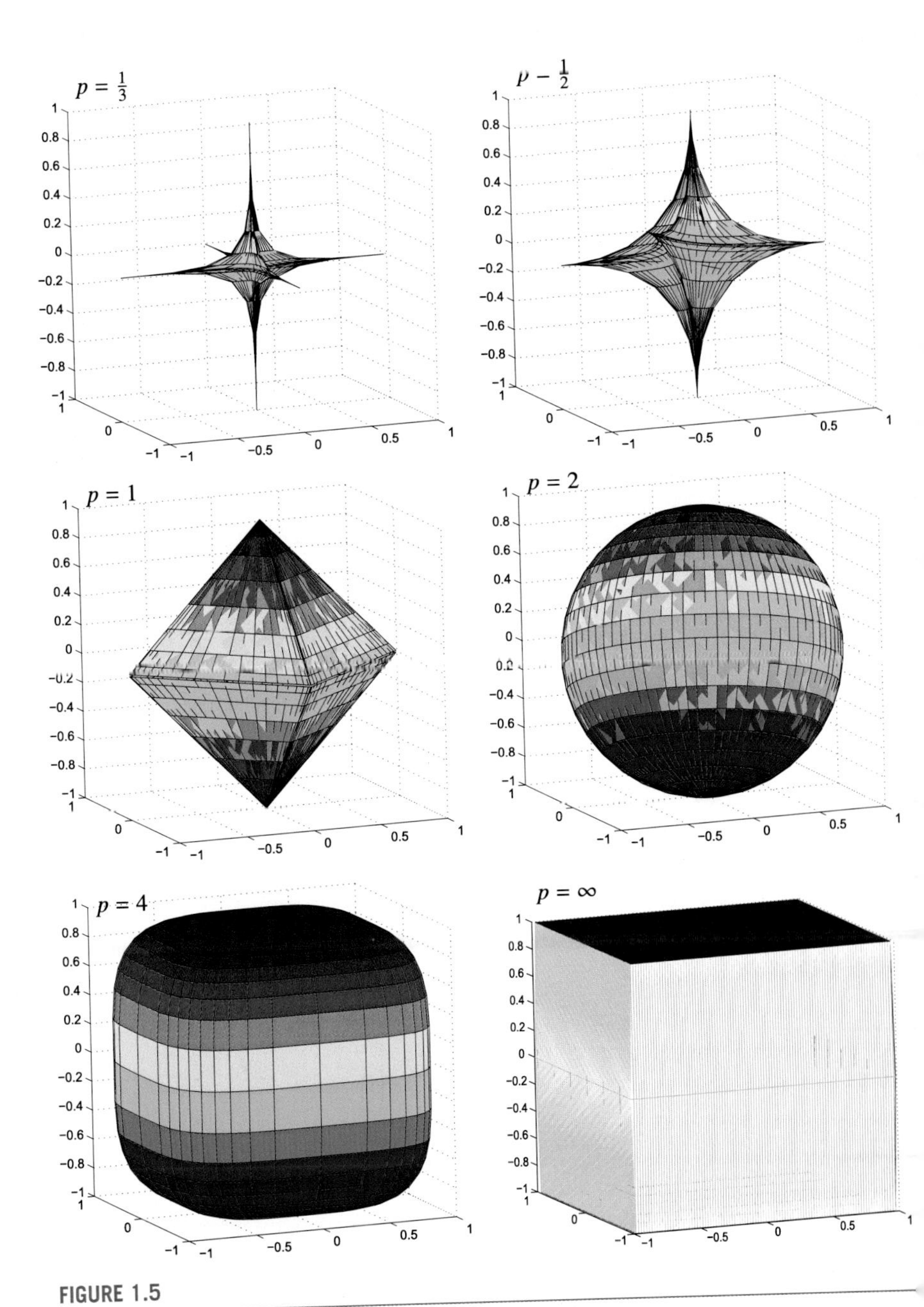

FIGURE 1.5

Spheres made by different l_p norms as distance criteria.

Here $supp(\boldsymbol{x}) = \{i : x_i \neq 0\}$ is the support of the $\boldsymbol{x}$, which is the set of the locations in $\boldsymbol{x}$ corresponding to non-zero elements. $|.|$ indicates the cardinality, the number of the elements.
Indeed, l_0 is the number of non-zero elements of the signal vector.

Here, we will have a quick review of the two concepts of basis and frame for a vectorial space [60–62].

Basis/dictionary

Considering a vector space $\mathbb{R}^n$, a set of n linearly independent vectors $\{\boldsymbol{\phi}_i | i = 1 \ldots, n\}$ is called a basis if its vectors span the vector space. In other words, every vector $\boldsymbol{x}$ of the vector space $\mathbb{R}^n$ can be uniquely represented by linearly combining the vectors in the basis. Mathematically, the set of independent vectors $\Phi = \{\boldsymbol{\phi}_i | i = 1 \ldots, n\}$ is a basis of $\mathbb{R}^n$ if

$$\forall \boldsymbol{x} \in \mathbb{R}^n, \quad \exists! \{c_i \; | i = 1, \ldots, n\} \Big(\boldsymbol{x} = \sum_{i=1}^{n} c_i \boldsymbol{\phi}_i, \quad \boldsymbol{\phi}_i \in \Phi\Big) \tag{1.6}$$

where $\exists!$ indicates uniquely existing. In some literature, the basis is called a '*dictionary*'. The expression of each vector by the basis vectors can be indicated as well by the following matrix equation:

$$\boldsymbol{x} = \Phi \boldsymbol{c}, \tag{1.7}$$

where Φ is a $n \times n$ matrix whose columns are $\boldsymbol{\phi}_i$ vectors. $\boldsymbol{c}$ is the vector of the coefficients c_i.

Orthonormal basis/dictionary

A basis $\Phi = \{\boldsymbol{\phi}_i | i = 1 \ldots, n\}$ is orthonormal, if its vectors $\boldsymbol{\phi}_i$ are orthogonal and unit. Mathematically,

$$\forall \boldsymbol{\phi}_i, \boldsymbol{\phi}_j \in \Phi, \qquad \langle \boldsymbol{\phi}_i, \boldsymbol{\phi}_j \rangle = \begin{cases} 1, & i = j, \\ 0, & i \neq j. \end{cases} \tag{1.8}$$

The importance of the orthonormal basis is its characteristic of easily obtainable basis coefficients c_i for a vector $\boldsymbol{x}$, where each $c_i = \langle \boldsymbol{x}, \boldsymbol{\phi}_i \rangle$, and in matrix notation

$$\boldsymbol{c} = \Phi^T \boldsymbol{x}. \tag{1.9}$$

Because if Φ is made of orthonormal columns, its inverse and transpose are the same, $\Phi^{-1} = \Phi^T$, and $\Phi^T \Phi = I$.

Frame/ over-complete dictionary

The concept of frame comes from a generalization of the basis by removing the condition of linearly independence amongst the vectors involved. This results in the involvement of some redundant vectors in addition to the d independent vectors from the vectorial space $\mathbb{R}^d$. Considering a set as $\{\boldsymbol{\phi}_i \mid i = 1, \ldots, n\}$ for the vectorial space $\mathbb{R}^d$ where $d < n$, this set is a frame if

$$\forall \boldsymbol{x} \in \mathbb{R}^d, \exists A, B, \quad 0 < A \leq B < \infty \quad A||\boldsymbol{x}||_2^2 \leq ||\Phi^T \boldsymbol{x}||_2^2 \leq B||\boldsymbol{x}||_2^2. \tag{1.10}$$

A and B are frame bounds. Pay attention to the following point as regards the frame bounds:

- The optimum A and B values fulfilling the above inequalities are, respectively, the maximum and minimum of them.
- The frame Φ is A-tight, if $A = B$.
- The frame Φ is Parseval, if $A = B = 1$.
- The frame Φ is equal-norm, if

$$\exists \alpha > 0, \forall \boldsymbol{\phi}_i \in \Phi, ||\boldsymbol{\phi}_i||_2 = \alpha.$$

- The frame Φ is unit-norm, if

$$\forall \boldsymbol{\phi}_i \in \Phi, ||\boldsymbol{\phi}_i||_2 = 1.$$

- Considering Φ as $d \times n$ matrix, then A and B are, respectively, smallest and biggest eigen-values of $\Phi\Phi^T$.
- $A > 0$ guarantees the linear independence of the rows.

In some literature, a frame is called an over-complete dictionary.

Alternate/dual frame

As a matter of fact, having a frame Φ, for a signal vector $\boldsymbol{x}$, there are infinitely many possible $\boldsymbol{c}$ where $\boldsymbol{x} = \Phi\boldsymbol{c}$. However, amongst these infinitely many possible $\boldsymbol{c}$ vectors of coefficients, there exists one $\boldsymbol{c}_{\min}$ which has the smallest l_2 norm. To obtain $\boldsymbol{c}_{\min}$, the alternate/dual frame of Φ, denoted $\widetilde{\Phi}$, is very useful. The alternate or dual frame is a frame fulfilling

$$\widetilde{\Phi}\Phi^T = \Phi\widetilde{\Phi}^T = I. \tag{1.11}$$

The name of canonical dual frame $\widetilde{\Phi} = (\Phi\Phi^T)^{-1}\Phi$ has been suggested for the same.[1] Since rows of Φ are independent from each other, $(\Phi\Phi^T)$ is reversible, and a canonical dual frame is feasible. Therefore, multiplying the two sides of $\Phi\boldsymbol{c} = \boldsymbol{x}$ by $\widetilde{\Phi}^T$

[1] It is also called the Moore–Penrose pseudoinverse.

we have

$$
\begin{aligned}
\widetilde{\Phi}^T \Phi \boldsymbol{c}_{\min} &= \quad \widetilde{\Phi}^T \boldsymbol{x}, \\
I \boldsymbol{c}_{\min} &= \quad ((\Phi\Phi^T)^{-1}\Phi)^T \boldsymbol{x}, \\
\boldsymbol{c}_{\min} &= \quad \Phi^T(\Phi\Phi^T)^{-1}\boldsymbol{x}.
\end{aligned} \tag{1.12}
$$

1.4 Sparsity

In most of the cases, a signal vector in a high dimensional vector space carries information expressible in much lower dimensions than the space the signal belongs to. Especially in the case of most natural signals, this issue can be observed where most of the signal duration contains low variance and few parts carry high variance information. That is exactly the sparsity where a few coefficients contain a big portion of all the energy of a big number of coefficients. As a result of this fact, in such a signal vector, there are k most dominant elements carrying the most impact. Such a signal is called k-sparse.

k-sparse signal

A signal vector $\boldsymbol{x}$ is k-sparse if it has at most k non-zero samples:

$$\boldsymbol{x} \text{ is } k\text{-sparse} \equiv ||\boldsymbol{x}||_0 \leq k. \tag{1.13}$$

It should be noted that $\boldsymbol{x} \neq 0$. Indeed the zero vector is out of the scope of the sparsity analysis. Compressive sensing does not require essentially the signal vector to be k-sparse at the current form, it is sufficient if it is k-sparse in some basis; an example is the case of a harmonic signal where the sparse basis can be provided by the Fourier transform. The condition of being k-sparse in some basis Φ means that while $\boldsymbol{x}$ may not be k-sparse at its current form, $\boldsymbol{x}\ \not{n}o\sigma_k$, it is transformed by Φ from a k-sparse signal. Mathematically

$$
\begin{aligned}
\boldsymbol{x} \quad &= \quad \Phi\widetilde{\boldsymbol{x}} \\
&\quad \widetilde{\boldsymbol{x}} \in \sum\nolimits_k,
\end{aligned} \tag{1.14}
$$

where $\sum_k$ is the set of all k-sparse vector signals; $\sum_k = \{\boldsymbol{x} \mid ||\boldsymbol{x}||_0 \leq k\}$. $\widetilde{\boldsymbol{x}}$ is a k-sparse representation of $\boldsymbol{x}$ in basis Φ.

k-sparsity is the simplest measure of sparsity as $l_0 \leq k$. It works sufficiently in compressive sensing.

Non-linearity of sparsity

As discussed, the sparsity of two signal vectors can hold on different dictionaries, but the linear combination of two k-sparse signal vectors is not essentially k-sparse except in the case that both of them are k-sparse in the same basis;

$$x, y \in \sum_k \quad \nRightarrow \quad x + y \in \sum_k, \tag{1.15}$$

where $\sum_k$ is the set of k-sparse signal vectors. But we can conclude that if two signal vectors are k-sparse then their addition is a $2k$-sparse vector signal:

$$x, y \in \sum_k \quad \Rightarrow \quad x + y \in \sum_{2k}, \tag{1.16}$$

where $\sum_k$, $\sum_{2k}$ are, respectively, the sets of k-sparse and $2k$-sparse signal vectors.

Sparsity and compressibility

For a compressible signal, a few elements of the signal vector elements can represent the signal vector. So the signal vector is represented by those few elements by keeping their values and locations, in other words the signal is compressed. For a compressible signal vector of x of length n, let us consider the compressed representation of the signal $\hat{x}$ by the k dominant elements. In this case when we compress the signal by representing these k elements on behalf of all n elements, the approximation error can be measured by the l_p norm as follows:

$$e_k(x)_p = \min_{\hat{x} \in \sum_k} \|x - \hat{x}\|_p. \tag{1.17}$$

As a matter of fact, as the signal is k-sparse, the above error is zero. This indicates the relation between the sparsity and compressibility where, as the signal is closer to k-sparsity, the compression error is closer to zero:

$$\text{if } x \in \sum_k \quad \Rightarrow \quad \forall p \; e_k(x)_p = 0. \tag{1.18}$$

1.5 Compressive sensing

When a signal segment represented by a vector is compressible, or in other words it has a sparse basis, it can be represented by much fewer samples instead of its uniformly spaced Nyquist rate sampled representation. The idea of compressive sensing (CS) is that since we have k dominant samples out of the n samples, we can reduce the sensed values to m values via a linear combination of n values. Instead of all n samples, one concludes to m samples where $m << n$, compressively sensing the signal.

Compressive sensing model

The standard formulation of compressive sensing is as follows. Having a signal vector x in the vector space of $\mathbb{R}^n$, this signal is compressively sensed and presented as a vector y in vector space of $\mathbb{R}^m$ much shorter in lengthy by the following linear operation where $m << n$:

$$y_{m\times 1} = A_{m\times n}x_{n\times 1}. \tag{1.19}$$

A is called a compressive sensing matrix. In standard CS, A is a matrix with fixed rows which previously have been determined, and in adaptive CS, A is depending on *a priori* measurements.

There are two fundamental questions in compressive sensing which have been the main research themes as follows:

1. What is an optimum compressive sensing matrix which maximally preserves the signal x information after transferring it to the short length measurement y?
2. How do we reconstruct the signal from the compressed sensed vector y?

The focus of this chapter is mainly on the first question as it is answered briefly in the following.

1.6 Essential properties of compressive sensing matrix

As the n-length signal vector x is compressively sensed by a $m \times n$ compressive sensing matrix as a much shorter m-length signal vector y, in order to maintain the recovery of x from y as a theoretical possibility, A should possess some properties. These properties are reviewed here.

1.6.1 Null space property (NSP)

After compressive sensing by the matrix A, we have the compressed sensed measurements y for different k-sparse signal vectors x. Let us consider the two sets: *(i)* $X = \{x\}$, the set of k-sparse n-length vectors x, and *(ii)* $Y = \{y\}$, the set of measurements by m-length vectors y by the compressive sensing matrix A. As depicted in Fig. 1.6, the theoretical expectation is that, as the resultant set Y is constituted by A matrix transfers from the members in X, each member of X can be approximately recovered by a reverse transfer to Δ. In order to have the theoretical possibility of the recovery of each signal vector x from its corresponding compressive sensed measure y, as a matter of fact there must be a bijection between the sets X and Y, as shown in Fig. 1.7. Mathematically,

$$\forall x, x' \in \sum_k, \ x \neq x' \quad \Rightarrow \quad Ax \neq Ax'. \tag{1.20}$$

The above condition is essential for A. To fulfill this necessary bijection in compressively sensing k-sparse signal vectors x, the sensing matrix A must have a property called the *Null Space Property,* as explained in the following.

Assuming the unwanted violation of the bijection of X and Y occurs, we have two different signal vectors of x_1 and x_2 where both are sensed by A as the same

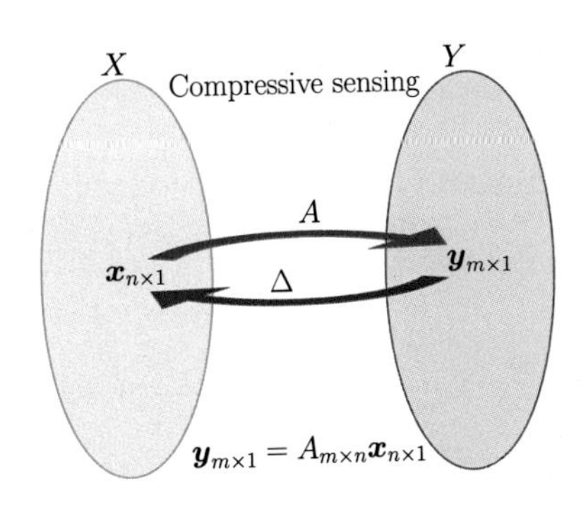

FIGURE 1.6

The theoretical expectation of the relations between the members of two sets: $X = \{x\}$, the set of k-sparse n-length vectors x, and $Y = \{y\}$, the set of measurements by m-length vectors y by the compressive sensing matrix A, and the possible recovery transfer Δ.

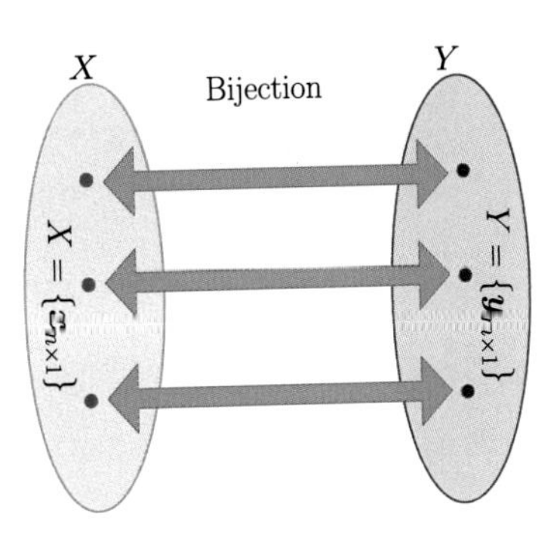

FIGURE 1.7

Bijection between two sets $X = \{x\}$, the set of k-sparse n-length vectors x, and $Y = \{y\}$, the set of measurements by m-length vectors y, by the compressive sensing matrix A is a theoretical requirement for possible recovery.

measurement y;

$$\exists x_1 \neq x_2 \quad | \quad y = Ax_1 = Ax_2. \tag{1.21}$$

Therefore, the difference vector $\delta = x_1 - x_2$ belongs to the null space of A as $A\delta = A(x_1 - x_2) = \mathbf{0}$. The null space of A is the set of signal vectors z where their transfer by A results in a zero vector:

$$\mathcal{N}(A) = \{z | Az = \mathbf{0}\}. \tag{1.22}$$

This is the starting point for finding an essential property for the compressive sensing matrix A which links the null space of A with the level of sparseness of signal vectors k. In the case of a wrong choice of A where $A\delta = A(x_1 - x_2) = \mathbf{0}$, since $\delta \neq \mathbf{0}$ some columns in A are linearly dependent. Now the question is how to avoid the unwanted condition of 1.21 by controlling the minimum number of linearly dependent columns of A, which is called spark(A).

Definition (Spark of a matrix)

The smallest number of columns of a matrix A which are linearly dependent is the 'spark of A'; we have

$$\text{spark}(A) = \min_{\boldsymbol{d} \neq \boldsymbol{0}} \|\boldsymbol{d}\|_0 \text{ s.t. } A\boldsymbol{d} = \boldsymbol{0}. \tag{1.23}$$

Some properties of the spark of a general matrix A are as follows:

- As a matter of fact, since $\boldsymbol{d} \neq \boldsymbol{0}$, spark$(A) \neq 0$.
- If A has dependent columns, spark$(A) \geq 2$ which means the number of dependent columns cannot be smaller than two.
- If A does not have dependency between columns, then its spark is defined as infinity.

Reference [63] presents the very useful theorem of '*The theorem of uniqueness*', which takes us to the property for A guaranteeing the bijection between X and Y by using the spark[2] of the A.

Theorem (uniqueness)

For a measured $\boldsymbol{y} \in \mathbb{R}^m$, there is not more than one signal vector $\boldsymbol{x} \in \sum_k$ such that $\boldsymbol{y} = A\boldsymbol{x}$ if and only if spark$(A) > 2k$.

Proof. Let us assume for a $\boldsymbol{y} \in \mathbb{R}^m$ that there does not exist more than one signal $\boldsymbol{x}$ such that $\boldsymbol{y} = A\boldsymbol{x}$, and as a contradiction the spark of A is less than or equal to $2k$. This implies that there exist maximally $2k$ linearly dependent columns in A. In other words, there exists a z in the null space of A which belongs to $\sum_{2k}$. As $z \in \sum_{2k}$, we can express it as $z = \boldsymbol{x}_1 - \boldsymbol{x}_2$ where $\boldsymbol{x}_1$ and $\boldsymbol{x}_2$ both are k-sparse vectors ($\boldsymbol{x}_1 \neq \boldsymbol{x}_2$). But as $z \in \mathcal{N}(A)$, it shows that $A(\boldsymbol{x}_1 - \boldsymbol{x}_2) = 0$, and thus $A\boldsymbol{x}_1 = A\boldsymbol{x}_2$. But this contracts the existence of not more than one vector $\boldsymbol{x}$ such that $\boldsymbol{y} = A\boldsymbol{x}$. Therefore, the spark of A is larger than $2k$, and the forward direction of the theorem is proved.

Now consider the spark of A to be higher than $2k$. Suppose that there exist two k-sparse signal vectors $\boldsymbol{x}_1$ and $\boldsymbol{x}_2$ that $\boldsymbol{y} = A\boldsymbol{x}_1 = A\boldsymbol{x}_2$. Then,

$$\begin{gathered} A(\boldsymbol{x}_1 - \boldsymbol{x}_2) = \boldsymbol{0}. \\ \text{Let } \boldsymbol{h} = \boldsymbol{x}_1 - \boldsymbol{x}_2, \text{ then } A\boldsymbol{h} = \boldsymbol{0}. \end{gathered} \tag{1.24}$$

Since $\boldsymbol{x}_1$ and $\boldsymbol{x}_2$ are k-sparse signal vectors, $\boldsymbol{h}$ is a $2k$-sparse vector signal. To fulfill $A\boldsymbol{h} = \boldsymbol{0}$ for a non-zero $\boldsymbol{h}$, it is required either that A has a set of linearly dependent

[2] The spark of the matrix might be confused with the rank of the matrix. Just to remind the reader, the rank of a matrix is the maximum number of independent columns of the matrix.

columns with the same number of columns as $\boldsymbol{h}$ non-zero elements, or that $\boldsymbol{h} = \mathbf{0}$. But $\boldsymbol{h}$ is $2k$-sparse while Spark$(A) > 2k$ which shows $\boldsymbol{h} = \mathbf{0}$. Therefore $\boldsymbol{x}_1 = \boldsymbol{x}_2$ contradicts the existence of more than one vector $\boldsymbol{x}$ and the backward direction of the theorem is proved too. □

The theory of uniqueness has been indicated in Corollary 1 in Ref. [63] in the other direction that we have already A and the theorem determining the minimum sparsity required for recovery.

Uniqueness: The measurement vector $\mathbf{y}$ *such that* $\mathbf{y} = A\mathbf{x}$ *necessarily represents the sparsest possible signal vector if* $||\boldsymbol{x}||_0 < Spark(A)/2$.

The property called the *Uniqueness theorem* can be expressed as follows in terms of the null space property.

Null Space Property: *For a measured* $\mathbf{y} \in \mathbb{R}^m$, *there is not more than one signal vector* $\boldsymbol{x} \in \sum_k$ *such that* $\mathbf{y} = A\boldsymbol{x}$, *if and only if A has the null space property of order* $2k$.

The matrix A has a null space of order $2k$ when its null space is without any $2k$-sparse vector;

$$\{\forall z \in \sum\nolimits_{2k} | Az = \mathbf{0}\} = \emptyset, \tag{1.25}$$

where $\emptyset$ is the empty set, $\sum_{2k}$ is the set of $2k$-sparse vectors. For an explanation of the NSP of order $2k$, we refer to the appendix.

In order to extend the conditions for semi-sparse compressible signal vectors, the null space property of order $2k$ should be extended to exclude $2k$-semi-sparse vectors in addition to k-sparse vectors. A $2k$-semi-sparse vector has at most $2k$ elements with dominant absolute values to be compared to other elements of the vector which are much smaller and trivial. Semi-sparsity entails the situation wherein the signal is not sparse exactly in the sense of having elements very close to zero which can be approximately considered as zero. In such a case, instead of having the definitions based on the number of zero elements, the equations and definitions will show how much smaller the summation of signal elements in a range is than a threshold. Fig. 1.8 illustrates sparsity versus semi-sparsity.

To this aim, a definition of NSP of order $2k$ is given where the null space of A excludes any semi-sparse signal. It is by using the $\mathbf{\Lambda}_\epsilon$, which is a semi-support which supports the dominant elements of the semi-sparse vector. Λ_ϵ elements are 1 for elements corresponding to the dominant elements of the semi-sparse vector, and 0 for the elements corresponding to other small trivial elements. Mathematically,

$$\mathbf{\Lambda}_\epsilon(i) = \begin{cases} 1 & \boldsymbol{x}(i) > \epsilon, \\ 0 & \boldsymbol{x}(i) \leq \epsilon, \end{cases} \qquad \epsilon > 0. \tag{1.26}$$

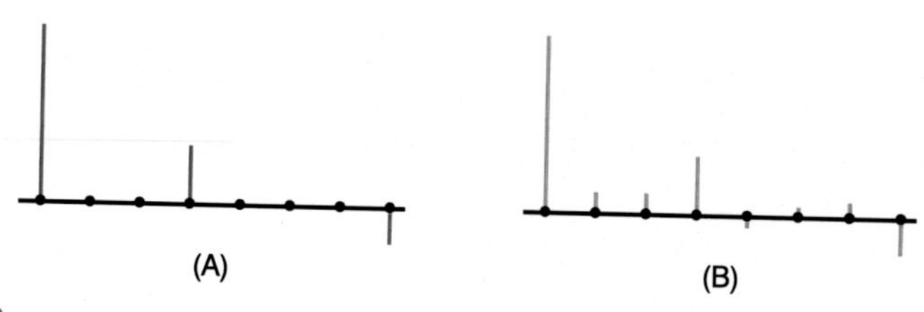

FIGURE 1.8

(A) A sparse signal vector, (B) a semi-sparse signal vector.

The complement of $\mathbf{\Lambda}_\epsilon$, denoted by $\mathbf{\Lambda}_\epsilon^c$, supports the trivial elements but not the dominant ones as follows:

$$\mathbf{\Lambda}_\epsilon^c(i) = \begin{cases} 0 & \boldsymbol{x}(i) > \epsilon, \\ 1 & \boldsymbol{x}(i) \leq \epsilon, \end{cases} \qquad \epsilon > 0. \tag{1.27}$$

Deploying the above-mentioned semi-support vector, the null space property of order $2k$ can be more restrictive by excluding $2k$-semi-sparse vectors in addition to $2k$-sparse vectors. In this way the extended null space property of order $2k$ is defined as follows.

Definition (Extended null space property)

The matrix A has the NSP of order 2k as

$$\exists c > 0, \forall z \in \mathcal{N}(A), \forall \mathbf{\Lambda}_\epsilon, |\mathbf{\Lambda}_\epsilon| \leq 2k, \quad \|z_{\mathbf{\Lambda}_\epsilon}\|_2 \leq c\frac{\|z_{\mathbf{\Lambda}_\epsilon^c}\|_1}{\sqrt{2k}}, \tag{1.28}$$

where $\mathbf{\Lambda}_\epsilon$ and $\mathbf{\Lambda}_\epsilon^c$ are defined in Eqs. (1.26) and (1.27).

It can be easily checked that if we consider a $2k$-semi-sparse/sparse vector having the above NSP of order $2k$, then the l_2 norm of $z_{\mathbf{\Lambda}_\epsilon}$, which is the vector made by the dominant elements of z, should be smaller than or equal to the l_1 norm of $z_{\mathbf{\Lambda}_\epsilon^c}$, the vector made by trivial elements of z. Such a thing is possible if and only if z is not a semi-sparse vector. Now we have a more general definition of NSP of order $2k$ which covers the compressible and semi-sparse signal vectors in addition to sparse signal vectors. The importance of the NSP condition in compressive sensing and recovery of sparse/semi-sparse signal vectors is clarified in the following. As a result of the theorem of uniqueness, the NSP property of A is a necessary condition for recovery. Now, the theorem of recovery given in the following indicates that if there is any recovery for compressive sensing by A, then A has essentially the NSP property.

The essence of the concept of recovery

Let us consider $\Delta : \mathbb{R}^m \to \mathbb{R}^n$, the recovery transfer for the compressed sensed vector signals by A. To guarantee the recovery of k-sparse vector signals by the transfer Δ from the sensed signal vectors by the compressive sensing matrix A, an error bound is used for the recovered vector signal difference from the original corresponding one using the evaluative error Eq. (1.17), and the following theorem holds [64].

Theorem (recovery)

Let the compressive sensing matrix and the corresponding vector transfer, respectively, be $A : \mathbb{R}^n \to \mathbb{R}^m$, *and* $\Delta : \mathbb{R}^m \to \mathbb{R}^n$; *if the recovery error is bounded by*

$$||\Delta(A\boldsymbol{x}) - \boldsymbol{x}||_2 \le c\frac{e_k(\boldsymbol{x})_1}{\sqrt{k}}, \tag{1.29}$$

then A has the null space property of order $2k$.

The preceding section indicates that the NSP for the compressive sensing matrix is essential to maintain the possibility of recovery. A recovery theorem indicates in addition that if there is any recovery from the compressively sensed vectors $\boldsymbol{y}$, then A has essentially the NSP of order $2k$. The proof of the theorem is in Ref. [64].

Maximum compression in compressive sensing (lower bound of m*)*

A very interesting result of the theorem of uniqueness follows.

For a compressive sensing matrix A of size $m \times n$ *which fulfills the conditions for measuring k-sparse signal vectors, the possible range for m is given by*

$$m \ge 2k. \tag{1.30}$$

This indicates that the maximum possible compressed sensed vector of a k-sparse signal vector by compressive sensing cannot be shorter than $2k$. Since m is the number of A rows, it can be simply obtained knowing that $\text{spark}(A) \in [2, m+1]$.

1.6.2 Restricted isometry property

The NSP property gives the necessary and sufficient conditions for the compressive sensing matrix but it is not strong enough for considerations of the noisy condition. To introduce robustness to the noisy condition, there is a necessary condition required for the sake of successful recovery of all sparse signal vectors. This concerns an isometry condition on the sensing matrix A, defined as follows [65].

Definition (Restricted isometry property (RIP))

The matrix A has the RIP of order k if

$$\forall \boldsymbol{x} \in \Sigma_k,\ \exists \delta_k \in (0,1),\quad (1-\delta_k)\|\boldsymbol{x}\|_2^2 \le \|A\boldsymbol{x}\|_2^2 \le (1+\delta_k)\|\boldsymbol{x}\|_2^2. \tag{1.31}$$

As is obvious from inequalities (1.31), this property of A guarantees that the norm-2 size of the measurements vectors $A\boldsymbol{x}$ by A would not go very far from the norm-2 size of signal vectors $\boldsymbol{x}$ due to noise to result in unsuitability; the oscillation range is limited to $\delta_k\|\boldsymbol{x}\|_2$ around $\|\boldsymbol{x}\|_2$.

Besides the robustness to noise, RIP helps a lot in the determination of the problem's dimensions, m, n, and k. Already n and k are acquired and we are indeed looking for the bounds of m which determine the lower dimension of the compressive sensing matrix. As an important result of the satisfaction of RIP, the lower band theorem for m has been given in Ref. [66] as follows.

Theorem (lower band of m (Theorem 3.5 of Ref. [66]))

Let $A_{m\times n}$ satisfy the restricted isometry property of order $k \leq \frac{n}{2}$ with constant $\delta \in (0, 1)$. Then the lower bound of m is given by

$$m \geq c_\delta k \log(\frac{n}{k}), \tag{1.32}$$

where the constant $c_\delta < 1$, and it depends only on δ as follows:

$$c_\delta \approx \frac{0.18}{\log\left(\sqrt{(1+\delta)/(1-\delta)}+1\right)}. \tag{1.33}$$

1.6.3 Coherence a simple way to check NSP

Until this point, one measure and two properties for the compressive sensing matrix A have been introduced: the Spark of A, the NSP, and the RIP, to evaluate A for possible recovery of k-sparse signal vectors from the measurement vectors by A. However, checking any of them in A requires a search over $\binom{n}{k}$ combinations of submatrices. Instead, the '*coherence*' measure of the matrix A has been suggested as a property that is easier to compute and applicable for a robust recovery.

Definition (Coherence of matrix A)

The coherence of A is defined as follows:

$$\mu(A) = \max_{1\leq i\leq j\leq n} \frac{|\langle \boldsymbol{a}_i, \boldsymbol{a}_j\rangle|}{||\boldsymbol{a}_i||_2||\boldsymbol{a}_j||_2}, \tag{1.34}$$

where $\boldsymbol{a}_i$ and $\boldsymbol{a}_j$ are any two columns of A.

The lower bound of the coherence of A is called the Welch bound and it is $\sqrt{\frac{n-m}{m(n-1)}}$, and the upper bound is 1. Since $m << n$, the lower bound is approximately $\frac{1}{\sqrt{m}}$. Thus

$$\mu(A) \in [\frac{1}{\sqrt{m}}, 1]. \tag{1.35}$$

Relation between coherence and spark of a matrix

It can be shown that for every matrix, between the spark of the matrix and its coherence there exists the following relation:

$$\text{Spark}(A) \geq 1 + \frac{1}{\mu(A)}. \tag{1.36}$$

From inequality (1.36), and the theorem of uniqueness, we have the following [63].

Theorem (Coherence approach to uniqueness)

For every measured signal vector $\mathbf{y} \in \mathbb{R}^m$*, there is maximally one k-sparse signal vector such that* $\mathbf{y} = A\mathbf{x}$*, if*

$$k < \frac{1}{2}\left(1 + \frac{1}{\mu(A)}\right). \tag{1.37}$$

In addition, this theorem gives *the upper limit of sparsity level* for the cases that we have an available compressive sensing matrix A. Applying the Welch bound in the theorem, in order to ensure the uniqueness, $k = O(\sqrt{m})$.

Coherence approach to RIP

Also there is a connection between coherence and RIP. Lemma 1.5 of Ref. [67] simplifies the evaluation of RIP in connection with $\mu(A)$ and k as follows.

RIP and coherence

For A with unit-norm columns and coherence $\mu(A)$, the RIP of order k is fulfilled with:

$$\forall k < \frac{1}{\mu}, \qquad \delta_k = k - 1. \tag{1.38}$$

Table 1.2 summarizes the theoretical foundations of compressive sensing including the basic properties, theorems, and relations.

1.7 Summary

This chapter gives a brief introduction to the theoretical fundamentals of compressive sensing. It clarifies the concept of information sparsity and presents a practical review to the measures of sparsity.

Compressive sensing comprises two main challenges: (i) how to construct a compressive sensing matrix, and (ii) how to recover the sparse signal from a much shorter segment as compressed measurement. Although this chapter focuses on the theoretical considerations in acquiring a sensing matrix, which is highly important in replying the first question, a theoretically well constructed sensing matrix guarantees the possibility of recovery as regards the second question, apart from its methodology. I

Table 1.2 Summary of the theoretical foundations of compressive sensing. The details of each concept has been provided inside the text.

Concept	Application	Description
Uniqueness	Recovery of $\boldsymbol{x}$	$\forall \boldsymbol{x}, \boldsymbol{x}' \in \sum_k, \boldsymbol{x} \neq \boldsymbol{x}' \quad \Rightarrow \quad A\boldsymbol{x} \neq A\boldsymbol{x}'$.
Uniqueness theorem	Uniqueness	spark$(A) > 2k$, a required property for A
Uniqueness theorem	Uniqueness	$\|\|\boldsymbol{x}\|\|_0 <$ Spark$(A)/2$. Having A, this inequality gives an upper bound of k
NSP order $2k$	Uniqueness	$\mathcal{N}(A) \cap \sum_{2k} = \emptyset$
Extended NSP of order $2k$	Semi-sparse recovery $\boldsymbol{x}$	$\exists c > 0, \forall \boldsymbol{z} \in \mathcal{N}(A), \forall \boldsymbol{\Lambda}_\epsilon$, $\|\boldsymbol{\Lambda}_\epsilon\| \leq 2k, \quad \|\boldsymbol{z}_{\boldsymbol{\Lambda}_\epsilon}\|_2 \leq c\frac{\|\boldsymbol{z}_{\boldsymbol{\Lambda}_\epsilon^c}\|_1}{\sqrt{2k}}$,
RIP of order k	Robustness to noise	$\forall \boldsymbol{x} \in \Sigma_k, \exists \delta_k \in (0,1)$, $(1-\delta_k)\|\|\boldsymbol{x}\|\|_2^2 \leq \|\|A\boldsymbol{x}\|\|_2^2 \leq (1+\delta_k)\|\|\boldsymbol{x}\|\|_2^2$
Lower bound of m	Maximum compression	$m \geq 2k$
Lower bound of m	Under RIP	$m \geq c_\delta k \log(\frac{n}{k})$, $\delta \in (0,1), c_\delta \approx \frac{0.18}{\log\left(\sqrt{(1+\delta)/(1-\delta)}+1\right)}$.
Coherence of A	Simpler property	$\mu(A) = \max_{1 \leq i \leq j \leq n} \frac{\|\langle \boldsymbol{a}_i, \boldsymbol{a}_j \rangle\|}{\|\|\boldsymbol{a}_i\|\|_2 \|\|\boldsymbol{a}_j\|\|_2}$ $\mu(A) \in [\frac{1}{\sqrt{m}}, 1]$
NSP via $\mu(A)$	Uniqueness	$k < \frac{1}{2}(1 + \frac{1}{\mu(A)})$
RIP via $\mu(A)$	Robustness to noise	$\forall k < \frac{1}{\mu}, \delta_k = k-1$, for A with unit-norm columns.

presents the required conditions for a compressive sensing matrix. Having addressed these, the theoretical background and the requirement behind the compressive sensing as treated in the literature have been provided to the reader. The chapter gives the theoretical requirement for the dimensions of the compressive sensing problem as well as the properties which should be fulfilled by a compressive sensing matrix.

Appendix 1.A

Null space property of order $2k$

The matrix A fulfills the null space property of order $2k$ if

$$\{\forall \boldsymbol{z} \in \sum\nolimits_{2k} \| A\boldsymbol{z} = \boldsymbol{0}\} = \emptyset, \tag{1.39}$$

where $\emptyset$ is the empty set, $\sum_{2k}$ is the set of $2k$-sparse vectors. The null space property can be represented using the null space of A:

$$\mathcal{N}(A) \cap \sum_{2k} = \emptyset, \tag{1.40}$$

where $\mathcal{N}(A)$ is the null space of A.

References

[1] Current world population, https://www.worldometers.info/world-population/, 2019.

[2] Digital 2019: Global internet use accelerates, 2019.

[3] H. Nyquist, Certain topics in telegraph transmission theory, Transactions of the American Institute of Electrical Engineers 47 (2) (1928) 617–644.

[4] C. Shannon, Communities in the presence of noise, proceeding of the institute of radio engineers, 1949.

[5] A.M. Bruckstein, D.L. Donoho, M. Elad, From sparse solutions of systems of equations to sparse modeling of signals and images, SIAM Review 51 (1) (2009) 34–81.

[6] R. DeVore, Nonlinear approximation, Acta Numerica 51 (1998) 150.

[7] M. Elad, Sparse and Redundant Representations: From Theory to Applications in Signal and Image Processing, Springer Science & Business Media, 2010.

[8] E.J. Candès, et al., Compressive sampling, in: Proceedings of the International Congress of Mathematicians, Madrid, Spain, vol. 3, 2006, pp. 1433–1452.

[9] E.J. Candès, J. Romberg, T. Tao, Robust uncertainty principles: exact signal reconstruction from highly incomplete frequency information, IEEE Transactions on Information Theory 52 (2) (2006) 489–509.

[10] E.J. Candes, J.K. Romberg, T. Tao, Stable signal recovery from incomplete and inaccurate measurements, Communications on Pure and Applied Mathematics: A Journal Issued by the Courant Institute of Mathematical Sciences 59 (8) (2006) 1207–1223.

[11] E.J. Candes, T. Tao, Near-optimal signal recovery from random projections: universal encoding strategies?, IEEE Transactions on Information Theory 52 (12) (2006) 5406–5425.

[12] D.L. Donoho, et al., Compressed sensing, IEEE Transactions on Information Theory 52 (4) (2006) 1289–1306.

[13] K. Melo, M. Khosravy, C. Duque, N. Dey, Chirp code deterministic compressive sensing: analysis on power signal, in: 4th International Conference on Information Technology and Intelligent Transportation Systems (ITITS 2019), 2019.

[14] A.A. Picorone, T.R. de Oliveira, R. Sampaio-Neto, M. Khosravy, M.V. Ribeiro, Channel characterization of low voltage electric power distribution networks for PLC applications based on measurement campaign, International Journal of Electrical Power & Energy Systems 116 (105) (2020) 554.

[15] E. Santos, M. Khosravy, M.A. Lima, A.S. Cerqueira, C.A. Duque, A. Yona, High accuracy power quality evaluation under a colored noisy condition by filter bank esprit, Electronics 8 (11) (2019) 1259.

[16] E. Santos, M. Khosravy, M.A. Lima, A.S. Cerqueira, C.A. Duque, Esprit associated with filter bank for power-line harmonics, sub-harmonics and inter-harmonics parameters estimation, International Journal of Electrical Power & Energy Systems 118 (105) (2020) 105731.

[17] M. Foth, R. Schroeter, J. Ti, Opportunities of public transport experience enhancements with mobile services and urban screens, International Journal of Ambient Computing and Intelligence (IJACI) 5 (1) (2013) 1–18.

[18] N. Gupta, M. Khosravy, N. Patel, T. Senjyu, A bi-level evolutionary optimization for coordinated transmission expansion planning, IEEE Access 6 (2018) 48455–48477.

[19] N. Gupta, M. Khosravy, K. Saurav, I.K. Sethi, N. Marina, Value assessment method for expansion planning of generators and transmission networks: a non-iterative approach, Electrical Engineering 100 (3) (2018) 1405–1420.
[20] M. Yamin, A.A.A. Sen, Improving privacy and security of user data in location based services, International Journal of Ambient Computing and Intelligence (IJACI) 9 (1) (2018) 19–42.
[21] C. Castelfranchi, G. Pezzulo, L. Tummolini, Behavioral implicit communication (BIC): communicating with smart environments, International Journal of Ambient Computing and Intelligence (IJACI) 2 (1) (2010) 1–12.
[22] C.E. Gutierrez, P.M.R. Alsharif, M. Khosravy, P.K. Yamashita, P.H. Miyagi, R. Villa, Main large data set features detection by a linear predictor model, AIP Conference Proceedings 1618 (2014) 733–737, AIP.
[23] S. Hemalatha, S.M. Anouncia, Unsupervised segmentation of remote sensing images using FD based texture analysis model and ISODATA, International Journal of Ambient Computing and Intelligence (IJACI) 8 (3) (2017) 58–75.
[24] M.H. Sedaaghi, R. Daj, M. Khosravi, Mediated morphological filters, in: 2001 International Conference on Image Processing, Proceedings, vol. 3, IEEE, 2001, pp. 692–695.
[25] M. Khosravy, N. Gupta, N. Marina, I.K. Sethi, M.R. Asharif, Morphological filters: an inspiration from natural geometrical erosion and dilation, in: Nature-Inspired Computing and Optimization, Springer, Cham, 2017, pp. 349–379.
[26] M. Khosravy, M.R. Asharif, K. Yamashita, A PDF-matched short-term linear predictability approach to blind source separation, International Journal of Innovative Computing, Information & Control 5 (11) (2009) 3677–3690.
[27] M. Khosravy, M.R. Asharif, K. Yamashita, A theoretical discussion on the foundation of Stone's blind source separation, Signal, Image and Video Processing 5 (3) (2011) 379–388.
[28] M. Khosravy, M.R. Asharif, K. Yamashita, A probabilistic short-length linear predictability approach to blind source separation, in: 23rd International Technical Conference on Circuits/Systems, Computers and Communications (ITC-CSCC 2008), Yamaguchi, Japan, 2008, pp. 381–384.
[29] M. Khosravy, M.R. Alsharif, K. Yamashita, A PDF-matched modification to Stone's measure of predictability for blind source separation, in: International Symposium on Neural Networks, Springer, Berlin, Heidelberg, 2009, pp. 219–222.
[30] M. Khosravy, Blind source separation and its application to speech, image and MIMO-OFDM communication systems, PhD thesis, University of the Ryukyus, Japan, 2010.
[31] M. Khosravy, M. Gupta, M. Marina, M.R. Asharif, F. Asharif, I. Sethi, Blind components processing a novel approach to array signal processing: a research orientation, in: 2015 International Conference on Intelligent Informatics and Biomedical Sciences, ICIIBMS, 2015, pp. 20–26.
[32] M. Khosravy, M.R. Asharif, M.H. Sedaaghi, Medical image noise suppression: using mediated morphology, IEICE, Technical Report 107 (461) (2008) 265–270.
[33] N. Dey, A.S. Ashour, A.S. Ashour, A. Singh, Digital analysis of microscopic images in medicine, Journal of Advanced Microscopy Research 10 (1) (2015) 1–13.
[34] M. Khosravy, M.R. Asharif, M.H. Sedaaghi, Medical image noise suppression using mediated morphology, IEICE Technical Report, IEICE (2008) 265–270.
[35] A.S. Ashour, S. Samanta, N. Dey, N. Kausar, W.B. Abdessalemkaraa, A.E. Hassanien, Computed tomography image enhancement using cuckoo search: a log transform based approach, Journal of Signal and Information Processing 6 (03) (2015) 244.
[36] M. Khosravy, N. Gupta, N. Marina, I. Sethi, M. Asharif, Brain action inspired morphological image enhancement, in: Nature-Inspired Computing and Optimization, Springer, Cham, 2017, pp. 381–407.
[37] M. Khosravy, N. Gupta, N. Marina, I. Sethi, M. Asharifa, Perceptual adaptation of image based on Chevreul–Mach bands visual phenomenon, IEEE Signal Processing Letters 24 (5) (2017) 594–598.
[38] G.V. Kale, V.H. Patil, A study of vision based human motion recognition and analysis, International Journal of Ambient Computing and Intelligence (IJACI) 7 (2) (2016) 75–92.
[39] B. Alenljung, J. Lindblom, R. Andreasson, T. Ziemke, User experience in social human–robot interaction, in: Rapid Automation: Concepts, Methodologies, Tools, and Applications, IGI Global, 2019, pp. 1468–1490.

[40] C.E. Gutierrez, M.R. Alsharif, K. Yamashita, M. Khosravy, A tweets mining approach to detection of critical events characteristics using random forest, International Journal of Next-Generation Computing 5 (2) (2014) 167–176.

[41] N. Kausar, S. Palaniappan, B.B. Samir, A. Abdullah, N. Dey, Systematic analysis of applied data mining based optimization algorithms in clinical attribute extraction and classification for diagnosis of cardiac patients, in: Applications of Intelligent Optimization in Biology and Medicine, Springer, 2016, pp. 217–231.

[42] M.H. Sedaaghi, M. Khosravi, Morphological ECG signal preprocessing with more efficient baseline drift removal, in: 7th IASTED International Conference, ASC, 2003, pp. 205–209.

[43] N. Dey, S. Samanta, X.-S. Yang, A. Das, S.S. Chaudhuri, Optimisation of scaling factors in electrocardiogram signal watermarking using cuckoo search, International Journal of Bio-Inspired Computation 5 (5) (2013) 315–326.

[44] M. Khosravy, M.R. Asharif, M.H. Sedaaghi, Morphological adult and fetal ECG preprocessing: employing mediated morphology, IEICE Technical Report, IEICE 107 (2008) 363–369.

[45] N. Dey, A.S. Ashour, F. Shi, S.J. Fong, R.S. Sherratt, Developing residential wireless sensor networks for ECG healthcare monitoring, IEEE Transactions on Consumer Electronics 63 (4) (2017) 442–449.

[46] M. Khosravi, M.H. Sedaaghi, Impulsive noise suppression of electrocardiogram signals with mediated morphological filters, in: 11th Iranian Conference on Biomedical Engineering, ICBME, 2004, pp. 207–212.

[47] N. Dey, S. Mukhopadhyay, A. Das, S.S. Chaudhuri, Analysis of P-QRS-T components modified by blind watermarking technique within the electrocardiogram signal for authentication in wireless telecardiology using DWT, International Journal of Image, Graphics and Signal Processing 4 (7) (2012) 33.

[48] T.W. Cabral, M. Khosravy, F.M. Dias, H.L.M. Monteiro, M.A.A. Lima, L.R.M. Silva, R. Naji, C.A. Duque, Compressive sensing in medical signal processing and imaging systems, in: Sensors for Health Monitoring, Elsevier, 2019, pp. 69–92.

[49] M. Khosravy, N. Punkoska, F. Asharif, M.R. Asharif, Acoustic OFDM data embedding by reversible Walsh–Hadamard transform, AIP Conference Proceedings 1618 (2014) 720–723, AIP.

[50] M. Baumgarten, M.D. Mulvenna, N. Rooney, J. Reid, Keyword-based sentiment mining using Twitter, International Journal of Ambient Computing and Intelligence (IJACI) 5 (2) (2013) 56–69.

[51] P. Sosnin, Precedent-oriented approach to conceptually experimental activity in designing the software intensive systems, International Journal of Ambient Computing and Intelligence (IJACI) 7 (1) (2016) 69–93.

[52] M. Khosravy, M.R. Alsharif, M. Khosravi, K. Yamashita, An optimum pre-filter for ICA based multi-input multi-output ofdm system, in: 2010 2nd International Conference on Education Technology and Computer, vol. 5, IEEE, 2010, pp. V5–129.

[53] F. Asharif, S. Tamaki, M.R. Alsharif, H. Ryu, Performance improvement of constant modulus algorithm blind equalizer for 16 qam modulation, International Journal on Innovative Computing, Information and Control 7 (4) (2013) 1377–1384.

[54] M. Khosravy, M.R. Alsharif, K. Yamashita, An efficient ICA based approach to multiuser detection in mimo OFDM systems, in: Multi-Carrier Systems & Solutions 2009, Springer, 2009, pp. 47–56.

[55] M. Khosravy, M.R. Alsharif, B. Guo, H. Lin, K. Yamashita, A robust and precise solution to permutation indeterminacy and complex scaling ambiguity in bss-based blind MIMO-OFDM receiver, in: International Conference on Independent Component Analysis and Signal Separation, Springer, 2009, pp. 670–677.

[56] M. Khosravy, A blind ica based receiver with efficient multiuser detection for multi-input multi-output OFDM systems, in: The 8th International Conference on Applications and Principles of Information Science (APIS), Okinawa, Japan, 2009, 2009, pp. 311–314.

[57] M. Khosravy, S. Kakazu, M.R. Alsharif, K. Yamashita, Multiuser data separation for short message service using ICA, SIP, IEICE Technical Report 109 (435) (2010) 113–117.

[58] S. Gupta, M. Khosravy, N. Gupta, H. Darbari, N. Patel, Hydraulic system onboard monitoring and fault diagnostic in agricultural machine, Brazilian Archives of Biology and Technology 62 (2019).

[59] S. Gupta, M. Khosravy, N. Gupta, H. Darbari, In-field failure assessment of tractor hydraulic system operation via pseudospectrum of acoustic measurements, Turkish Journal of Electrical Engineering & Computer Sciences 27 (4) (2019) 2718–2729.
[60] J. Kovacevic, A. Chebira, Life beyond bases: the advent of frames (part i), IEEE Signal Processing Magazine 24 (4) (2007) 86–104.
[61] J. Kovacevic, A. Chebira, Life beyond bases: the advent of frames (part ii), IEEE Signal Processing Magazine 24 (5) (2007) 115–125.
[62] D.T. Stoeva, On frames, dual frames, and the duality principle, Novi Sad Journal of Mathematics 45 (1) (2015) 183–200.
[63] D.L. Donoho, M. Elad, Optimally sparse representation in general (nonorthogonal) dictionaries via l1 minimization, Proceedings of the National Academy of Sciences 100 (5) (2003) 2197–2202.
[64] A. Cohen, W. Dahmen, R. DeVore, Compressed sensing and best k-term approximation, Journal of the American Mathematical Society 22 (1) (2009) 211–231.
[65] E. Candes, T. Tao, Decoding by linear programming, preprint, arXiv:math/0502327, 2005.
[66] M.A. Davenport, Random observations on random observations: Sparse signal acquisition and processing, PhD thesis, Rice University, 2010.
[67] Y.C. Eldar, G. Kutyniok, Compressed Sensing: Theory and Applications, Cambridge University Press, 2012.

CHAPTER

2

Recovery in compressive sensing: a review

Mahdi Khosravy[a], **Neeraj Gupta**[b], **Nilesh Patel**[c], **Carlos A. Duque**[c]

[a]*Media Integrated Communication Laboratory, Graduate School of Engineering, Osaka University, Suita, Osaka, Japan*

[b]*School of Engineering and Computer Science, Oakland University, Rochester, MI, United States*

[c]*Department of Electrical Engineering, Federal University of Juiz de Fora, Juiz de Fora, Brazil*

2.1 Introduction

In the era of the Internet of Things, and the increasing amount of information of all types, as well as the growing capability of modern devices in the processing of information of high quality, modern technology faces an increasing amount of information. These high levels of information quantity need: (i) to be initially sensed like signal acquisition, (ii) to be transmitted as in telecommunication systems, and (iii) to be processed. All these three tasks require data processing, storage memory, high transition bandwidth, and after all more power, especially for sensing. In the realm of signals and images, the compression was suggested and used wherein a signal/image was reshaped and saved in much lower size. The compressed signal/image was decompressed to the original size as it was needed for use. To avoid un-necessary sampling of the redundant information and bearing the side effect mentioned above, compressive sensing was suggested. It was pioneered by Candes, Romberg, and Tao and by Donoho [1–5]. Compressive sensing samples the signal by a much smaller number of samples than required by the Nyquist–Shannon theorem. It is based on an assumption of sparsity for the signal. Naturally, this assumption is true for most data forms of information in nature. Compressive sensing mainly is a challenge to (*i*) compressively measure a signal while its information content is kept preserved, (*ii*) to recover the original signal after compressive sensing. The compressive method has a great application potential and can be used in a wide range of applications, like location-based services [6], signal processing [7,8], smart environments [9], texture analysis [10], telecommunications [11–16], public transportation systems [17], acoustic OFDM [18], power line communications [19], power quality analysis [20–22], power system planning [23,24], human motion analysis [25], medical image processing [26–28], human–robot interaction [29], electrocardiogram processing [30–35], sentiment mining [36], text data processing [37], image enhancement [38, 39], image adaptation [40], software intensive systems [41], agriculture machinery [42,43], and data mining [44,45]. This chapter gives a quick review of both aspects of compressive sensing.

Compressive Sensing in Healthcare. https://doi.org/10.1016/B978-0-12-821247-9.00007-X

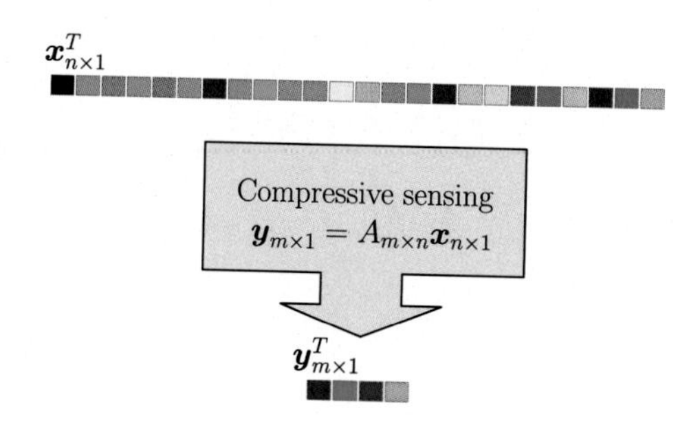

FIGURE 2.1

Compressive sensing.

2.1.1 Compressive sensing formulation

Compressive sensing is modeled as follows in its standard formulation. Consider $\boldsymbol{x}$ as a signal vector of length n belonging to the vector space of $\mathbb{R}^n$. Then $\boldsymbol{x}$ can be compressively sensed; in other words, it can be presented as a vector $\boldsymbol{y}$ of length m belonging to the vector space $\mathbb{R}^m$ as follows:

$$\boldsymbol{y}_{m\times1} = A_{m\times n}\boldsymbol{x}_{n\times1}. \tag{2.1}$$

Note that m is much shorter than n, $m << n$, which explains the term 'compressive sensing'. A is called a compressive sensing matrix. Fig. 2.1 shows the general schematic of compressive sensing wherein a long signal vector is sensed as a much shorter signal vector. As mentioned earlier two main questions of compressive sensing are as follows:

1. What conditions must a compressive sensing matrix obey to maximally preserve the information content of the signal?
2. Having the compressed sensed vector $\boldsymbol{y}$, how can the signal vector $\boldsymbol{x}$ be recovered from $\boldsymbol{y}$?

This chapter briefly answers these two main questions of compressive sensing with main focus on the second question.

2.2 Criteria required for a compressive sensing matrix

Not every matrix is suitable for use as a compressive sensing matrix. The matrix which compressively senses the signal should have several properties to keep its information content for recovery after being sensed. Here we briefly review these properties and the related theories.

2.2.1 Null space property

Clearly, the resultant measurement as the compressed sensed vector $\boldsymbol{y}$ should uniquely represent not more than one sparse signal vector $\boldsymbol{x}$. Formally

$$\forall \boldsymbol{x}, \boldsymbol{x}' \in \sum_k, \ \boldsymbol{x} \neq \boldsymbol{x}' \quad \Rightarrow \quad A\boldsymbol{x} \neq A\boldsymbol{x}', \tag{2.2}$$

where $\sum_k$ is the set of k-sparse vectors. A k-sparse vector is a vector with maximally k non-zero elements, while the rest of the elements are zero. This requires some properties for A.

Null space property of order k

The matrix A has the null space property of order k if

$$\{\forall z \in \sum_k | Az = 0\} = \emptyset, \tag{2.3}$$

where $\emptyset$ and $\sum_k$, respectively, indicate the empty set, and the set of k-sparse vectors. $\mathcal{N}(A) = \{z | Az = 0\}$ is called the null space of A. The null space property of order k can be also represented as follows:

$$\mathcal{N}(A) \cap \sum_k = \emptyset. \tag{2.4}$$

2.2.1.1 Uniqueness theorem [46]

To keep the measurements unique for their corresponding signal vectors, we should avoid that two different signal vectors, $\boldsymbol{x}_1$ and $\boldsymbol{x}_2$, are measured by A as the same $\boldsymbol{y}$;

$$\exists \boldsymbol{x}_1 \neq \boldsymbol{x}_2 \quad | \quad \boldsymbol{y} = A\boldsymbol{x}_1 = A\boldsymbol{x}_2. \tag{2.5}$$

Under such a condition the vector $\boldsymbol{\delta} = \boldsymbol{x}_1 - \boldsymbol{x}_2$ belongs to the null space of A. This is so because $A\boldsymbol{\delta} = A(\boldsymbol{x}_1 - \boldsymbol{x}_2) = 0$. It indicates linear dependency of some columns in A. The solution to the problem of avoiding such a condition is controlling the number of dependent columns in A.

Definition (Spark of a matrix)

The 'spark of A' is the minimum number of linearly dependent columns in A.
Clearly, spark(A) is at least 2.

Theorem (Uniqueness)

There is at most one k-sparse signal vector $\boldsymbol{x}$ *for a measured signal vector* $\boldsymbol{y} = A\boldsymbol{x}$, $\boldsymbol{y} \in \mathbb{R}^m$ *if and only if spark*$(A) > 2k$.

Maximum compression in compressive sensing

As a result of this theorem, in compressively sensing k-sparse signal vectors by A of size $m \times n$, m cannot be less than $2k$. This result is very similar to the result

of the Nyquist sampling theorem as regards the maximum frequency content of the sampled signal by a certain Nyquist rate but for maximum possible compression in compressive sensing signal vectors of a certain sparseness.
Besides, as a result Spark$(A) \in [2, m+1]$.

It has been treated as the theory of uniqueness in Corollary 1 of Ref. [46] in terms of

Uniqueness: *The equation* $\mathbf{y} = A\mathbf{x}$ *necessarily represents the sparsest measurement that is possible if* $||\mathbf{y}||_0 < Spark(A)/2$.

To extend the property for the compressible signal vectors, the null space property is defined as follows.

Definition (Null space property of order k**)**

The matrix A has the null space theory of order k as

$$\exists c > 0, \forall z \in \mathcal{N}(A), \forall \Lambda, |\Lambda| \leq k, \quad ||z_\Lambda||_2 \leq c\frac{||z_{\Lambda_c}||_1}{\sqrt{k}}, \tag{2.6}$$

where $\Lambda \subset \{1, 2, \ldots, n\}$ *is a set of indices representing a support for some of the indices of the vector.* Λ_c *is the complement set for* Λ *including the indices other than the ones indicated by* Λ.

Let us consider $W : \mathbb{R}^m \to \mathbb{R}^n$ as the recovery transfer. To ensure the recovery of k-sparse signals by the transfer W, an error bound is applied.

The approximation recovery error is measured by the l_p norm as follows:

$$e_k(\boldsymbol{x})_p = \min_{\hat{\boldsymbol{x}} \in \sum_k} ||\boldsymbol{x} - \hat{\boldsymbol{x}}||_p. \tag{2.7}$$

Applying the error bound, the following theorem is seen to hold [47].

Theorem (Best k**-term approximation)**

Let the compressive sensing matrix and the corresponding vector transfer, respectively, be $A : \mathbb{R}^n \to \mathbb{R}^m$, *and* $W : \mathbb{R}^m \to \mathbb{R}^n$, *if the recovery error is bounded as*

$$||W(Ax) - x||_2 \leq c\frac{e_k(\boldsymbol{x})_1}{\sqrt{k}}, \tag{2.8}$$

then A has the null space property of order 2k.

2.2.2 Restricted isometry property

The restricted isometry property (NSP) property provides the necessary and sufficient requirements for the compressive sensing matrix; however, it is not robust enough for consideration under the noise. To introduce robustness in the recovery of all sparse signal vectors to noise, a necessary condition is required. This entails an isometry condition on the sensing matrix A as follows [48].

Definition (Restricted isometry property)

The matrix A has the restricted isometry condition (RIP) of order k if

$$\forall \boldsymbol{x} \in \Sigma_k,\ \exists \delta_k \in (0,1)\,,\quad (1-\delta_k)||\boldsymbol{x}||_2^2 \le ||A\boldsymbol{x}||_2^2 \le (1+\delta_k)||\boldsymbol{x}||_2^2. \tag{2.9}$$

Determination of the problem's dimensions, which are m, n, and k, is the other advantage of RIP. We remind the reader that m and n are the dimensions of the compressive sensing matrix and the dimensions the vector spaces before and after compressive sensing, respectively, and k is the considered sparsity level of the signal vectors in the vector space $\mathbb{R}^n$. After obtaining n and k, the third dimension of the problem is m, that is, the lower dimension of the compressive sensing matrix; the ratio of $\frac{n}{m}$ determines the compression level in compressive sensing. The lower band theorem for m is found as a result of the RIP condition in Ref. [49].

Theorem (Lower bound of m)

If $A_{m\times n}$ satisfies the RIP of order $k \le \frac{n}{2}$ with constant $\delta \in (0,1)$, for the lower bound of m we have

$$m \ge c_\delta \log(\frac{n}{k}) \tag{2.10}$$

where the constant $c_\delta < 1$, and it depends only on δ as follows:

$$c_\delta \approx \frac{0.18}{\log\left(\sqrt{(1+\delta)/(1-\delta)}+1\right)}. \tag{2.11}$$

2.2.3 Coherence property

In addition to the three properties of Spark, NSP, and RIP for the compressive sensing matrix A, there is another property, which is much easier to compute than the above-mentioned three properties. This property is 'coherence'.

Definition (Coherence)

The coherence of the compressive sensing matrix A is as follows:

$$\mu(A) = \max_{1\le i\le j\le n} \frac{|\langle a_i, a_j\rangle|}{||a_i||_2||a_j||_2}, \tag{2.12}$$

where a_i, a_j are any two columns of A. The lower bound of the coherence of A is called the Welch bound and it is $\sqrt{\frac{n-m}{m(n-1)}}$, and the upper bound is 1. Since $m << n$

the lower bound is approximately $\frac{1}{\sqrt{m}}$. Thus,

$$\mu(A) \in [\frac{1}{\sqrt{m}}, 1]. \tag{2.13}$$

2.2.3.1 Coherence and spark of a matrix

For any matrix, the following relation holds between its spark and coherence:

$$\text{Spark}(A) \geq 1 + \frac{1}{\mu A}. \tag{2.14}$$

2.2.3.2 The upper bound of sparsity level

The inequality (2.14) together with the theorem of uniqueness results in a theorem [46] which gives an upper bound for the sparsity level that preserves the uniqueness.

Theorem (Upper bound of sparseness by coherence)

For every measured signal vector $\mathbf{y} \in \mathbb{R}^m$, there is maximally one k-sparse signal vector such that $\mathbf{y} = A\mathbf{x}$, if

$$k < \frac{1}{2}(1 + \frac{1}{\mu(A)}). \tag{2.15}$$

2.3 Recovery

After obtaining the compressed sensed signal vectors $\boldsymbol{y}$ from the original signal vectors $\boldsymbol{x}$, the main task in the context of compressive sensing will be recovering of the original signal vectors. Fig. 2.2 shows this important stage as the second crucial task in compressive sensing. Apart from the conditions and theory given in the former sections, let us again consider the compressive sensing process. A signal vector $\boldsymbol{x}$ of length n is linearly read by the matrix $A_{m\times n}$ as a much shorter signal vector $\boldsymbol{y}$ of length m, i.e. $m << n$. Now the question is how to recover $\boldsymbol{x}$ from $\boldsymbol{y}$ by using A. Let us look again at the formulation of compressive sensing with the variable of $\boldsymbol{x}$,

$$A_{m\times n}\boldsymbol{x}_{n\times 1} = \boldsymbol{y}_{n\times 1}. \tag{2.16}$$

As a set of linear equations it can be written as

$$\begin{aligned} a_{11}x_1 + a_{12}x_2 + \cdots + a_{1n}x_n &= y_1, \\ a_{21}x_1 + a_{22}x_2 + \cdots + a_{2n}x_n &= y_2, \\ &\vdots \\ a_{m1}x_1 + a_{m2}x_2 + \cdots + a_{mn}x_n &= y_m. \end{aligned} \tag{2.17}$$

Eqs. (2.16) and (2.17) represent a set of linear equations with multiple variables $x_1, x_2, \ldots x_n$ denoted by the signal vector $\boldsymbol{x}$. As a matter of fact for a set of linear equa-

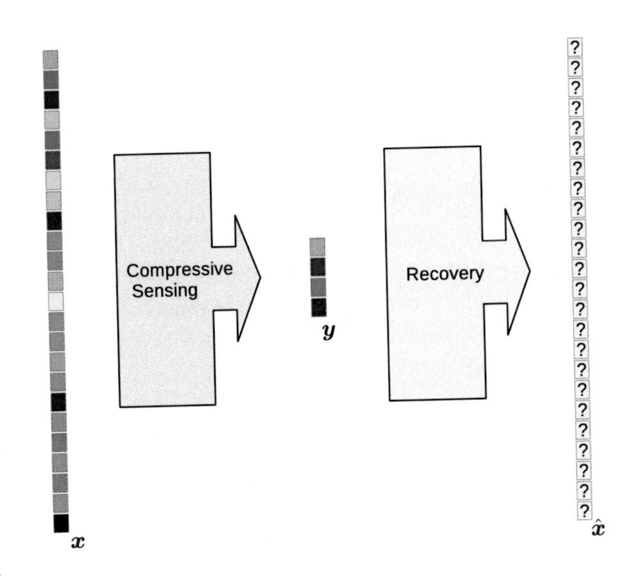

FIGURE 2.2

Schematic of recovery in compressive sensing.

tions, to have a certain answer for n variables, we must have n independent equations. In this set of m equations, not only we have less than n equations, which means the set of equations is incomplete, but the number of equations is much smaller than the requirement. In such a case, for a given A and $\boldsymbol{y}$ there are infinitely many possibilities for $\boldsymbol{x}$. So, how can we recover $\boldsymbol{x}$ after being compressively sensed as $\boldsymbol{y}$?

The answer to the question is quite interesting. Although there are infinitely many answers for the set of linear equations (2.17), there is only one solution that is maximally sparse. The solution to the compressive sensing problem is similar to blind source separation [50–55] in the same sense that, amongst infinitely many possible solutions for the set of linear equations of a mixture observation, there is only one solution with maximum independence between the recoveries. Therefore, the solution is via the search for the maximum sparse signal vector amongst all possible answers to the set of linear equations (2.17). As a matter of fact, the problem of recovery in compressive sensing is a search problem in a space of possible answers to an incomplete set of linear equations. This search problem can be solved through an optimization which looks for the best solution as the most sparse one.

The first fundamental way coming to mind for recovering the signal from the compressive sensed data is the search for the maximum sparsity via minimization of the l_1 norm as a measure of sparsity.

2.3.1 Recovery via minimization of l_1 norm

As in compressive sensing problem, we have the *a priori* knowledge of sparseness or semi-sparseness of the original signal vectors $\boldsymbol{x}$, after having the compressively sensed measurements as the $\boldsymbol{y}$; the signal vectors can be recovered by the following optimization problem;

$$\hat{\boldsymbol{x}} = \min_{z} ||\boldsymbol{z}||_0 \quad \text{subject to } \boldsymbol{z} \in \{\boldsymbol{z} : ||A\boldsymbol{z} - \boldsymbol{y}||_2 \leq \epsilon\}. \tag{2.18}$$

The inequality $||A\boldsymbol{z} - \boldsymbol{y}||_2 \leq \epsilon$ in the set covering $\boldsymbol{z}$ vectors is due to consideration of the noisy condition. In the case of the noise-free condition, the search space can be reduced to $\{\boldsymbol{z} : A\boldsymbol{z} = \boldsymbol{y}\}$. Note that here the optimization problems search for $\boldsymbol{x}$ itself as the sparse solution, while $\boldsymbol{x}$ can be sparse in another basis as $\boldsymbol{s}$ where $\boldsymbol{s} = \Phi\boldsymbol{x}$. Then the optimization problem gives $\boldsymbol{s}$ as the solution that should be transferred to $\boldsymbol{x}$ by going back to the basis of $\boldsymbol{x}$ rather than the basis ϕ.

Although the optimization problem given in Eq. (2.18) can effectively recover the vector signals, it is not practical to use because of the two following issues:

1. The norm zero as an objective function is non-convex.
2. It is non-linear and potentially very difficult to solve, and it is a NP-hard problem.

Although minimizing l_0 is an NP-hard problem maybe it is possible to deal with it by meta-heuristic evolutionary optimization techniques [56–67], as a common solution suggested by the literature is replacement of the norm-0 case by the norm-1 case, which is convex and feasible to solve. Therefore, the compressive sensing recovery is modified to

$$\hat{\boldsymbol{x}} = \min_{z} ||\boldsymbol{z}||_1 \quad \text{subject to } \boldsymbol{z} \in \{\boldsymbol{z} : ||A\boldsymbol{z} - \boldsymbol{y}||_2 \leq \epsilon\}. \tag{2.19}$$

The condition of $\boldsymbol{z} \in \{\boldsymbol{z} : ||A\boldsymbol{z} - \boldsymbol{y}||_2 \leq \epsilon\}$ only is a convex problem, and in the noise-free case this condition facilitates the problem as a linear programming problem [68].

In general, norm-1 as a cost function can be addressed by a replacement for any norm-p if $p < 1$. Reference [69] gives a very interesting review of norm-1 minimization for sparsity promotion as briefly given here. The minimization of norm-1 results in sparsity elevation was initially used in 1938 in Beurling's work, where a Fourier transform was interpolated from the partial observations [70]. In 1965, Logan deployed the norm-1 as a re-constructive property for the signals which are high-passed and their lower frequency content is missed. For the application of reconstruction of geophysics signals, in 1979, Taylor et al. suggested the use of norm-1 in an effective deconvolution [71] and later on in 1981, Levy et al. reconstructed the geophysics signals from a small portion of the spectrum of spike trains using minimization of norm-1 [72]. Donoho and Logan showed in 1992 that a band-limited signal corrupted by impulsive noise can be perfectly recovered under some conditions, as well as showing, in the case of wideband signal corruption by noise, that it can be stably recovered [73] through a search for a signal which is band-limited but with minimum norm-1 distance from the signal. In the 1990s, norm-1 minimization drew the attention of statisticians as a methodology for selection of the variables in a regression called Lasso [74], as well in signal and image processing for finding the sparsest form of the signal and image [75]. In 2001, Chen, Donoho, and Saunders

Table 2.1 Norm-1 minimization for sparsity promotion in the literature. The article citation numbers were taken on November 2019.

Literature	Citation	Year	Application
[70]	211	1938	Fourier transform extrapolation from partial observations
[76]	106	1965	Signal reconstruction after missing their low frequency contents
[71]	460	1979	Deconvolution
[72]	258	1981	Sparse spike train reconstruction by a portion of the spectrum with application in high-resolution reconstruction of geophysical signals
[73]	179	1992	Recovery of a band-limited signal under corruption by impulsive noise
[73]	179	1992	Recovery of a wideband signal under corruption by noise
[74]	30649	1996	Lasso a methodology for variable selection in regression
[75]	25247	1999	Searching for maximum sparse form signal/image
[68]	11714	2001	Atomic decomposition by basis pursuit

reviewed the Basis pursuit (BP) as an approach for signal decomposition via minimum norm-1 [68]. Table 2.1 summarizes the above review of norm-1 minimization in literature.

As mentioned and clarified earlier, the minimization of norm-1 is a strong tool for recovery of the sparse signal vector. Furthermore, the norm-1 minimization is a convex optimization problem with an accurate numerical solution. Although the most common form of the solution for this optimization problem is as the solution for the problem given in Eq. (2.19), there are other equivalent formulas for the same problem. A review to effective solutions to Eq. (2.19) has been given in Ref. [77].

2.3.2 Greedy algorithms

The greedy algorithm solves the problem by iterative methods wherein the signal coefficients and support are iteratively approximated and the process continues until the error of some convergence decreases to less than a certain level. The main concept for this iterative search is 'greediness'. In greedy search methods, at each iteration, there is a stage of decision making according to some criteria. Here we present a brief review of these methods.

2.3.2.1 Pursuits

Greedy pursuits are algorithms for which their history pre-dates compressive sensing as they date back to the introduction of 'projection pursuit' [78]. They are composed of two steps: (i) selecting elements, and (ii) updating the coefficients. A very common initialization is just consideration of zero elements as $\hat{\boldsymbol{x}}_0 = \boldsymbol{0}$. Due to initializing by zero elements, the initial residual error is $\boldsymbol{y}$ itself as $\boldsymbol{r}_0 = \boldsymbol{y} - \boldsymbol{A}\hat{\boldsymbol{x}}_0$. Also, the initial set of non-zero elements (the support set) is an empty set; $T = \emptyset$. After initializing the $\hat{\boldsymbol{x}}_0$, then at each iteration, the elements are updated as the support set is shaped and its elements are updated. This process continues until the residual error becomes less than a certain level of value. Algorithm 1 describes the general process of a greedy solution to compressive sensing recovery. Also, Fig. 2.3 depicts the iteration loop in a general greedy algorithm.

Input: y, A, and stopping criteria
$i = 1$;
while $i = i + 1$, *until stopping criteria not met* **do**
  Calculate $\boldsymbol{g}_i = A^T \boldsymbol{r}_i$;
  Select elements (columns from A) using the magnitude of the elements of $\boldsymbol{g}_i$;
  Revised the estimation of $\hat{\boldsymbol{x}}_i$, by decreasing the cost function

$$F(\hat{\boldsymbol{x}}_i) = ||\boldsymbol{y} - A\hat{\boldsymbol{x}}_i||_2^2 \tag{2.20}$$

end
$\boldsymbol{r}_i = \boldsymbol{y} - A\hat{\boldsymbol{x}}_i$;
Output: $\boldsymbol{r}_i, \hat{\boldsymbol{x}}_i$;

Algorithm 1: General algorithm of a greedy pursuit solver for compressive sensing recovery.

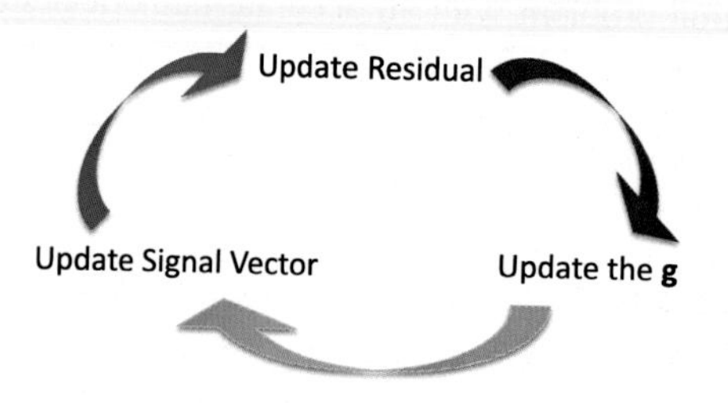

FIGURE 2.3

The iteration loop in general greedy algorithm.

2.3.2.2 Matching pursuit

Matching pursuit (MP) [79] is also known as a 'pure greedy algorithm'. MP is a simple pursuit algorithm. At each time, it selects one column of A such as the jth column as A_j, and each iteration updates the elements by the same selected column in an incremental way. In each iteration to minimize $||\boldsymbol{y} - A\hat{\boldsymbol{x}}_i||_2^2$, the approximation of $\hat{\boldsymbol{x}}_{i,j^{[i]}}$ is updated with respect to the selected coefficients as

$$\hat{\boldsymbol{x}}_{i,j^{[i]}} = \hat{\boldsymbol{x}}_{i-1,j^{[i]}} + \frac{\boldsymbol{g}_i[j^{[i]}]}{||A_{j^{[i]}}||_2^2} \tag{2.21}$$

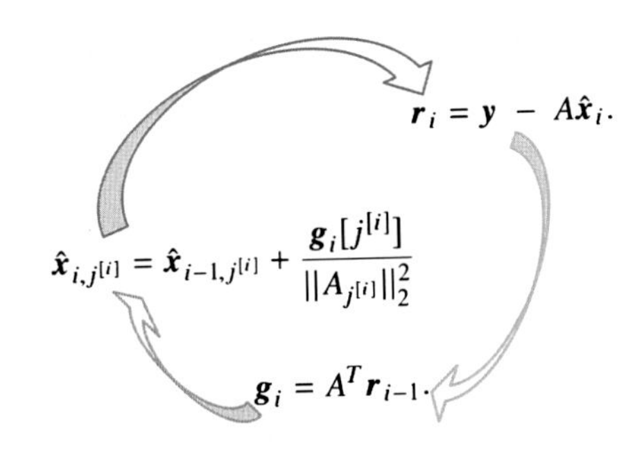

FIGURE 2.4

The iteration loop in the matching pursuit algorithm.

where $g_i[j^{[i]}]$ is the updating parameter by the jth column of A at ith iteration. As we have an updated $\hat{x}_i$ at the ith iteration, it can be used for updating the residual value as

$$r_i = y - A\hat{x}_i. \tag{2.22}$$

Then the updated residual is used for calculating the next updating parameter of the signal vector:

$$g_i = A^T r_{i-1}. \tag{2.23}$$

Having an updated g_i, the estimation of the signal vector is updated, then the updated signal vector is used for updating the residual, and so on. This iterative loop is shown in Fig. 2.4. Also, Algorithm 2 outlines the matching pursuit approach in compressive sensing recovery.

Input: y, A, and stopping criteria
$r_0 = y$, $\hat{x}_0 = \mathbf{0}$;
$i = 1$;
while $i = i + 1$, *until stopping criteria not met* **do**
  Calculate $g_i = A^T r_{i-1}$;
  $j^{[i]} = \text{argmax}_j \frac{|g_{j,i}|}{||A_j||_2}$;
  $\hat{x}_{i,j^{[i]}} = \hat{x}_{i-1,j^{[i]}} + \frac{|g_{j^{[i]},i}|}{||A_j^{[i]}||_2^2}$;
  $r_i = r_{i-1} - \frac{A_j g_{j^{[i]},i}}{||A_j^{[i]}||_2^2}$;
end
Output: $r_i, \hat{x}_i$;

Algorithm 2: Matching pursuit algorithm for compressive sensing recovery.

2.3.2.3 Orthogonal matching pursuit

The orthogonal matching pursuit (OMP) [79] or orthogonal greedy algorithm is more complicated than MP. Thc OMP starts the search by finding a column of A with maximum correlation with measurements $\boldsymbol{y}$ at the first step and thereafter at each iteration it searches for the column of A with maximum correlation with the current residual. In each iteration, the estimation of the signal vector is updated by the highly correlated column of A as A_j as explained in detail in the following. Algorithm 3 illustrates the OMP for recovery in compressive sensing. The input variables to the algorithm are the measurement vector $\boldsymbol{y}$, and the compressive sensing matrix A. The residual vector is taken initially as $\boldsymbol{y}$, as at each ith iteration it is calculated from $\boldsymbol{r}_i = \boldsymbol{y} - A\hat{\boldsymbol{x}}_i$. Since the estimated signal vector is initially $\mathbf{0}$, we have $\boldsymbol{r}_0 = \boldsymbol{y}$. There is an effective parameter as support vector Λ_i which at each iteration is updated. Λ_i determines which columns of A are to be considered and which columns to be set as $\mathbf{0}$, while the signal vector estimation is updated by the following equation:

$$\begin{aligned} \hat{\boldsymbol{x}}_i|_{\Lambda_i^c} &:= \mathbf{0}, \\ \hat{\boldsymbol{x}}_i|_{\Lambda_i} &:= A^\dagger \boldsymbol{y}. \end{aligned} \tag{2.24}$$

Λ_i is a support vector where it is initially empty, $\Lambda_0 = \emptyset$, then at each iteration it is updated by the updating parameter of each iteration $\boldsymbol{g}_i$:

$$\Lambda_i = \Lambda_{i-1} \cup j[i]. \tag{2.25}$$

Input: y, A
Initialize: $\hat{x}_0 = \mathbf{0}, \boldsymbol{r}_0 = y, \boldsymbol{\Lambda}_0 = \emptyset$;
$i = 1$;
while $i = i + 1$, *until stopping criteria fulfilled* **do**
  Calculate the updating vector from the residual:
  $g_i = A^T \boldsymbol{r}_{i-1}$;
  Find j of the column in A with highest autocorrelation with residual vector:
  $j_i = \operatorname{argmax}_j \frac{|g_i[j]|}{||A_j||_2}$;
  Update the support vector by addition of j index:
  $\Lambda_i = \Lambda_{i-1} \cup j[i]$;
  Update the signal vector estimation:
  $\hat{\boldsymbol{x}}_i|_{\Lambda_i} := A^\dagger \boldsymbol{y}$;
  $\hat{\boldsymbol{x}}_i|_{\Lambda_i^c} := \mathbf{0}$;
  Update the residual signal vector:
  $\boldsymbol{r}_i = \boldsymbol{y} - A\hat{\boldsymbol{x}}_i$;
end
Output: $\hat{x}$;

Algorithm 3: Orthogonal matching pursuit (OMP) algorithm for compressive sensing recovery.

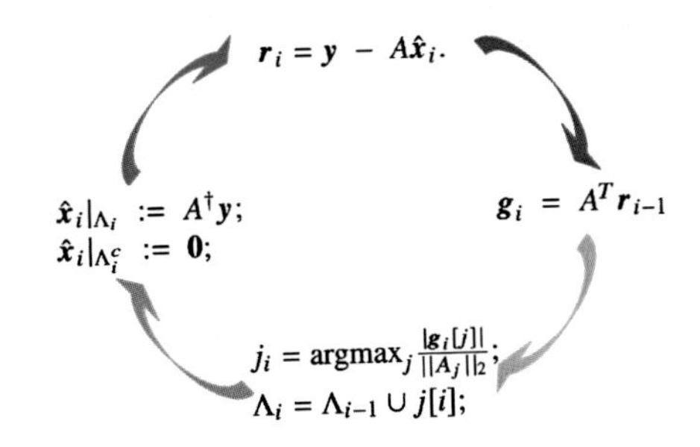

FIGURE 2.5

The iteration loop in the matching pursuit algorithm.

$j[i]$ is the index of the column of A which has the maximum correlation with the current residual $\boldsymbol{r}_i$, and it is determined as the element of $\boldsymbol{g}$ which gives the maximum value,

$$j_i = \text{argmax}_j \frac{|g_i[j]|}{||A_j||_2}. \tag{2.26}$$

The current residual vector $\boldsymbol{r}_i$ gives the current updating vector $\boldsymbol{g}_i$ as

$$\boldsymbol{g}_i = A^T \boldsymbol{r}_{i-1}. \tag{2.27}$$

The above equations are in the iterative loop as shown in Fig. 2.5. The corresponding vectors are iteratively updated till a stopping criterion is met or the number of iterations passes a limit. Note that despite MP, in updating the $\hat{\boldsymbol{x}}$ all the elements of the current support vector are in use. Also, despite MP, OMP never does re-selection of an element. Indeed, at each iteration, the current residual is orthogonal to the elements already adjusted. Interestingly, when $\boldsymbol{x}$ is exactly k-sparse without any noise, OMP recovers it exactly in k iterations.

The OMP implementation steps are as follows.

Step 1: Taking the compressive sensing matrix A, and the measurement vector $\boldsymbol{y}$, initializing the approximation of the signal vector $\hat{\boldsymbol{x}}$ equivalent to $\boldsymbol{0}$.

Step 2: Calculating/updating the residual $\boldsymbol{r}_i = \boldsymbol{y} - A\hat{\boldsymbol{x}}_i$; as a matter of fact the initial value for $\boldsymbol{r}$ is $\boldsymbol{y}$.

Step 3: Calculate the updating vector $\boldsymbol{g}$ from the residual: $\boldsymbol{g}_i = A^T \boldsymbol{r}_{i-1}$.

Step 4: Using the updating vector, finding $j_{\max}$, the index of the column in A which maximally correlates with the residual.

Step 5: Updating the support vector Λ_i by using $j_{\max}$.

Step 6: Shaping the $A^\dagger$ by using the support vector Λ_i.

Step 7: Updating $\hat{\boldsymbol{x}}$ in the direction of minimizing $||\boldsymbol{y} - A\hat{\boldsymbol{x}}||_2$. Updating is via Eqs. (2.24) where $\dagger$ is the notation for the pseudo-inverse operator.

Step 8: If the criterion for stopping is fulfilled put out the last update of $\hat{\boldsymbol{x}}$, otherwise go to Step 2.

2.3.2.4 Iterative hard thresholding

There are a variety of greedy pursuit solvers for compressive sensing problems. However, we here focus just on the main ones. Another greedy recovery technique is 'iterative hard thresholding' (ITH) which is even more directive in updating the signal vector estimation than OMP [79]. The methodology of ITH starts from an initial value of the $\mathbf{0}$ vector for $\hat{\boldsymbol{x}}$, then in a very straightforward manner it updates the signal vector by a gradient descent algorithm as given by the formula

$$\hat{\boldsymbol{x}}_i = H_k\big(\hat{\boldsymbol{x}}_{i-1} + A^T(\boldsymbol{y} - A\hat{\boldsymbol{x}}_{i-1})\big) \tag{2.28}$$

where H_k is the hard thresholding operator. It keeps the k elements of its operand with the biggest amplitude without change and sets all the others to zero. Indeed ITH is based on the *a priori* knowledge that $\boldsymbol{x}$ is a k-sparse signal vector. Algorithm 4 outlines the algorithm of ITH in compressive sensing recovery.

Input: $\boldsymbol{y}$, A, sparsity level k
Initialize: $\hat{\boldsymbol{x}}_0 = \mathbf{0}$;
$i = 1$;
while $i = i + 1$, *until stopping criteria fulfilled* **do**
| $\hat{\boldsymbol{x}}_i = H_k\big(\hat{\boldsymbol{x}}_{i-1} + A^T(\boldsymbol{y} - A\hat{\boldsymbol{x}}_{i-1})\big)$;
end
Output: $\hat{\boldsymbol{x}}$;

Algorithm 4: Iterative hard thresholding algorithm for compressive sensing recovery.

2.4 Summary

This chapter gives a brief to the recovery solutions in compressive sensing. First, it simply explains the idea of compressive sensing in a simplified manner for a reader of moderate mathematical knowledge. Compressive sensing comprises two main challenges: (i) how to construct a compressive sensing matrix, and (ii) how to recover the sparse signal from a much shorter segment as compressed measurement. The chapter goes through the recovery via the two main categories of norm-1 minimization and greedy pursuits. The norm-1 minimization pre-dates compressive sensing, and the chapter reviews it in a systematic way. It leads to its application as a recovery solution in compressive sensing. The main greedy pursuits for signal vector recovery in compressive sensing as matching pursuit (MP), orthogonal matching pursuit (OMP), and iterative hard thresholding are explained. The related algorithms are given and clarified in detail. Also the iteration loop of each methodology is depicted to give the reader a clear view of the implementation of compressively sensed signal recovery by these techniques.

References

[1] E.J. Candès, et al., Compressive sampling, in: Proceedings of the International Congress of Mathematicians, Madrid, Spain, vol. 3, 2006, pp. 1433–1452.

[2] E.J. Candès, J. Romberg, T. Tao, Robust uncertainty principles: exact signal reconstruction from highly incomplete frequency information, IEEE Transactions on Information Theory 52 (2) (2006) 489–509.

[3] E.J. Candes, J.K. Romberg, T. Tao, Stable signal recovery from incomplete and inaccurate measurements, Communications on Pure and Applied Mathematics: A Journal Issued by the Courant Institute of Mathematical Sciences 59 (8) (2006) 1207–1223.

[4] E.J. Candes, T. Tao, Near-optimal signal recovery from random projections: universal encoding strategies?, IEEE Transactions on Information Theory 52 (12) (2006) 5406–5425.

[5] D.L. Donoho, et al., Compressed sensing, IEEE Transactions on Information Theory 52 (4) (2006) 1289–1306.

[6] M. Yamin, A.A.A. Sen, Improving privacy and security of user data in location based services, International Journal of Ambient Computing and Intelligence (IJACI) 9 (1) (2018) 19–42.

[7] M.H. Sedaaghi, R. Daj, M. Khosravi, Mediated morphological filters, in: 2001 International Conference on Image Processing, Proceedings, vol. 3, IEEE, 2001, pp. 692–695.

[8] M. Khosravy, N. Gupta, N. Marina, I.K. Sethi, M.R. Asharif, Morphological filters: an inspiration from natural geometrical erosion and dilation, in: Nature-Inspired Computing and Optimization, Springer, Cham, 2017, pp. 349–379.

[9] C. Castelfranchi, G. Pezzulo, L. Tummolini, Behavioral implicit communication (BIC): communicating with smart environments, International Journal of Ambient Computing and Intelligence (IJACI) 2 (1) (2010) 1–12.

[10] S. Hemalatha, S.M. Anouncia, Unsupervised segmentation of remote sensing images using FD based texture analysis model and ISODATA, International Journal of Ambient Computing and Intelligence (IJACI) 8 (3) (2017) 58–75.

[11] M. Khosravy, M.R. Alsharif, M. Khosravi, K. Yamashita, An optimum pre-filter for ICA based multi-input multi-output OFDM system, in: 2010 2nd International Conference on Education Technology and Computer, vol. 5, IEEE, 2010, pp. V5–129.

[12] F. Asharif, S. Tamaki, M.R. Alsharif, H. Ryu, Performance improvement of constant modulus algorithm blind equalizer for 16 QAM modulation, International Journal on Innovative Computing, Information and Control 7 (4) (2013) 1377–1384.

[13] M. Khosravy, M.R. Alsharif, K. Yamashita, An efficient ICA based approach to multiuser detection in MIMO OFDM systems, in: Multi-Carrier Systems & Solutions 2009, Springer, 2009, pp. 47–56.

[14] M. Khosravy, M.R. Alsharif, B. Guo, H. Lin, K. Yamashita, A robust and precise solution to permutation indeterminacy and complex scaling ambiguity in BSS-based blind MIMO-OFDM receiver, in: International Conference on Independent Component Analysis and Signal Separation, Springer, 2009, pp. 670–677.

[15] M. Khosravy, A blind ICA based receiver with efficient multiuser detection for multi-input multi-output OFDM systems, in: The 8th International Conference on Applications and Principles of Information Science (APIS), Okinawa, Japan, 2009, 2009, pp. 311–314.

[16] M. Khosravy, S. Kakazu, M.R. Alsharif, K. Yamashita, Multiuser data separation for short message service using ICA, SIP, IEICE Technical Report 109 (435) (2010) 113–117.

[17] M. Foth, R. Schroeter, J. Ti, Opportunities of public transport experience enhancements with mobile services and urban screens, International Journal of Ambient Computing and Intelligence (IJACI) 5 (1) (2013) 1–18.

[18] M. Khosravy, N. Punkoska, F. Asharif, M.R. Asharif, Acoustic OFDM data embedding by reversible Walsh–Hadamard transform, AIP Conference Proceedings 1618 (2014) 720–723.

[19] A.A. Picorone, T.R. de Oliveira, R. Sampaio-Neto, M. Khosravy, M.V. Ribeiro, Channel characterization of low voltage electric power distribution networks for PLC applications based on measurement campaign, International Journal of Electrical Power & Energy Systems 116 (105) (2020) 554.

[20] E. Santos, M. Khosravy, M.A. Lima, A.S. Cerqueira, C.A. Duque, A. Yona, High accuracy power quality evaluation under a colored noisy condition by filter bank ESPRIT, Electronics 8 (11) (2019) 1259.
[21] E. Santos, M. Khosravy, M.A. Lima, A.S. Cerqueira, C.A. Duque, Esprit associated with filter bank for power-line harmonics, sub-harmonics and inter-harmonics parameters estimation, International Journal of Electrical Power & Energy Systems 118 (105) (2020) 731.
[22] K. Melo, M. Khosravy, C. Duque, N. Dey, Chirp code deterministic compressive sensing: analysis on power signal, in: 4th International Conference on Information Technology and Intelligent Transportation Systems (ITITS 2019), IOS Press, 2020, pp. 125–134.
[23] N. Gupta, M. Khosravy, N. Patel, T. Senjyu, A bi-level evolutionary optimization for coordinated transmission expansion planning, IEEE Access 6 (2018) 48455–48477.
[24] N. Gupta, M. Khosravy, K. Saurav, I.K. Sethi, N. Marina, Value assessment method for expansion planning of generators and transmission networks: a non-iterative approach, Electrical Engineering 100 (3) (2018) 1405–1420.
[25] G.V. Kale, V.H. Patil, A study of vision based human motion recognition and analysis, International Journal of Ambient Computing and Intelligence (IJACI) 7 (2) (2016) 75–92.
[26] M. Khosravy, M.R. Asharif, M.H. Sedaaghi, Medical image noise suppression: using mediated morphology, IEICE, Technical Report 107 (461) (2008) 265–270.
[27] N. Dey, A.S. Ashour, A.S. Ashour, A. Singh, Digital analysis of microscopic images in medicine, Journal of Advanced Microscopy Research 10 (1) (2015) 1–13.
[28] M. Khosravy, M.R. Asharif, M.H. Sedaaghi, Medical image noise suppression using mediated morphology, in: IEICE Tech. Rep., IEICE, 2008, pp. 265–270.
[29] B. Alenljung, J. Lindblom, R. Andreasson, T. Ziemke, User experience in social human–robot interaction, in: Rapid Automation: Concepts, Methodologies, Tools, and Applications, IGI Global, 2019, pp. 1468–1490.
[30] M.H. Sedaaghi, M. Khosravi, Morphological ECG signal preprocessing with more efficient baseline drift removal, in: 7th IASTED International Conference, ASC, 2003, pp. 205–209.
[31] N. Dey, S. Samanta, X.-S. Yang, A. Das, S.S. Chaudhuri, Optimisation of scaling factors in electrocardiogram signal watermarking using cuckoo search, International Journal of Bio-Inspired Computation 5 (5) (2013) 315–326.
[32] M. Khosravy, M.R. Asharif, M.H. Sedaaghi, Morphological adult and fetal ECG preprocessing: employing mediated morphology, IEICE Technical Report, IEICE 107 (2008) 363–369.
[33] N. Dey, A.S. Ashour, F. Shi, S.J. Fong, R.S. Sherratt, Developing residential wireless sensor networks for ECG healthcare monitoring, IEEE Transactions on Consumer Electronics 63 (4) (2017) 442–449.
[34] M. Khosravi, M.H. Sedaaghi, Impulsive noise suppression of electrocardiogram signals with mediated morphological filters, in: 11th Iranian Conference on Biomedical Engineering, ICBME, 2004, pp. 207–212.
[35] N. Dey, S. Mukhopadhyay, A. Das, S.S. Chaudhuri, Analysis of P-QRS-T components modified by blind watermarking technique within the electrocardiogram signal for authentication in wireless telecardiology using DWT, International Journal of Image, Graphics and Signal Processing 4 (7) (2012) 33.
[36] M. Baumgarten, M.D. Mulvenna, N. Rooney, J. Reid, Keyword-based sentiment mining using Twitter, International Journal of Ambient Computing and Intelligence (IJACI) 5 (2) (2013) 56–69.
[37] C.E. Gutierrez, P.M.R. Alsharif, M. Khosravy, P.K. Yamashita, P.H. Miyagi, R. Villa, Main large data set features detection by a linear predictor model, AIP Conference Proceedings 1618 (2014) 733–737.
[38] A.S. Ashour, S. Samanta, N. Dey, N. Kausar, W.B. Abdessalemkaraa, A.E. Hassanien, Computed tomography image enhancement using cuckoo search: a log transform based approach, Journal of Signal and Information Processing 6 (03) (2015) 244.
[39] M. Khosravy, N. Gupta, N. Marina, I. Sethi, M. Asharif, Brain action inspired morphological image enhancement, in: Nature-Inspired Computing and Optimization, Springer, Cham, 2017, pp. 381–407.
[40] M. Khosravy, N. Gupta, N. Marina, I. Sethi, M. Asharifa, Perceptual adaptation of image based on Chevreul–Mach bands visual phenomenon, IEEE Signal Processing Letters 24 (5) (2017) 594–598.

[41] P. Sosnin, Precedent-oriented approach to conceptually experimental activity in designing the software intensive systems, International Journal of Ambient Computing and Intelligence (IJACI) 7 (1) (2016) 69–93.
[42] S. Gupta, M. Khosravy, N. Gupta, H. Darbari, N. Patel, Hydraulic system onboard monitoring and fault diagnostic in agricultural machine, Brazilian Archives of Biology and Technology 62 (2019).
[43] S. Gupta, M. Khosravy, N. Gupta, H. Darbari, In-field failure assessment of tractor hydraulic system operation via pseudospectrum of acoustic measurements, Turkish Journal of Electrical Engineering & Computer Sciences 27 (4) (2019) 2718–2729.
[44] C.E. Gutierrez, M.R. Alsharif, K. Yamashita, M. Khosravy, A tweets mining approach to detection of critical events characteristics using random forest, International Journal of Next-Generation Computing 5 (2) (2014) 167–176.
[45] N. Kausar, S. Palaniappan, B.B. Samir, A. Abdullah, N. Dey, Systematic analysis of applied data mining based optimization algorithms in clinical attribute extraction and classification for diagnosis of cardiac patients, in: Applications of Intelligent Optimization in Biology and Medicine, Springer, 2016, pp. 217–231.
[46] D.L. Donoho, M. Elad, Optimally sparse representation in general (nonorthogonal) dictionaries via l1 minimization, Proceedings of the National Academy of Sciences 100 (5) (2003) 2197–2202.
[47] A. Cohen, W. Dahmen, R. DeVore, Compressed sensing and best k-term approximation, Journal of the American Mathematical Society 22 (1) (2009) 211–231.
[48] E. Candes, T. Tao, Decoding by linear programming, preprint, arXiv:math/0502327, 2005.
[49] M.A. Davenport, Random observations on random observations: Sparse signal acquisition and processing, PhD thesis, Rice University, 2010.
[50] M. Khosravy, Blind source separation and its application to speech, image and MIMO-OFDM communication systems, PhD thesis, University of the Ryukyus, Japan, 2010.
[51] M. Khosravy, M. Gupta, M. Marina, M.R. Asharif, F. Asharif, I. Sethi, Blind components processing a novel approach to array signal processing: a research orientation, in: 2015 International Conference on Intelligent Informatics and Biomedical Sciences, ICIIBMS, 2015, pp. 20–26.
[52] M. Khosravy, M.R. Alsharif, K. Yamashita, A PDF-matched modification to stone's measure of predictability for blind source separation, in: International Symposium on Neural Networks, Springer, Berlin, Heidelberg, 2009, pp. 219–222.
[53] M. Khosravy, M.R. Asharif, K. Yamashita, A probabilistic short-length linear predictability approach to blind source separation, in: 23rd International Technical Conference on Circuits/Systems, Computers and Communications (ITC-CSCC 2008), Yamaguchi, Japan, ITC-CSCC, 2008, pp. 381–384.
[54] M. Khosravy, M.R. Asharif, K. Yamashita, A theoretical discussion on the foundation of Stone's blind source separation, Signal, Image and Video Processing 5 (3) (2011) 379–388.
[55] M. Khosravy, M.R. Asharif, K. Yamashita, A PDF-matched short-term linear predictability approach to blind source separation, International Journal of Innovative Computing, Information & Control 5 (11) (2009) 3677–3690.
[56] M. Khosravy, N. Gupta, N. Patel, T. Senjyu, Frontier Applications of Nature Inspired Computation, Springer, 2020.
[57] N. Gupta, M. Khosravy, O.P. Mahela, N. Patel, Plant biology-inspired genetic algorithm: superior efficiency to firefly optimizer, in: Applications of Firefly Algorithm and Its Variants, Springer, 2020, pp. 193–219.
[58] C. Moraes, E. De Oliveira, M. Khosravy, L. Oliveira, L. Honório, M. Pinto, A hybrid bat-inspired algorithm for power transmission expansion planning on a practical Brazilian network, in: Applied Nature-Inspired Computing: Algorithms and Case Studies, Springer, 2020, pp. 71–95.
[59] M. Khosravy, N. Gupta, N. Patel, T. Senjyu, C.A. Duque, Particle swarm optimization of morphological filters for electrocardiogram baseline drift estimation, in: Applied Nature-Inspired Computing Algorithms and Case Studies, Springer, 2020, pp. 1–21.
[60] G. Singh, N. Gupta, M. Khosravy, New crossover operators for real coded genetic algorithm (RCGA), in: 2015 International Conference on Intelligent Informatics and Biomedical Sciences (ICIIBMS), IEEE, 2015, pp. 135–140.

[61] N. Gupta, N. Patel, B.N. Tiwari, M. Khosravy, Genetic algorithm based on enhanced selection and log-scaled mutation technique, in: Proceedings of the Future Technologies Conference, Springer, 2018, pp. 730–748.
[62] N. Gupta, M. Khosravy, N. Patel, I. Sethi, Evolutionary optimization based on biological evolution in plants, Procedia Computer Science 126 (2018) 146–155.
[63] J. Kaliannan, A. Baskaran, N. Dey, A.S. Ashour, M. Khosravy, R. Kumar, ACO based control strategy in interconnected thermal power system for regulation of frequency with HAE and UPFC unit, in: International Conference on Data Science and Application (ICDSA-2019), in: LNNS, Springer, 2019.
[64] N. Gupta, M. Khosravy, N. Patel, O. Mahela, G. Varshney, Plants genetics inspired evolutionary optimization: a descriptive tutorial, in: Frontier Applications of Nature Inspired Computation, Springer, 2020, pp. 53–77.
[65] M. Khosravy, N. Gupta, N. Patel, O. Mahela, G. Varshney, Tracing the points in search space in plants biology genetics algorithm optimization, in: Frontier Applications of Nature Inspired Computation, Springer, 2020, pp. 180–195.
[66] N. Gupta, M. Khosravy, N. Patel, S. Gupta, G. Varshney, Evolutionary artificial neural networks: comparative study on state of the art optimizers, in: Frontier Applications of Nature Inspired Computation, Springer, 2020, pp. 302–318.
[67] N. Gupta, M. Khosravy, N. Patel, S. Gupta, G. Varshney, Artificial neural network trained by plant genetics-inspired optimizer, in: Frontier Applications of Nature Inspired Computation, Springer, 2020, pp. 266–280.
[68] S.S. Chen, D.L. Donoho, M.A. Saunders, Atomic decomposition by basis pursuit, SIAM Review 43 (1) (2001) 129–159.
[69] Y.C. Eldar, G. Kutyniok, Compressed Sensing: Theory and Applications, Cambridge University Press, 2012.
[70] A. Beurling, Sur les intégrales de Fourier absolument convergentes et leur application à une transformation fonctionelle, in: Ninth Scandinavian Mathematical Congress, 1938, pp. 345–366.
[71] H.L. Taylor, S.C. Banks, J.F. McCoy, Deconvolution with the l1 norm, Geophysics 44 (1) (1979) 39–52.
[72] S. Levy, P.K. Fullagar, Reconstruction of a sparse spike train from a portion of its spectrum and application to high-resolution deconvolution, Geophysics 46 (9) (1981) 1235–1243.
[73] D.L. Donoho, B.F. Logan, Signal recovery and the large sieve, SIAM Journal on Applied Mathematics 52 (2) (1992) 577–591.
[74] R. Tibshirani, Regression shrinkage and selection via the lasso, Journal of the Royal Statistical Society, Series B (Methodological) 58 (1) (1996) 267–288.
[75] S. Mallat, A Wavelet Tour of Signal Processing, Elsevier, 1999.
[76] B.F. Logan, Properties of high-pass signals, PhD thesis, Columbia University, 1965.
[77] S.R. Becker, E.J. Candès, M.C. Grant, Templates for convex cone problems with applications to sparse signal recovery, Mathematical Programming Computation 3 (3) (2011) 165.
[78] J.H. Friedman, J.W. Tukey, A projection pursuit algorithm for exploratory data analysis, IEEE Transactions on Computers 100 (9) (1974) 881–890.
[79] S.G. Mallat, Z. Zhang, Matching pursuits with time-frequency dictionaries, IEEE Transactions on Signal Processing 41 (12) (1993) 3397–3415.

CHAPTER

3 A descriptive review to sparsity measures

Mahdi Khosravy[a], Naoko Nitta[a], Neeraj Gupta[b], Nilesh Patel[b], Noboru Babaguchi[b]
[a]*Media Integrated Communication Laboratory, Graduate School of Engineering, Osaka University, Suita, Osaka, Japan*
[b]*Computer Science and Engineering, Oakland University, Rochester, MI, United States*

3.1 Introduction

Compressive sensing [1–5] as a recent approach of sensing data with much less sensory data than required by the Shannon–Nyquist sampling theorem brings about several considerable advantages to information–communication systems as: (*i*) much fewer required sensory devices, (*ii*) a much higher data rate, (*iii*) a much milder requirement for data storage, (*iv*) a much higher data rate transmission on a same bandwidth (or the same data rate transmission but on a much lower bandwidth), and (*v*) the energy efficiency at sensory level.

Compressive sensing works based on the fact that the data which carries information is of a sparse nature, and the sparsity of the data is a measure of the informative value. Therefore, the data samples are linearly mixed and sensed as a much smaller number of samples by a much smaller number of sensors, and thereafter can be saved in a storage memory of much lower size, or being transferred on much lower communication bandwidth (or on the same commutation bandwidth but with the much higher data rate). The original data samples of the higher volume are retrieved from the compressively sensed data by consideration of the fact that, amongst infinitely many possible solutions to an incomplete set of linear equations, the one with the highest sparsity has informative value and can represent the original data.

Compressive sensing has been already deployed in a wide range of applications as fast magnetic resonance imaging (MRI) [6], accelerated dynamic MRI [7], [8], CT scan reconstruction [9], compressive radar imaging [10], multiplexing of PET detectors [11], medical ultrasound [12], background subtraction [13], image fusion [14], image denoising [15], compressive MUSIC [16], spectral compressive sensing [17], wide-band spectrum sensing [18], multi-user detection [19], sparse channel estimation [20], cognitive radio [21], OFDM [22], network data acquisition [23], cognitive IoT [24], undersampled AFM imaging [25], confocal microscopy [26], medical signal and imaging [27], ECG compression [28], energy-efficient ECG compression [29], wireless ECG sensor [30], wearable ECG [31], wireless fetal ECG [32], scalp EEG [33], MRI [6,7], medical CT scan [9], medical physiological imaging [11]

Compressive Sensing in Healthcare. https://doi.org/10.1016/B978-0-12-821247-9.00008-1

Table 3.1 Compressive sensing applications.

Application field	Topic	Literature
Imaging systems	MRI	[6]
	accelerated dynamic MRI	[7]
	CT scan reconstruction	[9]
	compressive radar imaging	[10]
	multiplexing of PET detectors	[11]
	medical ultrasound	[12]
Image processing	background subtraction	[13]
	image fusion	[14]
	image denoising	[15]
Spectrum estimation	compressive MUSIC	[16]
	spectral compressive sensing	[17]
	wide-band spectrum sensing	[18]
Telecommunications	multi-user detection	[19]
	sparse channel estimation	[20]
	cognitive radio	[21]
	OFDM	[22]
Internet of things	network data acquisition	[23]
	cognitive IoT	[24]
Microscopy	undersampled AFM imaging	[25]
	confocal microscopy	[26]
Biomedical engineering	medical signal and imaging	[27]
	ECG compression	[28]
	energy-efficient ECG compression	[29]
	wireless ECG sensor	[30]
	wearable ECG	[31]
	wireless fetal ECG	[32]
	scalp EEG	[33]
	MRI	[6,7]
	medical CT scan	[9]
	Medical physiological imaging	[11]

Table 3.1 lists some selected applications of compressive sensing found in the literature.

Despite the applications of compressive sensing being well developed, still there is a great potential for application in a variety of fields as text data processing [34], signal processing [35,36], blind signal processing [37–42], power line communications [43], power system planning [44,45], medical image processing [46–48], image enhancement [49,50], power quality analysis [51,52], image adaptation [53], data mining [54,55], electrocardiogram processing [27,56–61], acoustic OFDM [62], telecommunications [63–68], agriculture machinery [69,70], public transportation systems [71], human motion analysis [72], location based services [73], human-robot interaction [74], sentiment mining [75], software intensive systems [76], texture analysis [77], and smart environments [78].

The fundamental theory behind the compressive sensing originates from the concept of sparsity as a measure of data non-uniformity. Since recovery in compressive sensing is via maximization of sparsity, this chapter reviews the defined formulas for measuring sparsity occurring in the literature. Each formula and its approach to sparsity is interpreted.

3.2 Compressive sensing

Most of the data of natural types have a sparse form where most of the samples are zero or close to zero, and few samples have considerable value. Considering the data in signal vector form, then the signal vector is sparse if most of the vector elements are zero and few non-zero, and it is semi-sparse when most of the elements are close to zero and few elements have a considerable non-zero value. Fig. 3.1 shows a sparse and a semi-sparse signal vector. Logically, sparse data can be sampled by just the dominant values and their corresponding locations, in this way the data will be compressed. This is due to sparsity which can be extended to the semi-sparse case by ignoring the close-to-zero values in counting the non-zero element.

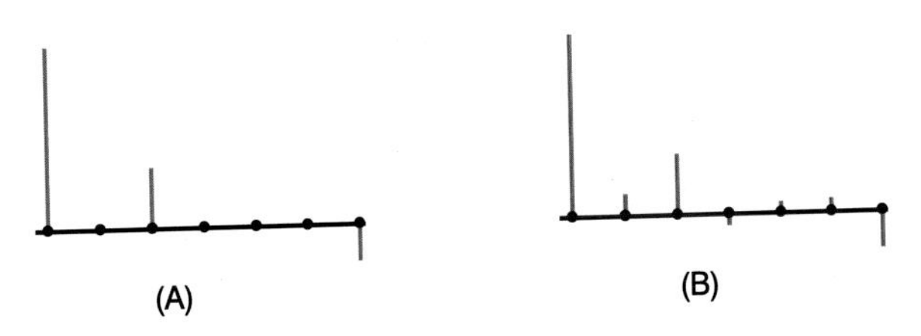

FIGURE 3.1

An example of (A) a sparse signal vector, and (B) a semi-sparse signal vector.

Relying upon the sparsity concept, the compressive sensing directly compresses the data and collects the data at the sensory level where, instead of equally spaced uniform sensing the data based on the Shannon–Nyquist theorem, long data vector values are linearly mixed and sensed as a short length data vector. Fig. 3.2 shows how a long signal vector $\boldsymbol{x}$ of length n is sensed by an $n \times m$ sensing matrix A as a much shorter signal vector $\boldsymbol{y}$ of length m where $m << n$.

Mathematically, a sparse signal vector $\boldsymbol{x}$ from the vector space $\mathbb{R}^n$ is compressively measured as $\boldsymbol{y}$ belonging to vector space of $\mathbb{R}^m$ as follows:

$$y_{m\times 1} = A_{m\times n}\boldsymbol{x}_{n\times 1}, \tag{3.1}$$

where A is the compressive matrix of size $m \times n$ and $m << n$. In the viewpoint of the recovery of $\boldsymbol{x}$ from the measurements $\boldsymbol{y}$, since $m << n$, Eq. (3.1) represents an

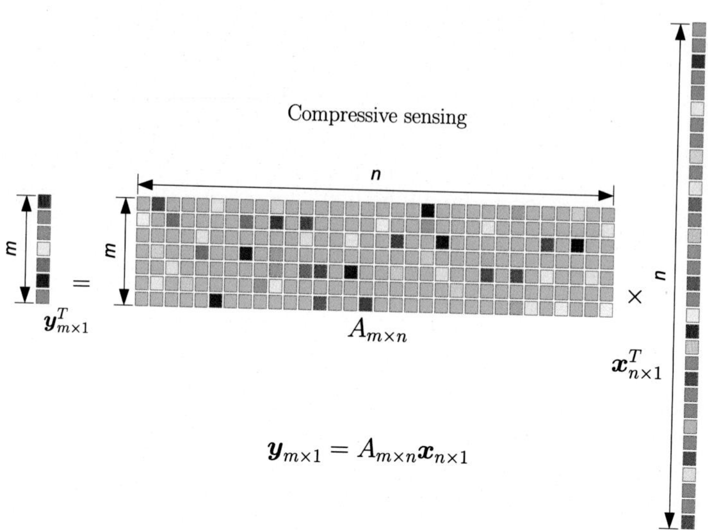

FIGURE 3.2

Compressive sensing model.

incomplete set of linear equations as follows:

$$\begin{aligned}
a_{11}x_1 + a_{12}x_2 + \cdots + a_{1n}x_n &= y_1, \\
a_{21}x_1 + a_{22}x_2 + \cdots + a_{2n}x_n &= y_2, \\
&\vdots \\
a_{m1}x_1 + a_{m2}x_2 + \cdots + a_{mn}x_n &= y_m.
\end{aligned} \tag{3.2}$$

Eqs. (3.2) are composed of a set of equations with multiple variables, $x_1, x_2, \ldots, x_n$, as the elements of the signal vector $\boldsymbol{x}$. For having a solution for n variables, we should have n independent equations. But this set has m equations which is less than the requirement and therefore, there are infinitely many possible solutions, and the problem sounds ill-posed.

Indeed, the problem is not ill-posed, and the key solution to this condition is in the sparsity of the $\boldsymbol{x}$ vectors. Although there are infinitely many solutions, the true solution is the one with maximum sparsity. Therefore, the recovery can be done by searching for the maximum over the search space of possible solutions. This is feasible via an optimization problem of the search for maximization of an available cost function as a measure of the sparsity of the recovery approximation as follows:

$$\begin{aligned}
\underset{\boldsymbol{x}}{\operatorname{argmax}} \quad & \mathcal{S}(\boldsymbol{x}) \\
\text{subjected to} \quad & A\boldsymbol{x} = \boldsymbol{y},
\end{aligned} \tag{3.3}$$

where $\mathcal{S}(.)$ is an increasing measure of signal vector sparsity. As it is observable, the key to the compressive sensing recovery is an optimization process through a measure of sparsity. Because of this key role of the sparsity measure, the focus of this chapter is on functions presented in the literature for measuring the sparsity.

3.3 Sparsity

In English vocabularies, 'sparsity' refers to a situation wherein something is not spread uniformly. For example, as in the case of hair loss, the hairs get gradually more sparse, or in the case of deforestation, some land areas get empty of trees, and so on. In the case of signals, images and in general data, sparsity concerns non-uniformity of the values spread along the time–spatial dimensions. As a matter of fact, as data is inherited from more information content, its non-uniformity is higher. Therefore, there is a direct relationship between being informative and sparsity. For most of the naturally obtained data, such as speech and images, due to the sparsity of their content, there is a capacity for compression by transforming from the sparse basis to a non-sparse basis. As explained earlier this matter has been effectively employed in compressive sensing, and maximization of the sparsity measure is a way to recover the original shape of data. k-sparsity is a measure of sparseness that has been used in most of the compressive sensing literature.

k-sparse signal

A signal vector $\boldsymbol{x}$ is measured as k-sparse, if it maximally contains k non-zero samples. k-sparsity can be phrased in terms of norm-0 as a signal vector is k-sparse if

$$\|\boldsymbol{x}\|_0 \leq k, \tag{3.4}$$

where $\|.\|_0$ indicates the norm-0 case. k-sparsity is a decreasing measure of sparsity in the sense that when a signal is k-sparse, the higher k indicates less sparsity. In the literature, there are other measures of sparsity too. The sequence of the chapter reviews all the measures of the sparsity and organizes the corresponding measures in a way such each of them represents an increasing measure of sparseness.

3.4 Review to sparsity measures

As k-sparsity indicates the number of non-zero elements of the signal as a measure of sparsity, the lower value indicates the higher sparsity. In the literature, there are a variety of measures for the sparsity of a signal vector where some of them are increasing, which means that their higher value means that we have a higher sparsity. Here most of these measures are listed and organized in increasing form of formulas that are usable for maximization of sparsity. Hurley and Rickard list six properties from the literature that a measure of sparsity should have [79]. In the sense that if we have an increasing measure of sparsity $S(.)$, increasing the sparsity of $\boldsymbol{x}$ via any of the six properties must be effectively quantified by the measure $\mathcal{S}(\boldsymbol{x})$. These properties are as follows:

- Robin Hood
- Scaling
- Rising tide
- Cloning
- Bill Gates
- Babies

Although these properties are not in the focus of this chapter, a brief description of each of them is given here. As an important aspect of sparsity is in a distribution where most of the values are concentrated at a few samples, there are lots of samples with a low value or even no value. The properties are based on financial concepts, as the values given to the samples have been resembled as the wealth of individuals and sparsity as a measure of economic inequity. The first four properties are called Dalton's laws by Ref. [79], as taken from a century-old paper of Dalton [80]. The last two are from Ref. [81].

Robin Hood: Dalton's first law

An important aspect of sparsity is having a distribution with most of the values concentrated for few samples and lots of samples with low values, and this reminds one of the inequality in the wealth distribution. A way to decrease this inequality is taking from the rich and giving to the poor, or the Robin Hood way of stealing from the rich and donating to the poor. Dalton's first law indicates that a Robin Hood act should decrease the sparsity, and if we apply the same effect to a signal vector by giving a percentage of the high values to the low values, the sparsity measure should show a lower value. A sparsity measure must follow the sparsity changes by the effect of the Robin Hood property.

Scaling: Dalton's second law

Dalton's second law of sparsity indicates that a sparsity measure should not be variant to scaling. This means that as the wealth of all identities is equally multiplied, the inequality of poor and rich does not change. In other words, the concepts of 'rich' and 'poor' are relative and comparative to other identities. This property as seen from the viewpoint of signal/image sparsity means that if we scale the signal/image, the relative concentrations in the distribution of the values have the same form, however the values are scaled. Mathematically, this property is written as follows:

$$\mathcal{S}(\alpha \boldsymbol{x}) = \mathcal{S}(\boldsymbol{x}), \tag{3.5}$$

where α is a positive multiplier, and $\boldsymbol{x}$ is the signal vector.

Rising tide: Dalton's third law

Dalton's third law indicates that if there is inequality of wealth, giving an additional equal amount of wealth to all society members, the inequality of wealth decreases. To be observable there should be at least a difference between two members' wealth in a society. Giving an additional equal wealth to all will make the available difference more negligible. As in the case of sparsity, if we have a signal vector with some difference in its elements, then giving a bias to the signal values should decrease the measure of sparsity. Mathematically, this property is indicated as follows:

$$\exists(i,j), i \neq j, \boldsymbol{x}(i) \neq \boldsymbol{x}(j) \quad \rightarrow \quad \mathcal{S}(\boldsymbol{x}+\alpha) < \mathcal{S}(\boldsymbol{x}), \tag{3.6}$$

where α is a positive value. A measure of sparsity must follow the scaling property.

Cloning: Dalton's fourth law

The fourth law of Dalton indicates that if two societies suffer the same level of inequality of wealth, putting the two societies as one society together does not change the wealth inequality. Extending to the concept of sparsity, this means that if we append two signal vectors with the same level of sparsity to each other and make a longer signal vector, then the resultant signal vector has the same sparsity level. This effect also should be reflected in a measure of sparsity.

Bill Gates

Bill Gates is an economic phenomenon as an individual with a huge amount of wealth compared with other individuals. As one of the society members gains huge amounts of wealth, the wealth distribution becomes very sparse. Therefore, giving to a data sample a value many times higher than the others highly increases the sparseness. This effect also should be reflected in a measure of sparsity.

Babies

When there is some wealth in society, adding members with zero wealth increases the sparsity of wealth. Also, the sparsity of data is the same, as we have a non-zero signal vector, adding zero elements increases its sparsity. This effect also should be reflected in a measure of sparsity.

Reference [79] reviews 16 measures of sparsity from the literature and compares them with satisfactory results as regards the above-mentioned properties. These measures and their definitions are presented in Tables 3.2 and 3.3. However, the range of their values has not been set inside a range that can indicate the level of sparsity between certain possible minimum and maximum sparsity levels. Although these measures can be used for sparsity maximization, they cannot be solely used as a sparsity level indicator. Each of them, in order to be used as a sparsity level indicator, should be modified to be in a certain range from zero considering the value of the maximum sparsity. This section gives a brief review of these basic measures. The measures have been manipulated to have an increasing relation with sparsity.

Table 3.2 Measures of sparsity in the literature for a signal vector $\boldsymbol{x} = \{x_i\}$ (part 1).

Measure		Definition	Ref.
$\mathcal{S}_{l_0}(\boldsymbol{x})$	$=$	$-\|\boldsymbol{x}\|_0, \quad \boldsymbol{x} \neq \boldsymbol{0}$	[82,83]
$\mathcal{S}_{l_{0,\epsilon}}(\boldsymbol{x})$	$=$	$-\|\boldsymbol{x}\|_{0,\epsilon} = n - \#\{i, \|x_i\| > \epsilon\} \quad \boldsymbol{x} \neq \boldsymbol{0}$	[84]
$\mathcal{S}_{\tanh_{a,b}}(\boldsymbol{x})$	$=$	$-\sum_{i=1}^{n} \tanh\left((a \times \|x_i\|)^b\right) \quad \boldsymbol{x} \neq \boldsymbol{0}$	[82]
$\mathcal{S}_{\log}(\boldsymbol{x})$	$=$	$-\sum_{i=1}^{n} \log(1 + x_i^2) \quad \boldsymbol{x} \neq \boldsymbol{0}$	[81]
$\mathcal{S}_{l_1}(\boldsymbol{x})$	$=$	$-\sum_{i=1}^{n} \|x_i\| \quad \boldsymbol{x} \neq \boldsymbol{0}$	[82]
$\mathcal{S}_{l_p}(\boldsymbol{x})$	$=$	$-\left(\sum_{i=1}^{n} \|x_i\|^p\right)^{\frac{1}{p}} \quad \boldsymbol{x} \neq \boldsymbol{0}$	[85,86]
$\mathcal{S}_{\frac{l_2}{l_1}}(\boldsymbol{x})$	$=$	$\frac{\sqrt{\sum_{i=1}^{n} \|x_i\|^2}}{\sum_{i=1}^{n} \|x_i\|} \quad \boldsymbol{x} \neq \boldsymbol{0}$	[82]
κ^4	$=$	$\frac{\sum_{i=1}^{n} x_i^4}{\left(\sum_{i=1}^{n} x_i^2\right)^2} \quad \boldsymbol{x} \neq \boldsymbol{0}$	[87]
$\mathcal{S}_{u_\theta}$	$=$	$-\min_{i,j} \|x_i - x_j\|,$ $\frac{\|i-j\|}{n} \geq \theta,$ $x_1 \leq x_2 \leq \cdots \leq x_n, \ \boldsymbol{x} \neq \boldsymbol{0}.$	[82]
H_G	$=$	$-\sum_{i=1}^{n} \log x_i^2 \quad \boldsymbol{x} \neq \boldsymbol{0}$	[88,89]
H_S	$=$	$-\sum_{i=1}^{n} \tilde{x}_i \log \tilde{x}_i \quad \boldsymbol{x} \neq \boldsymbol{0}$ $\tilde{x}_i = {x_i}^2 / \|\boldsymbol{x}\|_2^2$	[88,89]
H'_S	$=$	$-\sum_{i=1}^{n} \|x_i\| \log x_i^2 \quad \boldsymbol{x} \neq \boldsymbol{0}$	[88,89]
$\mathcal{S}_{l^p_-}$	$=$	$\sum_{i, x_i \neq 0} x_i^p$ $p < 0, \ \boldsymbol{x} \neq \boldsymbol{0}$	[88,89]

Measure $\mathcal{S}_{l_0}$

We start from $\mathcal{S}_{l_0}$ which uses the norm-0, which indicates the number of non-zero elements. As a matter of fact, increasing the zero elements and having fewer non-zero elements results in more sparsity. However, this increment in the number of zero elements can be up to the level having at least one non-zero element. If all elements are zero, then the sparsity is dropped to its minimum level where all signal elements have an equal value. Therefore the number of zero elements is an increasing measure of sparsity:

$$\mathcal{S}_{l_0} = -\|\boldsymbol{x}\|_0, \qquad \boldsymbol{x} \neq \mathbf{0}. \tag{3.7}$$

$\|\boldsymbol{x}\|_0$ is interestingly comparable with Shannon entropy as follows [82]:

$$H(x) = -\int f(x)\log(f(x))dx, \tag{3.8}$$

where $f(x)$ is the density function of x. This measure is widely used in independent component analysis [37–42]. Although its formulation looks different from l_0, when it is applied to practical data, they become very similar.

Measure $\mathcal{S}_{l_{0,\epsilon}}$

The logic of this measure is the same as the former measure, using l_0, but instead of counting on zero elements, it considers the measure of being close to zero, ϵ. Indeed, the sparsity measure of $l_{0,\epsilon}$ takes into account the noise measure:

$$\mathcal{S}_{l_{0,\epsilon}}(\boldsymbol{x}) = -\#\{i, |x_i| > \epsilon\} \qquad \boldsymbol{x} \neq \mathbf{0}. \tag{3.9}$$

The quantity ϵ depends on the variance of noise but not the variance of $\boldsymbol{x}$. According to Ref. [84], determination of ϵ is an open problem. A solution to overcome this issue is using some approximate measure instead of $\mathcal{S}_{l_0^{\epsilon}}$. The two following measures concern two approximations.

Measure $\mathcal{S}_{\mathbf{tanh}_{a,b}}$

A solution to overcome the problem of the determination of ϵ is using an approximate measure, using tanh instead of $\mathcal{S}l_{0,\epsilon}$:

$$\mathcal{S}_{\tanh_{a,b}}(\boldsymbol{x}) = -\sum_{i=1}^{n} \tanh\big((a \times |x_i|)^b\big) \qquad \boldsymbol{x} \neq \mathbf{0}, \tag{3.10}$$

where a and b are positive constants and $b > 1$, e.g. $b = 2, 3$, or 4. The replacement of tanh instead of a threshold set of ϵ leads to the output being close to 1 when x_i has a big value.

Measure $\mathcal{S}_{\mathbf{log}}$

Another solution is using an approximate measure, using the logarithm function:

$$\mathcal{S}_{\log}(\boldsymbol{x}) = -\sum_{i=1}^{n} \log(1 + x_i^2) \qquad \boldsymbol{x} \neq \mathbf{0}. \tag{3.11}$$

Measure $\mathcal{S}_{l_1}$

Due to using norm-0 in Eq. (3.7), its maximization is an NP-hard problem which requires meta-heuristic evolutionary algorithms [90–101]. Also, the sensitivity of norm-0 to noise addition can lead to an incorrect presentation of a sparse signal as non-sparse. The norm-1 has been suggested as an effective replacement for norm-0 [82]:

$$\mathcal{S}_{l_1}(\boldsymbol{x}) = -\sum_{i=1}^{n} |x_i| \qquad \boldsymbol{x} \neq \mathbf{0}. \tag{3.12}$$

Measure $\mathcal{S}_{l_p}$

One very reason of using norm-1 instead of norm-0 is its potential for linear programming. In general, the replacement by norm-p has been suggested in the literature:

$$\mathcal{S}_{l_p}(\boldsymbol{x}) = -\Big(\sum_{i=1}^{n} |x_i|^p\Big)^{\frac{1}{p}} \qquad \boldsymbol{x} \neq \mathbf{0}. \tag{3.13}$$

Measure $\mathcal{S}_{\frac{l_2}{l_1}}$

Measure $\mathcal{S}_{\frac{l_2}{l_1}}$ is an interesting measure wherein the ratio of norm-2 to norm-1 is taken into account as the sparsity. We have

$$\mathcal{S}_{\frac{l_2}{l_1}}(\boldsymbol{x}) = \frac{\sqrt{\sum_{i=1}^{n} |x_i|^2}}{\sum_{i=1}^{n} |x_i|} \qquad \boldsymbol{x} \neq \mathbf{0}, \tag{3.14}$$

Here, we try to present a geometrical point of view of this measure via 3-dimensional Cartesian space.

The ratio of l_2 distance to l_1 distance from the center

The numerator is clearly the Cartesian vector size and the denominator is summation of the vector elements. Let us consider $\boldsymbol{x} = [x_1, x_2, x_3]^T$. Fig. 3.3A depicts $\boldsymbol{x}$ in the space $\mathbb{R}^3$. The depicted signal vector shown in Fig. 3.3A has low sparsity, as all its elements x_1, x_2, and x_3 have significant values. In this case as is clearly observable, $\|\boldsymbol{x}\|_2$ as the distance of $\boldsymbol{x}$ from the center is much smaller than $\|\boldsymbol{x}\|_1 = |x_1| + |x_2| + |x_3|$ as the summation of the absolute values of the elements. The measure $\mathcal{S}_{\frac{l_2}{l_1}}(\boldsymbol{x})$ is much smaller than one. Let us increase the sparsity of $\boldsymbol{x}$ by setting one of the elements close to zero. As depicted in Fig. 3.3B by setting x_3 close to zero, the relative difference between $\|\boldsymbol{x}\|_1$ and $\|\boldsymbol{x}\|_2$ is reduced, and $\|\boldsymbol{x}\|_2$ comes very close to $\|[x_1, x_2]^T\|_2$. Now, let us increase the sparsity of $\boldsymbol{x}$ close to its maximum by setting all elements except one close to zero. As depicted in Fig. 3.3C, x_2 keeps a significant value and the other elements are set close to zero. In this case, $\|\boldsymbol{x}\|_1$ and

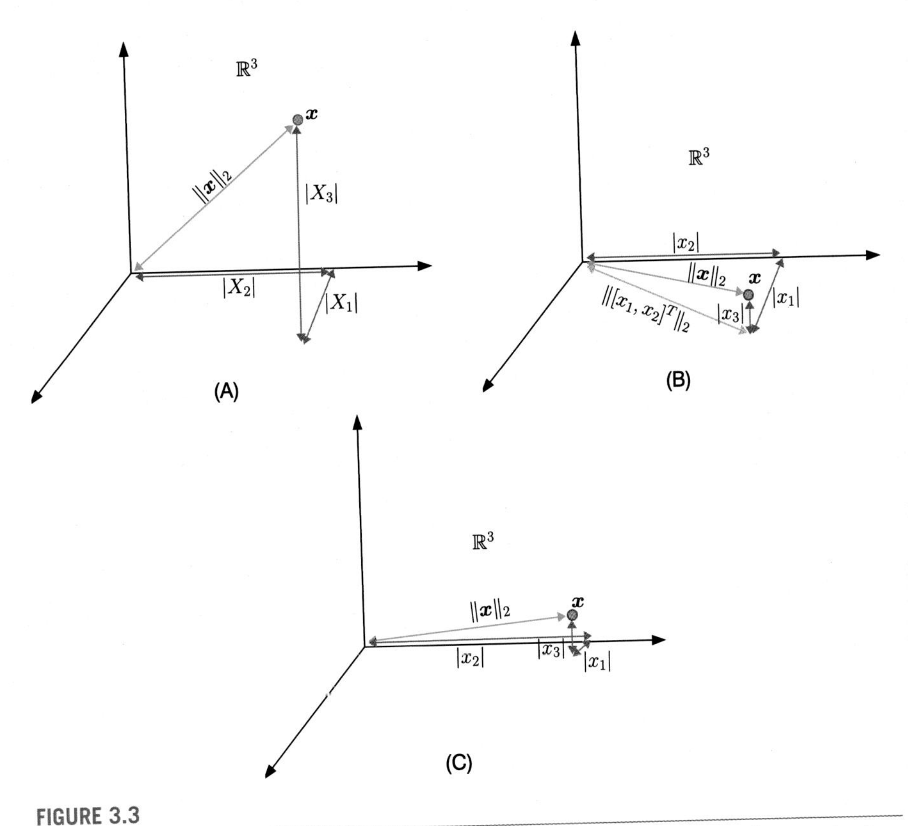

FIGURE 3.3

A geometrical view of the measure $S_{\frac{l_2}{l_1}}(\boldsymbol{x})$ as the ratio of the point distance from center to the summation of its coordinates when (A) it is non-sparse signal vector, (B) it is more sparse by setting one coordinate close to zero, and (C) it is very sparse by having just one coordinate far from zero, and the rest are very close to zero.

$\|\boldsymbol{x}\|_2$ are almost equal and $S_{\frac{l_2}{l_1}}(\boldsymbol{x})$ is very close to one. If we set all the elements to zero except one of them, it can be seen that this ratio will have exactly one as its maximum value.

Measure κ^4

The measure of kurtosis is another measure of sparsity:

$$\kappa^4 = \frac{\sum_{i=1}^{n} x_i^4}{\left(\sum_{i=1}^{n} x_i^2\right)^2} \qquad \boldsymbol{x} \neq \mathbf{0}, \tag{3.15}$$

On quick inspection, it can be observed that this measure is a modification of $S_{\frac{l_2}{l_1}}$ wherein, instead of each element of $\boldsymbol{x}$, its squared term has been used.

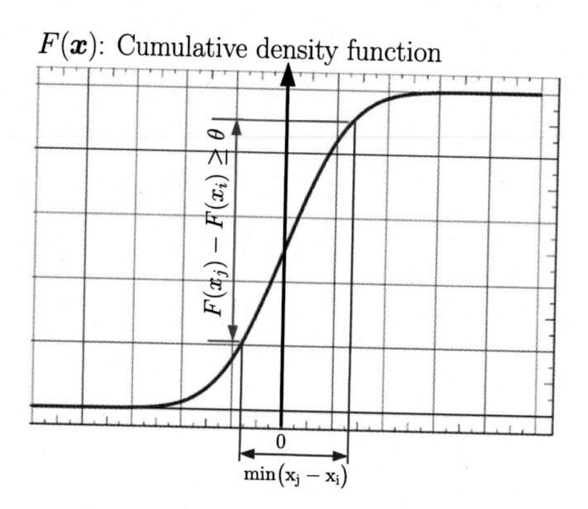

FIGURE 3.4

Graphical presentation of the u_θ measure of sparsity of $\boldsymbol{x}$ by its cumulative density function $F(\boldsymbol{x})$.

Measure $\mathcal{S}_{u_\theta}$

Another measure of sparseness [82] is $\mathcal{S}_{u_\theta}$. It is a measure of the minimum variation in data values around zero for the highest density of data values. Consider the $\boldsymbol{x}$ signal vector of elements x_i which are univariate random variables with cumulative density function (CDF) of $F(\boldsymbol{x})$. The highest density of data values is determined by a parameter θ, which is the probability mass of the data values x_i located in a minimum variation range around zero:

$$\begin{aligned} u_\theta = & \min_{x_i,\, x_j} (x_j - x_i) \\ \text{s.t. } & F(x_j) - F(x_i) \geq \theta,\ x_i \leq 0 \leq x_j. \end{aligned} \tag{3.16}$$

For example, if $\theta = 0.9$, then u_θ gives the minimum range around zero wherein 90 percent of the signal vector values are located. Fig. 3.4 graphically presents the measure u_θ of a univariate random variable x_i by its cumulative density function $F(\boldsymbol{x})$. To deploy u_θ in a measure of sparsity, Ref. [82] suggests an empirical approach to CDF as explained in the following. The suggested empirical CDF is by sorting all the elements x_i in ascending order. After sorting, the index of the data i resembles the density of the data, and the mass density in an interval from i to j is $\frac{|i-j|}{n}$ while n is the data size. Using the above-mentioned empirical CDF, the measure of sparsity of $\boldsymbol{x} = [x_1, x_2, \ldots, x_i, \ldots, x_n]^T$ based on u_θ is as follows:

$$\begin{aligned} \mathcal{S}_{u_\theta}(\boldsymbol{x}) = & -\min_{i,j}(x_j - x_i), \\ & \tfrac{|i-j|}{n} \geq \theta, \\ & x_1 \leq x_2 \leq \cdots \leq x_n, \\ & x_i \leq 0 \leq x_j,\ \boldsymbol{x} \neq \boldsymbol{0}. \end{aligned} \tag{3.17}$$

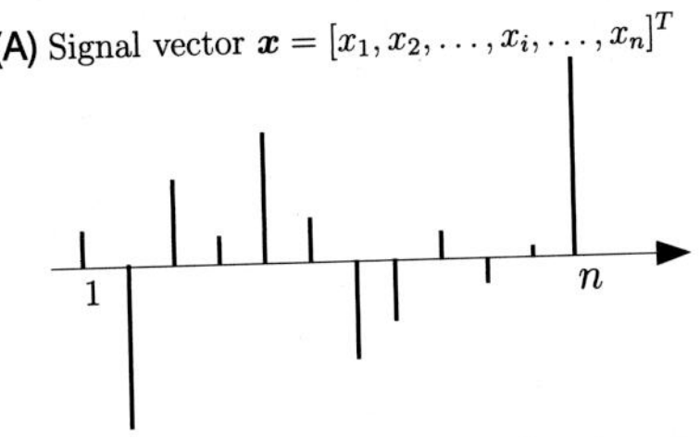

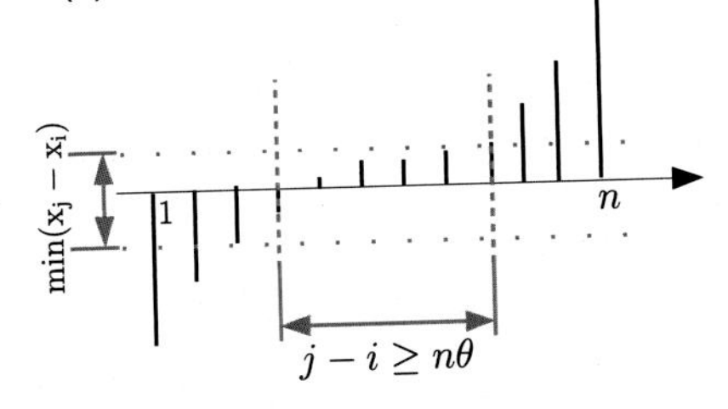

FIGURE 3.5

Graphical presentation of the S_{u_θ} measure of sparsity via an empirical CDF.

Note that the x_i are sorted in ascending order. For clarity, let us show the index of the sorted data i on the horizontal axis and the corresponding data value x_i on the vertical axis, as depicted by Fig. 3.5. S_{u_θ} is a measure of the minimum range of the data values variation for the range of data values containing the highest density as determined by the parameters θ.

The θ value is acquired depending on the application. It can be 0.5, 0.75, 0.9, and even 0.99.

Measures H_G, H_S, H'_S, and l^p_-

Diversity measurement through entropy is another approach to measuring the sparsity [88,89], which leads to H_S, H'_S, H_G, and l^p_- as follows. The Gaussian entropy measure of diversity is by the 'logarithm of the energy' as

$$H_G = -\sum_{i=1}^{n} \log x_i^2, \qquad \boldsymbol{x} \neq \boldsymbol{0}. \tag{3.18}$$

Reference [88] gives an effective algorithm for the maximization of the Gaussian entropy measure of diversity for sparsity maximization.

The diversity measure by the Shannon entropy H_S is as follows:

$$H_S = -\sum_{i=1}^{n} \tilde{x}_i \log \tilde{x}_i \qquad \boldsymbol{x} \neq \boldsymbol{0}, \tag{3.19}$$

where $\tilde{x}_i = |x_i^2|/||\boldsymbol{x}||_2^2$. There are modified versions of the Shannon entropy by replacement of $\tilde{x}_i$ by other terms [89] to improve the convexity of the measure. H_S' is a modified diversity measure of the Shannon entropy as $\tilde{x}_i := |x_i|$:

$$H_S' = -\sum_{i=1}^{n} |x_i| \log x_i^2 \qquad \boldsymbol{x} \neq \boldsymbol{0}. \tag{3.20}$$

References [88,89] extended the l_p measure for negative exponents by

$$\mathcal{S}_{l_-^p} = \sum_{i, x_i \neq 0} x_i^p, \tag{3.21}$$

where $p < 0$.

Hoyer measure

TheHoyer measure is a sparsity measure which by normalization by $\frac{l_2}{l_1}$ is obtained [102] as follows:

$$\text{Hoyer}(\boldsymbol{x}) = \frac{\left(\sqrt{n} - \frac{\|\boldsymbol{x}\|_1}{\|\boldsymbol{x}\|_2}\right)}{\left(\sqrt{n} - 1\right)} \qquad \boldsymbol{x} \neq \boldsymbol{0}, \tag{3.22}$$

where $\|.\|_1$ and $\|.\|_2$ are, respectively, l_1 and l_2 norms, and n is the signal vector length.

pq-mean measure

The pq-mean is a general measure of sparsity. In this chapter, we have modified it in terms of the increasing proportionality with sparsity. It is indeed the fraction of p-norm mean to q-norm mean, written as follows:

$$pq - \text{mean}(\boldsymbol{x}) = \frac{\left(\frac{1}{n}\sum_{i=1}^{n} x_i^p\right)^{\frac{1}{p}}}{\left(\frac{1}{n}\sum_{i=1}^{n} x_i^q\right)^{\frac{1}{q}}} \qquad p > q, \boldsymbol{x} \neq \boldsymbol{0}, \tag{3.23}$$

where n is the signal vector length. Reference [103] uses this measure with $q = 2$ and $0 < q < 1$. Also, the measures $S_{\frac{l_2}{l_1}}$, and κ^4 are special cases of the same measure, where in $\mathcal{S}_{\frac{l_2}{l_1}}$, p and q are acquired as 2 and 1, and in κ^4, they are taken as 4 and 2, respectively.

Measure $\mathcal{S}_{\text{Gini}}$

Finally, the Gini index as a measure of sparsity comes from economics as a measure of wealth inequality [80,104]. Indeed, the Gini index gives the *inequality in wealth* and as a sparsity measure it is given by Ref. [81]. The Gini index measure of sparsity

Table 3.3 Measures of sparsity in literature for a signal vector $\boldsymbol{x} = \{x_i\}$ (part 2).

$\text{Hoyer}(\boldsymbol{x})$	=	$(\sqrt{n} - \frac{\|\boldsymbol{x}\|_1}{\|\boldsymbol{x}\|_2})(\sqrt{n}-1)^{-1}$ $\quad \boldsymbol{x} \neq \boldsymbol{0}$,	[102]
$pq-\text{mean}(\boldsymbol{x})$	=	$\frac{(\frac{1}{n}\sum_{i=1}^{n} x_i^p)^{\frac{1}{p}}}{(\frac{1}{n}\sum_{i=1}^{n} x_i^q)^{\frac{1}{q}}}$ $\quad p > q,\ \boldsymbol{x} \neq \boldsymbol{0}$, n is the signal vector length. Special case in Ref. [103]: $p = 2$ and $0 < q < 1$.	[103]
$S_{\text{Gini}}(\boldsymbol{x})$	=	$1 - 2\frac{1}{n}\sum_{i=1}^{n} \frac{\tilde{\boldsymbol{x}}(i)}{\|\boldsymbol{x}\|_1}(\frac{n-i-\frac{1}{2}}{n})$, $\quad \boldsymbol{x} \neq \boldsymbol{0}$, $\tilde{\boldsymbol{x}}$ is obtained by sorting elements of $\boldsymbol{x}$ in ascending order.	[81]

for the signal vector $\boldsymbol{x}$ is as follows:

$$S_{\text{Gini}}(\boldsymbol{x}) = 1 - 2\frac{1}{n}\sum_{i=1}^{n} \frac{\tilde{\boldsymbol{x}}(i)}{||\boldsymbol{x}||_1}(\frac{n-i-\frac{1}{2}}{n}) \qquad \boldsymbol{x} \neq \boldsymbol{0}, \tag{3.24}$$

where $\tilde{\boldsymbol{x}}$ is obtained by sorting the elements of $\boldsymbol{x}$ in ascending order.

It is worth mentioning that according to the evaluative review by Hurley and Rickard [79] just the Gini index and pq-mean with $p \leq 1$, $q > 1$ satisfy all the six properties that a sparse measure should possess.

3.5 Summary

Compressive sensing is a recent advanced technique for below-Nyquist rate sampling of signals and images. It has a variety of applications decreasing the data volume, decreasing the required storage memory, increasing the data rate, reducing the sensory devices, increasing the economic aspects of imaging systems, reducing the data acquisition time, accelerating the imaging process, reducing the required dose of medical markers in imaging, reducing the power required for sensors, etc. As sparsity has the main role in compressive sensing's theoretical foundations, a measure of sparsity plays a highly important role in the recovery process of compressive sensing. Although k-sparsity is the main measure of sparsity used in compressive sensing theory, there are other sparsity measures too. This chapter briefly reviews the different measures of sparsity presented in the literature, while putting all the sparsity measure formulas in increasing relation with sparsity to be applicable for sparsity maximization. Furthermore, the chapter briefly lists six laws that a sparsity measure should obey.

References

[1] E.J. Candès, et al., Compressive sampling, in: Proceedings of the International Congress of Mathematicians, Madrid, Spain, vol. 3, 2006, pp. 1433–1452.
[2] E.J. Candès, J. Romberg, T. Tao, Robust uncertainty principles: exact signal reconstruction from highly incomplete frequency information, IEEE Transactions on Information Theory 52 (2) (2006) 489–509.
[3] E.J. Candes, J.K. Romberg, T. Tao, Stable signal recovery from incomplete and inaccurate measurements, Communications on Pure and Applied Mathematics: A Journal Issued by the Courant Institute of Mathematical Sciences 59 (8) (2006) 1207–1223.
[4] E.J. Candes, T. Tao, Near-optimal signal recovery from random projections: universal encoding strategies?, IEEE Transactions on Information Theory 52 (12) (2006) 5406–5425.
[5] K. Melo, M. Khosravy, C. Duque, N. Dey, Chirp code deterministic compressive sensing: analysis on power signal, in: 4th International Conference on Information Technology and Intelligent Transportation Systems (ITITS 2019), 2019.
[6] R. Chartrand, Fast algorithms for nonconvex compressive sensing: MRI reconstruction from very few data, in: 2009 IEEE International Symposium on Biomedical Imaging: From Nano to Macro, IEEE, 2009, pp. 262–265.
[7] R. Otazo, E. Candes, D.K. Sodickson, Low-rank plus sparse matrix decomposition for accelerated dynamic MRI with separation of background and dynamic components, Magnetic Resonance in Medicine 73 (3) (2015) 1125–1136.
[8] K. Khare, C.J. Hardy, K.F. King, P.A. Turski, L. Marinelli, Accelerated MR imaging using compressive sensing with no free parameters, Magnetic Resonance in Medicine 68 (5) (2012) 1450–1457.
[9] G.-H. Chen, J. Tang, S. Leng, Prior image constrained compressed sensing (PICCS): a method to accurately reconstruct dynamic CT images from highly undersampled projection data sets, Medical Physics 35 (2) (2008) 660–663.
[10] R. Baraniuk, P. Steeghs, Compressive radar imaging, in: 2007 IEEE Radar Conference, IEEE, 2007, pp. 128–133.
[11] P.D. Olcott, G. Chinn, C.S. Levin, Compressed sensing for the multiplexing of pet detectors, in: 2011 IEEE Nuclear Science Symposium Conference Record, IEEE, 2011, pp. 3224–3226.
[12] Z. Chen, A. Basarab, D. Kouamé, Compressive deconvolution in medical ultrasound imaging, IEEE Transactions on Medical Imaging 35 (3) (2015) 728–737.
[13] V. Cevher, A. Sankaranarayanan, M.F. Duarte, D. Reddy, R.G. Baraniuk, R. Chellappa, Compressive sensing for background subtraction, in: European Conference on Computer Vision, Springer, 2008, pp. 155–168.
[14] X. Li, S.-Y. Qin, Efficient fusion for infrared and visible images based on compressive sensing principle, IET Image Processing 5 (2) (2011) 141–147.
[15] C.A. Metzler, A. Maleki, R.G. Baraniuk, From denoising to compressed sensing, IEEE Transactions on Information Theory 62 (9) (2016) 5117–5144.
[16] J.M. Kim, O.K. Lee, J.C. Ye, Compressive music: revisiting the link between compressive sensing and array signal processing, IEEE Transactions on Information Theory 58 (1) (2012) 278–301.
[17] M.F. Duarte, R.G. Baraniuk, Spectral compressive sensing, Applied and Computational Harmonic Analysis 35 (1) (2013) 111–129.
[18] Y.L. Polo, Y. Wang, A. Pandharipande, G. Leus, Compressive wide-band spectrum sensing, in: 2009 IEEE International Conference on Acoustics, Speech and Signal Processing, IEEE, 2009, pp. 2337–2340.
[19] B. Wang, L. Dai, Y. Zhang, T. Mir, J. Li, Dynamic compressive sensing-based multi-user detection for uplink grant-free NOMA, IEEE Communications Letters 20 (11) (2016) 2320–2323.
[20] C.R. Berger, Z. Wang, J. Huang, S. Zhou, Application of compressive sensing to sparse channel estimation, IEEE Communications Magazine 48 (11) (2010) 164–174.
[21] S.K. Sharma, E. Lagunas, S. Chatzinotas, B. Ottersten, Application of compressive sensing in cognitive radio communications: a survey, IEEE Communications Surveys and Tutorials 18 (3) (2016) 1838–1860.

[22] L. Dai, Z. Wang, Z. Yang, Compressive sensing based time domain synchronous ofdm transmission for vehicular communications, IEEE Journal on Selected Areas in Communications 31 (9) (2013) 460–469.

[23] S. Li, L. Da Xu, X. Wang, Compressed sensing signal and data acquisition in wireless sensor networks and Internet of things, IEEE Transactions on Industrial Informatics 9 (4) (2012) 2177–2186.

[24] Z. Li, B. Chang, S. Wang, A. Liu, F. Zeng, G. Luo, Dynamic compressive wide-band spectrum sensing based on channel energy reconstruction in cognitive Internet of things, IEEE Transactions on Industrial Informatics 14 (6) (2018) 2598–2607.

[25] T. Arildsen, C.S. Oxvig, P.S. Pedersen, J. Østergaard, T. Larsen, Reconstruction algorithms in undersampled AFM imaging, IEEE Journal of Selected Topics in Signal Processing 10 (1) (2015) 31–46.

[26] P. Ye, J.L. Paredes, G.R. Arce, Y. Wu, C. Chen, D.W. Prather, Compressive confocal microscopy, in: 2009 IEEE International Conference on Acoustics, Speech and Signal Processing, IEEE, 2009, pp. 429–432.

[27] T.W. Cabral, M. Khosravy, F.M. Dias, H.L.M. Monteiro, M.A.A. Lima, L.R.M. Silva, R. Naji, C.A. Duque, Compressive sensing in medical signal processing and imaging systems, in: Sensors for Health Monitoring, Elsevier, 2019, pp. 69–92.

[28] L.F. Polania, R.E. Carrillo, M. Blanco-Velasco, K.E. Barner, Compressed sensing based method for ECG compression, in: 2011 IEEE International Conference on Acoustics, Speech and Signal Processing (ICASSP), IEEE, 2011, pp. 761–764.

[29] H. Mamaghanian, N. Khaled, D. Atienza, P. Vandergheynst, Compressed sensing for real-time energy-efficient ECG compression on wireless body sensor nodes, IEEE Transactions on Biomedical Engineering 58 (9) (2011) 2456–2466.

[30] A.M. Dixon, E.G. Allstot, D. Gangopadhyay, D.J. Allstot, Compressed sensing system considerations for ECG and EMG wireless biosensors, IEEE Transactions on Biomedical Circuits and Systems 6 (2) (2012) 156–166.

[31] H. Ding, H. Sun, K.-m. Hou, Abnormal ECG signal detection based on compressed sampling in wearable ECG sensor, in: 2011 International Conference on Wireless Communications and Signal Processing (WCSP), IEEE, 2011, pp. 1–5.

[32] Z. Zhang, T.-P. Jung, S. Makeig, B.D. Rao, Compressed sensing for energy-efficient wireless telemonitoring of noninvasive fetal ECG via block sparse Bayesian learning, IEEE Transactions on Biomedical Engineering 60 (2) (2012) 300–309.

[33] A.M. Abdulghani, A.J. Casson, E. Rodriguez-Villegas, Compressive sensing scalp EEG signals: implementations and practical performance, Medical & Biological Engineering & Computing 50 (11) (2012) 1137–1145.

[34] C.E. Gutierrez, P.M.R. Alsharif, M. Khosravy, P.K. Yamashita, P.H. Miyagi, R. Villa, Main large data set features detection by a linear predictor model, AIP Conference Proceedings 1618 (2014) 733–737.

[35] M.H. Sedaaghi, R. Daj, M. Khosravi, Mediated morphological filters, in: 2001 International Conference on Image Processing, Proceedings, vol. 3, IEEE, 2001, pp. 692–695.

[36] M. Khosravy, N. Gupta, N. Marina, I.K. Sethi, M.R. Asharif, Morphological filters: an inspiration from natural geometrical erosion and dilation, in: Nature-Inspired Computing and Optimization, Springer, Cham, 2017, pp. 349–379.

[37] M. Khosravy, M.R. Asharif, K. Yamashita, A PDF-matched short-term linear predictability approach to blind source separation, International Journal of Innovative Computing, Information & Control 5 (11) (2009) 3677–3690.

[38] M. Khosravy, M.R. Asharif, K. Yamashita, A theoretical discussion on the foundation of Stone's blind source separation, Signal, Image and Video Processing 5 (3) (2011) 379–388.

[39] M. Khosravy, M.R. Asharif, K. Yamashita, A probabilistic short-length linear predictability approach to blind source separation, in: 23rd International Technical Conference on Circuits/Systems, Computers and Communications (ITC-CSCC 2008), Yamaguchi, Japan, ITC-CSCC, 2008, pp. 381–384.

[40] M. Khosravy, M.R. Alsharif, K. Yamashita, A PDF-matched modification to Stone's measure of predictability for blind source separation, in: International Symposium on Neural Networks, Springer, Berlin, Heidelberg, 2009, pp. 219–222.
[41] M. Khosravy, Blind source separation and its application to speech, image and MIMO-OFDM communication systems, PhD thesis, University of the Ryukyus, Japan, 2010.
[42] M. Khosravy, M. Gupta, M. Marina, M.R. Asharif, F. Asharif, I. Sethi, Blind components processing a novel approach to array signal processing: a research orientation, in: 2015 International Conference on Intelligent Informatics and Biomedical Sciences, ICIIBMS, 2015, pp. 20–26.
[43] A.A. Picorone, T.R. de Oliveira, R. Sampaio-Neto, M. Khosravy, M.V. Ribeiro, Channel characterization of low voltage electric power distribution networks for PLC applications based on measurement campaign, International Journal of Electrical Power & Energy Systems 116 (105) (2020) 554.
[44] N. Gupta, M. Khosravy, N. Patel, T. Senjyu, A bi-level evolutionary optimization for coordinated transmission expansion planning, IEEE Access 6 (2018) 48455–48477.
[45] N. Gupta, M. Khosravy, K. Saurav, I.K. Sethi, N. Marina, Value assessment method for expansion planning of generators and transmission networks: a non-iterative approach, Electrical Engineering 100 (3) (2018) 1405–1420.
[46] M. Khosravy, M.R. Asharif, M.H. Sedaaghi, Medical image noise suppression: using mediated morphology, IEICE, Technical Report 107 (461) (2008) 265–270.
[47] N. Dey, A.S. Ashour, A.S. Ashour, A. Singh, Digital analysis of microscopic images in medicine, Journal of Advanced Microscopy Research 10 (1) (2015) 1–13.
[48] M. Khosravy, M.R. Asharif, M.H. Sedaaghi, Medical image noise suppression using mediated morphology, in: IEICE Tech. Rep., IEICE, 2008, pp. 265–270.
[49] A.S. Ashour, S. Samanta, N. Dey, N. Kausar, W.B. Abdessalemkaraa, A.E. Hassanien, Computed tomography image enhancement using cuckoo search: a log transform based approach, Journal of Signal and Information Processing 6 (03) (2015) 244.
[50] M. Khosravy, N. Gupta, N. Marina, I. Sethi, M. Asharif, Brain action inspired morphological image enhancement, in: Nature-Inspired Computing and Optimization, Springer, Cham, 2017, pp. 381–407.
[51] E. Santos, M. Khosravy, M.A. Lima, A.S. Cerqueira, C.A. Duque, ESPRIT associated with filter bank for power-line harmonics, sub-harmonics and inter-harmonics parameters estimation, International Journal of Electrical Power & Energy Systems 118 (105) (2020) 731.
[52] E. Santos, M. Khosravy, M.A. Lima, A.S. Cerqueira, C.A. Duque, A. Yona, High accuracy power quality evaluation under a colored noisy condition by filter bank ESPRIT, Electronics 8 (11) (2019) 1259.
[53] M. Khosravy, N. Gupta, N. Marina, I. Sethi, M. Asharifa, Perceptual adaptation of image based on Chevreul–Mach bands visual phenomenon, IEEE Signal Processing Letters 24 (5) (2017) 594–598.
[54] C.E. Gutierrez, M.R. Alsharif, K. Yamashita, M. Khosravy, A tweets mining approach to detection of critical events characteristics using random forest, International Journal of Next-Generation Computing 5 (2) (2014) 167–176.
[55] N. Kausar, S. Palaniappan, B.B. Samir, A. Abdullah, N. Dey, Systematic analysis of applied data mining based optimization algorithms in clinical attribute extraction and classification for diagnosis of cardiac patients, in: Applications of Intelligent Optimization in Biology and Medicine, Springer, 2016, pp. 217–231.
[56] M.H. Sedaaghi, M. Khosravi, Morphological ECG signal preprocessing with more efficient baseline drift removal, in: 7th IASTED International Conference, ASC, 2003, pp. 205–209.
[57] N. Dey, S. Samanta, X.-S. Yang, A. Das, S.S. Chaudhuri, Optimisation of scaling factors in electrocardiogram signal watermarking using cuckoo search, International Journal of Bio-Inspired Computation 5 (5) (2013) 315–326.
[58] M. Khosravy, M.R. Asharif, M.H. Sedaaghi, Morphological adult and fetal ECG preprocessing: employing mediated morphology, IEICE Technical Report, IEICE 107 (2008) 363–369.
[59] N. Dey, A.S. Ashour, F. Shi, S.J. Fong, R.S. Sherratt, Developing residential wireless sensor networks for ECG healthcare monitoring, IEEE Transactions on Consumer Electronics 63 (4) (2017) 442–449.

[60] M. Khosravi, M.H. Sedaaghi, Impulsive noise suppression of electrocardiogram signals with mediated morphological filters, in: 11th Iranian Conference on Biomedical Engineering, ICBME, 2004, pp. 207–212.
[61] N. Dey, S. Mukhopadhyay, A. Das, S.S. Chaudhuri, Analysis of P-QRS-T components modified by blind watermarking technique within the electrocardiogram signal for authentication in wireless telecardiology using DWT, International Journal of Image, Graphics and Signal Processing 4 (7) (2012) 33.
[62] M. Khosravy, N. Punkoska, F. Asharif, M.R. Asharif, Acoustic OFDM data embedding by reversible Walsh–Hadamard transform, AIP Conference Proceedings 1618 (2014) 720–723.
[63] M. Khosravy, M.R. Alsharif, M. Khosravi, K. Yamashita, An optimum pre-filter for ICA based multi-input multi-output OFDM system, in: 2010 2nd International Conference on Education Technology and Computer, vol. 5, IEEE, 2010, pp. V5–129.
[64] F. Asharif, S. Tamaki, M.R. Alsharif, H. Ryu, Performance improvement of constant modulus algorithm blind equalizer for 16 QAM modulation, International Journal on Innovative Computing, Information and Control 7 (4) (2013) 1377–1384.
[65] M. Khosravy, M.R. Alsharif, K. Yamashita, An efficient ICA based approach to multiuser detection in MIMO OFDM systems, in: Multi-Carrier Systems & Solutions 2009, Springer, 2009, pp. 47–56.
[66] M. Khosravy, M.R. Alsharif, B. Guo, H. Lin, K. Yamashita, A robust and precise solution to permutation indeterminacy and complex scaling ambiguity in BSS-based blind MIMO-OFDM receiver, in: International Conference on Independent Component Analysis and Signal Separation, Springer, 2009, pp. 670–677.
[67] M. Khosravy, A blind ICA based receiver with efficient multiuser detection for multi-input multi-output OFDM systems, in: The 8th International Conference on Applications and Principles of Information Science (APIS), Okinawa, Japan, 2009, 2009, pp. 311–314.
[68] M. Khosravy, S. Kakazu, M.R. Alsharif, K. Yamashita, Multiuser data separation for short message service using ICA, SIP, IEICE Technical Report 109 (435) (2010) 113–117.
[69] S. Gupta, M. Khosravy, N. Gupta, H. Darbari, N. Patel, Hydraulic system onboard monitoring and fault diagnostic in agricultural machine, Brazilian Archives of Biology and Technology 62 (2019).
[70] S. Gupta, M. Khosravy, N. Gupta, H. Darbari, In-field failure assessment of tractor hydraulic system operation via pseudospectrum of acoustic measurements, Turkish Journal of Electrical Engineering & Computer Sciences 27 (4) (2019) 2718–2729.
[71] M. Foth, R. Schroeter, J. Ti, Opportunities of public transport experience enhancements with mobile services and urban screens, International Journal of Ambient Computing and Intelligence (IJACI) 5 (1) (2013) 1–18.
[72] G.V. Kale, V.H. Patil, A study of vision based human motion recognition and analysis, International Journal of Ambient Computing and Intelligence (IJACI) 7 (2) (2016) 75–92.
[73] M. Yamin, A.A.A. Sen, Improving privacy and security of user data in location based services, International Journal of Ambient Computing and Intelligence (IJACI) 9 (1) (2018) 19–42.
[74] B. Alenljung, J. Lindblom, R. Andreasson, T. Ziemke, User experience in social human–robot interaction, in: Rapid Automation: Concepts, Methodologies, Tools, and Applications, IGI Global, 2019, pp. 1468–1490.
[75] M. Baumgarten, M.D. Mulvenna, N. Rooney, J. Reid, Keyword-based sentiment mining using Twitter, International Journal of Ambient Computing and Intelligence (IJACI) 5 (2) (2013) 56–69.
[76] P. Sosnin, Precedent-oriented approach to conceptually experimental activity in designing the software intensive systems, International Journal of Ambient Computing and Intelligence (IJACI) 7 (1) (2016) 69–93.
[77] S. Hemalatha, S.M. Anouncia, Unsupervised segmentation of remote sensing images using FD based texture analysis model and ISODATA, International Journal of Ambient Computing and Intelligence (IJACI) 8 (3) (2017) 58–75.
[78] C. Castelfranchi, G. Pezzulo, L. Tummolini, Behavioral implicit communication (BIC): communicating with smart environments, International Journal of Ambient Computing and Intelligence (IJACI) 2 (1) (2010) 1–12.

[79] N. Hurley, S. Rickard, Comparing measures of sparsity, IEEE Transactions on Information Theory 55 (10) (2009) 4723–4741.
[80] H. Dalton, The measurement of the inequality of incomes, The Economic Journal 30 (119) (1920) 348–361.
[81] S. Rickard, M. Fallon, The Gini index of speech, in: Proceedings of the 38th Conference on Information Science and Systems (CISS'04), 2004.
[82] J. Karvanen, A. Cichocki, Measuring sparseness of noisy signals, in: 4th International Symposium on Independent Component Analysis and Blind Signal Separation, Citeseer, 2003, pp. 125–130.
[83] D.L. Donoho, M. Elad, V.N. Temlyakov, Stable recovery of sparse overcomplete representations in the presence of noise, IEEE Transactions on Information Theory 52 (1) (2005) 6–18.
[84] G. Rath, C. Guillemot, J.-J. Fuchs, Sparse approximations for joint source-channel coding, in: 2008 IEEE 10th Workshop on Multimedia Signal Processing, IEEE, 2008, pp. 481–485.
[85] D.L. Donoho, Sparse components of images and optimal atomic decompositions, Constructive Approximation 17 (3) (2001) 353–382.
[86] N. Saito, B.M. Larson, B. Bénichou, Sparsity vs. statistical independence from a best-basis viewpoint, in: Wavelet Applications in Signal and Image Processing VIII, vol. 4119, International Society for Optics and Photonics, 2000, pp. 474–486.
[87] B.A. Olshausen, D.J. Field, Sparse coding of sensory inputs, Current Opinion in Neurobiology 14 (4) (2004) 481–487.
[88] B.D. Rao, K. Kreutz-Delgado, An affine scaling methodology for best basis selection, IEEE Transactions on Signal Processing 47 (1) (1999) 187–200.
[89] K. Kreutz-Delgado, B.D. Rao, Measures and algorithms for best basis selection, in: Proceedings of the 1998 IEEE International Conference on Acoustics, Speech and Signal Processing, ICASSP'98 (Cat. No. 98CH36181), vol. 3, IEEE, 1998, pp. 1881–1884.
[90] J. Kaliannan, A. Baskaran, N. Dey, A.S. Ashour, M. Khosravy, R. Kumar, ACO based control strategy in interconnected thermal power system for regulation of frequency with HAE and UPFC uni, in: International Conference on Data Science and Application (ICDSA-2019), in: LNNS, Springer, 2019.
[91] M. Khosravy, N. Gupta, N. Patel, T. Senjyu, Frontier Applications of Nature Inspired Computation, Springer, 2020.
[92] N. Gupta, M. Khosravy, O.P. Mahela, N. Patel, Plant biology-inspired genetic algorithm: superior efficiency to firefly optimizer, in: Applications of Firefly Algorithm and Its Variants, Springer, 2020, pp. 193–219.
[93] C. Moraes, E. De Oliveira, M. Khosravy, L. Oliveira, L. Honório, M. Pinto, A hybrid bat-inspired algorithm for power transmission expansion planning on a practical Brazilian network, in: Applied Nature-Inspired Computing: Algorithms and Case Studies, Springer, 2020, pp. 71–95.
[94] M. Khosravy, N. Gupta, N. Patel, T. Senjyu, C.A. Duque, Particle swarm optimization of morphological filters for electrocardiogram baseline drift estimation, in: Applied Nature-Inspired Computing: Algorithms and Case Studies, Springer, 2020, pp. 1–21.
[95] G. Singh, N. Gupta, M. Khosravy, New crossover operators for real coded genetic algorithm (RCGA), in: 2015 International Conference on Intelligent Informatics and Biomedical Sciences (ICIIBMS), IEEE, 2015, pp. 135–140.
[96] N. Gupta, N. Patel, B.N. Tiwari, M. Khosravy, Genetic algorithm based on enhanced selection and log-scaled mutation technique, in: Proceedings of the Future Technologies Conference, Springer, 2018, pp. 730–748.
[97] N. Gupta, M. Khosravy, N. Patel, I. Sethi, Evolutionary optimization based on biological evolution in plants, Procedia Computer Science 126 (2018) 146–155.
[98] N. Gupta, M. Khosravy, N. Patel, O. Mahela, G. Varshney, Plants genetics inspired evolutionary optimization: a descriptive tutorial, in: Frontier Applications of Nature Inspired Computation, Springer, 2020.
[99] M. Khosravy, N. Gupta, N. Patel, O. Mahela, G. Varshney, Tracing the points in search space in plants biology genetics algorithm optimization, in: Frontier Applications of Nature Inspired Computation, Springer, 2020.

[100] N. Gupta, M. Khosravy, N. Patel, S. Gupta, G. Varshney, Artificial neural network trained by plant genetics-inspired optimizer, in: Frontier Applications of Nature Inspired Computation, Springer, 2020.
[101] N. Gupta, M. Khosravy, N. Patel, S. Gupta, G. Varshney, Evolutionary artificial neural networks: comparative study on state of the art optimizers, in: Frontier Applications of Nature Inspired Computation, Springer, 2020.
[102] P.O. Hoyer, Non-negative matrix factorization with sparseness constraints, Journal of Machine Learning Research 5 (Nov) (2004) 1457–1469.
[103] A.M. Bronstein, M.M. Bronstein, M. Zibulevsky, Y.Y. Zeevi, Sparse ICA for blind separation of transmitted and reflected images, International Journal of Imaging Systems and Technology 15 (1) (2005) 84–91.
[104] C. Gini, Measurement of inequality of incomes, The Economic Journal 31 (121) (1921) 124–126.

CHAPTER

Compressive sensing in practice and potential advancements

4

Ayan Banerjee, Sandeep K.S. Gupta
Impact Lab, Arizona State University, Tempe, AZ, United States

4.1 Introduction

Technological advances in wearables and mobile sensing is creating opportunities for mobile computing applications to gather context data from humans and their surroundings at a remarkable volume. This will generate data at an alarming rate leading to the "data deluge" problem that has serious implications in system design issues such as real time computation, communication and storage. The fundamental sampling theory requires data to be sampled at twice the maximum desired resolution for accurate reconstruction of information. However, in practice in most domains data is *sparse*, i.e., relevant information in data is contained in much fewer elements than the number of collected samples. For example, images in the visual spectrum may be represented using much fewer data points in the Discrete Wavelet Transform (DWT) or Discrete Cosine Transform (DCT) linear transform domain without loss of information. The theory of Compressive Sensing (CS) [1–6] was developed from this in-the-field observation. CS takes advantage of this compressed representation of information in data and recovers original data with much fewer samples than the fundamental Nyquist rate.

As shown in Fig. 4.1 the CS theory argues that a k-sparse n-dimensional signal can be accurately recovered with $m << n$ samples collected using a $m \times n$ sensing matrix $\boldsymbol{\Theta}$ instead of an $n \times n$ identity matrix, where n is the Nyquist mandated number of samples. The measurement vector for the sensor $y = \boldsymbol{\Theta} x$ has much fewer samples m than n; hence the sensor that uses CS is much simpler than a sensor sensing at Nyquist rate. This leads to significant improvements in power consumption, form factor, and storage capacity. Since the original signal x is k-sparse, the measurement signal y is basically a linear combination of at most k non-zero elements of x. If the position of the k non-zero elements were known, then finding the sensing matrix and then subsequently recovering x would be a trivial problem. However, since the positions are not known, we have to search every possible $m \times k$ submatrices of $\boldsymbol{\Theta}$ to find the best k sparse approximation of x. This problem is NP hard and has combinatorial complexity. However, with random sampling, it can be proven that any $m \times k$ submatrix of $\boldsymbol{\Theta}$ has full rank and hence can encode every k sparse vector in the

Compressive Sensing in Healthcare. https://doi.org/10.1016/B978-0-12-821247-9.00009-3

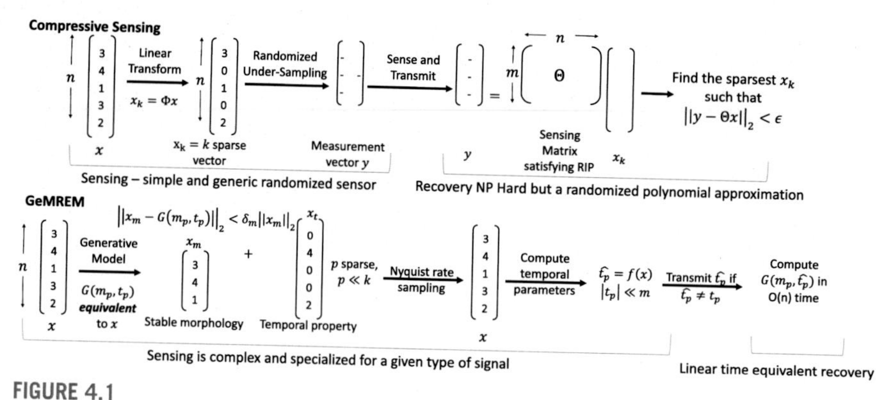

FIGURE 4.1

Fundamental methods of compressive sensing and GeMREM.

measurement vector. Further, the recovery of x from y can be performed in polynomial time. Compressive sensing thus allows us to recover relevant information from signals using simpler sensors, by exploiting the sparsity of signals in linear domain, and solving an NP-hard recovery problem in polynomial time by using randomized sampling. Significant research in this area has led to several important inventions including the single pixel camera [7], simple and efficient sensors in photonics (EARS project NSF ECCS #1443936), contactless physiological sensing (EasySense project NSF IIS #1231754), robust sensing for cognitive radio, ultra wideband wireless communication (NSF CCF # 0910765) and sensors for medical imaging, hyper-spectral imaging, and bio-informatics (NSF ECCS # 1418710) and in medical imaging and sensing [8].

In many practical scenarios signals may not be sparse or even compressible (i.e., approximated using a sparse signal) in any linear transformed domain. Such signals often have a unique shape property that is repeated at random intervals. In this document, we will refer to these signals as *quasi-periodic* signals. For example, signals from the human body such as electro-cardiogram (ECG), or brain signals, or chaotic signals are not sufficiently sparse in any linear transform domain for CS to be beneficial [9,10]. The complex shape characteristics of these signals can be modeled using non-linear generative models. The models encode significant information from the raw data, which are used for decision making. As such for many applications the raw data is less significant than these models. Furthermore, in many practical applications, two data sets or signals with high sample-by-sample error may contain the same relevant information, and are equivalent. Hence, essentially a generative model is a *nonlinear sparsifying operator*, which sparsifies a signal with respect to relevant information content. This is typically true in case of physiological sensing, where the diagnosis of a disease is performed based on ranges of metrics derived from the signal and not the exact signal value. Two signals differing sample-by-sample may not be diagnostically different. Thus, instead of recovering with the goal of sample-by-sample accuracy, it is often sufficient to recover an *equivalent* signal, which has the same *relevant information* as the original signal sampled at the Nyquist rate.

Compressive sensing employs random sensing following a pre-specified distribution. Such random sensing through time is based on the fact that the signal is sparse in some domain that is a linear transformation of the time domain. A side effect of such sampling is that during reconstruction the associated error is distributed uniformly over recovered time domain samples. This effect may not be appropriate for many applications. For example, in satellite imagery of distant planetary objects, an image may contain Region of Interests (ROI). These are particular areas of the image that are more interesting than other regions. For example, an image may contain different landforms such as sand dunes, or water channels. However, the signatures of these landforms are only seen in a very limited part of the image. In case of uniform reconstruction error distribution, it is highly probable that errors are distributed on the signatures of these landforms. Hence, usage of compressive sensing on such images can result in potential loss of interesting landform features.

With this observation in-the-field, other methods can be explored that may attempt to move away from the random sampling methodology in compressive sensing. The need for better performance in practice is *selective random sampling*. This can be achieved in several ways however, in this chapter we will discuss two broad techniques.

a) Model-based sampling. Non-linear models of signals can be used, such that the model parameters act as a low dimensional representation of the signal. Compressive sensing in the model parameter domain can then lead to even more savings without compromising the accuracy on relevant information. However, the problem is that model-based compression techniques most typically reduce communication load and require sensing at Nyquist rate to operate in an error free manner. For example, Generative Model-based Resource Efficient Modeling (GeMREM), is an effective model-based communication mechanism for continuous monitoring of physiological signals in a body sensor network. It considers that the relevant information in a quasi-periodic signal x can be encoded in a generative model $G(m_p, t_p)$ having two parameters: a) morphology m_p, which characterizes the unique shape property of the signal, and b) temporal properties t_p, which characterizes the underlying temporal variations. The generative model provides a signal that is equivalent to x with respect to certain relevant information. Further, m_p is relatively stable and does not change over time. A sensor samples the signal x at Nyquist rate and compares it with a generative model. Thus, the sensor for GeMREM is much more complex than a sensor sensing at Nyquist rate or executing CS. The sensor then transmits only the changes in t_p to the recovery algorithm. Using t_p and previously learned m_p the equivalent signal $G(m_p, t_p)$ can be recovered in linear time, $O(n)$. With such *non-linear sparsifying operators* and the idea of *equivalent reconstruction*, GeMREM was found to achieve 40 and 300 times communication compression for ECG and photoplethysmogram signals, respectively. Such compressions are much greater than the in-the-field sensing compression performance of CS algorithms.

b) Region of interest aware sensing. In this method, the raw data can be divided into ROIs, and the sparsity parameters of each of these regions can be varied based on the reconstruction accuracy needs. For example in case of satellite imagery, the entire image can be divided into blocks of ROIs. The ROIs can then be sampled at different rates and individually reconstructed to recover the entire image.

In this chapter, we will focus on: a) the practical manifestation of CS efficiency and accuracy on various medical monitoring examples, b) the practical implications of the performance bounds suggested by CS, and c) advances that can potentially improve the efficacy and accuracy of CS by leaps and bounds. In summary, CS exploits sparsity in linear domain, reduces sensing frequency, and recovers randomly undersampled signals by minimizing sample by sample error. Although the sensor device is very simple and has significant sensing power savings, the recovery is complex, and the data compression ratio is much less than what can be exploited. However, if we can potentially move away from random sampling and utilize the idea of selective random sampling, then we can potentially improve the reconstruction efficacy of compressive sensing. This chapter will show the challenges in compressive sensing and its potential advancements.

4.2 Compressive sensing theory

Considering a compressible signal $x \in \mathbb{R}^n$ has a sparse representation $\alpha \in \mathbb{R}^n$ on a certain orthogonal basis $\boldsymbol{\Phi} \in \mathbb{R}^{n \times n}$, given as $x = \boldsymbol{\Phi}\alpha$, α is a k-sparse vector that contains only k non-zero elements. Given a random sensing matrix $\boldsymbol{\Theta} \in \mathbb{R}^{m \times n}$, the measurement signal $y \in \mathbb{R}^m$ can be expressed as

$$y = \boldsymbol{\Theta}x = \boldsymbol{\Theta}\boldsymbol{\Phi}\alpha. \tag{4.1}$$

It is proven that as long as the sensing matrix $\boldsymbol{\Theta}$ satisfies the Restricted Isometry Property (RIP) of order $2k$, the signal information α can be well preserved by the random encoding (RE) scheme in Eq. (4.1) [11,12]. This holds true even if the sensing matrix $\boldsymbol{\Theta}$ is a under-determined matrix ($m < n$), which represents a dimensionality reduction from $\mathbb{R}^n$ to $\mathbb{R}^m$. Therefore, the random sample y is a compressed representation of the signal coefficient α that is encoded by $\boldsymbol{A} = \boldsymbol{\Theta}\boldsymbol{\Phi}$. It is also proven that a sensing matrix randomly generated from Bernoulli distributions can easily satisfy the RIP of order $2k$ for $m = O(k \cdot \log(n/k))$ [11,12]. This indicates that the undersampling ratio (m/n) or compression ratio (n/m) achievable by compressive sensing is proportional to the signal sparsity k on the chosen basis $\boldsymbol{\Phi}$.

To recover the sparse coefficient α (the original signal x can then be reconstructed as $x = \boldsymbol{\Phi}\alpha$), we need to solve the under-determined linear equation in (4.1). It has been proven that by utilizing the sparsity condition as prior knowledge, α can be exactly recovered by solving the sparse approximation (SA) problem (ℓ_0 pseudo-norm minimization) defined as

$$\min_{\alpha} \|\alpha\|_0, \quad \text{subject to } \|y - \boldsymbol{\Theta}\boldsymbol{\Phi}\alpha\|_2^2 \leq \zeta, \tag{4.2}$$

where ϵ is an error tolerance term to enhance the reconstruction robustness considering that the random sample is contaminated by an additive noise. The SA problem is to find the sparsest vector out of the solution space of $\|y - \boldsymbol{\Theta}\boldsymbol{\Phi}\alpha\|_2^2 \leq \zeta$. Although the SA problem in (4.2) is NP hard, its solution can be robustly estimated by either heuristic methods such as orthogonal matching pursuit (OMP) [13,14] or linear programming through the relaxation to a basis pursuit (ℓ_1-norm minimization) problem [15–17].

4.3 Example compressive sensing implementations

The theory of compressive sensing can be applied to several examples. In this chapter, we will consider applications from two broad application domains which are inherently very different: a) physiological signal monitoring, and b) planetary image acquisition from satellites and spacecrafts.

4.3.1 Compressive sensing in physiological signal monitoring

Considering the example of cardiac monitoring we can see an application of compressive sensing. ECG morphology has two important characteristics: a) a sharp R peak, and b) four blunt and low amplitude waves. CS for ECG monitoring will need two types of orthonormal basis vectors: a) wavelet basis for capturing the sharp R peak, and b) discrete cosine basis for capturing the blunt waves. The method of using multiple basis can be realized by concatenating the basis matrices of each of the transforms to form a $n \times 2m$ matrix A. For each of the transforms we take the m greatest coefficients. The optimization problem (4.2) is then solved using the homotopy driven ℓ_1-penalized least squares regression method (LASSO homotopy) [18]. In an execution, m samples of ECG are taken at random in a second and is passed through to the base station. For every $2m$ samples, the base station uses the dual basis and estimates $2n$ ECG samples using the LASSO homotopy algorithm. It then stitches $2n$ consecutive samples to form the ECG time series.

In the field application results

In this subsection, we consider three medical monitoring databases and apply compressive sensing to evaluate its efficacy and reconstruction accuracy in practice.

Databases:

- **ICU database of ECG signals:** ECG data is collected using Holter monitors and GeMREM-based sensors for 25 patients in the ICU of St Luke's hospital for a period of 24 hours. Holter monitors gave raw data at 125 Hz, while GeMREM sensors represented the signal using the generative models.
- **Home use database:** This is an ongoing study funded by NIH, where ECG data is collected using Holter monitors and GeMREM sensors from 100 patients in free living condition at home. Data is collected for a duration of 24 hours.

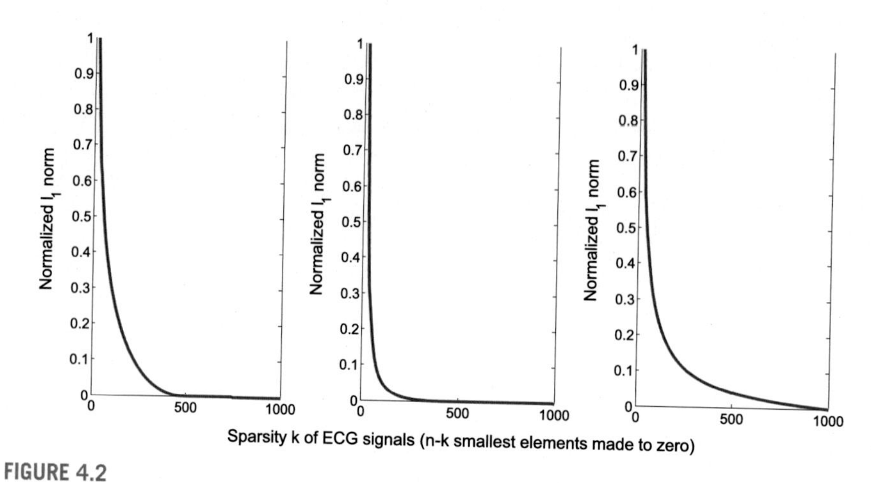

FIGURE 4.2

CS for ECG: error bound of recovery algorithm vs. sparsity assumption of k.

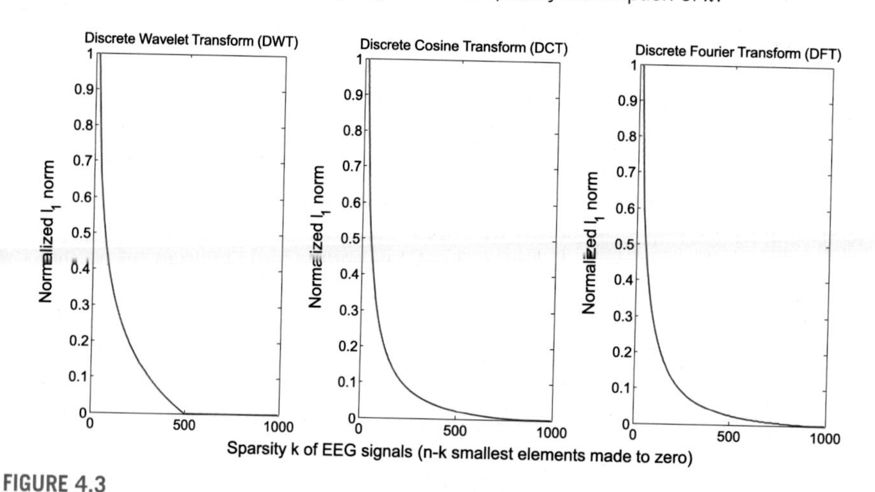

FIGURE 4.3

CS for EEG: error bound of recovery algorithm vs. sparsity assumption of k.

- **IMPACT Photoplethysmogram (PPG) data:** GeMREM was also designed for compressive data transmission from PPG sensors. A total of 30 minute worth of PPG data from 10 volunteers were collected.
- **Electroencephalogram data:** This is an online database of 106 subjects from Physionet [60]. Each subject has 108 channels of EEG data obtain through medical grade sensors. Each EEG signal is of duration of approximately 30 min.

A large scale experiment was carried out in the IMPACT Lab with ECG and electro-encephalograms (EEG) signals from 50 subjects to determine the upper bound of recovery error with respect to varying sparsity. The signals are not sparse in time domain and hence three linear transformed domains were considered Discrete Wavelet Transforms (DWT), Discrete Fourier Transform (DFT), and Discrete Cosine Transform (DCT). Even in transformed domains the signals were not exactly sparse or compressible. Hence, the best k sparse approximation was used and the upper bound of the error is plotted against different k. Figs. 4.2 and 4.3 show that when k is reduced the error increases exponentially. Hence, in practice we can only get a compression of the order of 1.5 or 2.

4.3.2 Compressive sensing in THEMIS imaging

In this example, we consider the images acquired by the Thermal Emission Imaging System (THEMIS) on the Mars Odyssey spacecraft. There are three types of landforms that are interesting: a) sand dunes, b) valley, and c) polar ice caps.

Data set. The data sets used in this example are the Infrared (IR) (band 9) and Visible (band 3) images collected by THEMIS. The images considered are the Brightness Temperature Records (BTR) of a scene in Kelvin. In the visible range, Apparent Brightness Record (ABR) images are considered that describe the calibrated scene radiance. THEMIS images of dunes and valley network at different spatial locations are considered.

The sparsity domain selected is the discrete cosine transform (DCT) and wavelet domain. For an image x of size N, with dimensions $R \times C$, only $S << N$ samples in the sparse domains are required for reconstructing the image. Traditionally compression algorithm operate after sensing, where an image x is first sensed at the Nyquist rate and then compressed to a smaller size. However, such techniques involve acquiring all N samples or observations, which is followed by compression. Compressive sensing provides an alternative methodology to directly sense lesser number of samples to obtain a compressed representation y, having length M, where $S < M << N$. CS allows sub-Nyquist sampling of an image, which is followed by a reconstruction step to obtain the original scene. This can be performed with a sensing matrix and the measurement step is expressed as

$$y = \boldsymbol{\Phi} x, \tag{4.3}$$

where $\boldsymbol{\Phi}$ is a random sensing matrix of dimension $M \times N$. The M samples of x are acquired randomly following different types of distributions. This can lead to a reduction of memory requirement, however, due to a reconstruction step the computational complexity is increases which also leads to an increase in image acquisition time. It is important to ensure that the sensing matrix $\boldsymbol{\Phi}$ satisfies the Restricted Isometry Property (RIP) to achieve accurate reconstruction by solving an under-determined set of equations.

There are several methods to solve the reconstruction problem for images. One potential way is to use l_1-minimization. The problem can be formulated as

$$\hat{x} = \min\{\|x\|_1 : y = \boldsymbol{\Phi} x\}, \tag{4.4}$$

where $\hat{x}$ is the reconstructed image. Different types of solutions can be used for this optimization including homotopy [1–4] or other basis pursuit strategies [17,19].

In-the-field application results

In a small scale experiment we have applied CS techniques on 15 THEMIS images and the results are shown in [20]. The reconstruction error was measured using common metrics such as Peak Signal-to-Noise Ratio (PSNR), which is expressed as

$$\text{PSNR} = 10\log_{10}\frac{Q^2}{MSE}, \tag{4.5}$$

where Q is the dynamic range of pixel values, which is 255 for 8-bit images and the Mean Squared Error (MSE) is given as

$$\text{MSE} = \frac{1}{N}\sum_{i=1}^{N}(x(i) - \hat{x}(i))^2. \tag{4.6}$$

PSNR is calculated in decibels and represents the information loss from $\hat{x}$ due to compression, where higher PSNR values represent better image quality.

For different types of landforms the PSNR obtained is less than 30 dB for a compression ratio of 2. If the compression ratio is increased PSNR drastically reduces. For example, for a compression ratio of 10 the PSNR drops to below 20 dB.

4.4 Review of CS literature

In recent years, several attempts have been made to adapt CS techniques for preserving shape features as summarized in Table 4.1. The basic method of random sampling in time domain, then recovery by solving an over determined set of linear equations (as shown in Fig. 4.1) has been applied with different forms of linear transforms for many physiological signals [1–6]. In some efforts related to ECG signal recovery [21,22], researchers have considered the ECG signal to be bock sparse, i.e., the signal can be divided into blocks, many of which are zero while of some of them have non-zero values. The data samples in the blocks that are non-zero are modeled as parameterized gaussian random variables. These parameters are then estimated using Type II maximum likelihood minimization. While these techniques increase the recovery accuracy, the compression ratio achieved is at most 2.5. The second approach, proposes the notion of model-based compressive sensing and considers that in the transformed domain, the non-zero items of the sparse representation fits into a given mathematical model [23,24]. Such an assumption reduces the complexity of recovery algorithms since they now have to only search values that fit a given model. A recent effort [25] shows potential for utilizing such models to increase recovery accuracy at higher compression ratios for ECG (≈10). The work uses a personalized basis learned from an hour long trace of ECG signal to recover compressively sensed data.

Table 4.1 Overview of compressive sensing research.

CS techniques	Signal properties	Sparsity domain	Recovery algorithm	Error metric
Random sampling [1–6,26–31]	Signals sparse in any linear domain, transform coefficients are generic	Linear domain such as DWT, DCT, gabor	Convex relaxation, greedy algorithm, Bergman iterative algorithm,	ℓ_2 minimization
Block sparsity exploitation [21,22,32–35]	Signals are non-zero in blocks such as ECG	time domain	Type II maximum likelihood minimization, non-convex optimization	ℓ_2 minimization
Model-based CS [23,24,36–39]	Coefficients in the transformed domain follow specific models	Linear transforms	Iterative greedy algorithms, personalized learned basis	ℓ_2 minimization
Nonlinear basis pursuit [40]	Signal is sparse in a non-linear domain	polynomial sparsity models	Iterative greedy using numerical solvers	ℓ_2 minimization
Lossy CS [32,41], where only required information is preserved	Signal not sparse in any domain	Linear transforms	Combinatorial sublinear algorithms, or linear optimization	emotion classification accuracy
Manifold embedding [42–44]	High dimensional data	Linear	Application specific	ℓ_2 minimization

The third approach assumes that the signal is sparse in a non-linear domain. They propose a non-linear basis pursuit algorithm that approximates the non-linear transformation using Taylor series expansion and uses numerical solutions to recover the original signal. The fourth approach reduces the dimensionality of the signal using manifold embedding. All these efforts improve the recovery accuracy but do not necessarily increase compression ratio. Recently there have been made efforts to model signals using high dimensional continuous curves or manifolds which can be effectively represented as a non-linear continuous function of fewer parameters.

Nearly all these techniques consider recovery of signal for reducing sample by sample error. A recently concluded study at the IMPACT Lab [9] show that although ECG signals recovered using compressive sensing had low mean square error, they had significant distortions in the morphology of the signal beats, which can potentially result in wrong diagnosis [45]. The idea of diagnostic equivalence is not limited to physiological signals, but also applies to signals in other domain. For example, in case of audio classification, relative distribution of power spectrum in the frequency transform, and not the exact power values, is important for classification of different musical instruments.

4.4.1 Practical manifestations of theoretical bounds

Linear sparsity and recovery error bound. In CS, a k-sparse signal $\alpha \in \mathbb{R}^n$, can be recovered by strategically sensing only $m << n$ samples $y \in \mathbb{R}^m$ using a sensing matrix $A \in \mathbb{R}^n \times \mathbb{R}^m$, such that $y = A\alpha$ and using a recovery algorithm $\Delta(y)$. The aim is to minimize the l_2 norm of the error between recovered signal and original signal, $\|\Delta(A\alpha) - \alpha\|_2$. For signals which are not sparse, the fundamental theorem of compressive sensing [46] gives an upper bound on the error of recovery as

$$\|\Delta(A\alpha) - \alpha\|_2 \leq C\frac{\sigma_k(\alpha)_1}{\sqrt{k}}, \tag{4.7}$$

where $\sigma_k(\alpha)_1$ is the best k-sparse approximation of α with respect to the l_1 norm and C is a constant.

Manifold-based CS and shape preservation. In compressive sensing theory, the idea of manifold embedding can potentially be used to preserve shape features of high dimensional data such as facial image or brain scan [42–44]. An S-manifold $\mathbb{M}$ is a topological space in $\mathbb{R}^n$ such that there is a homeomorphism f from the manifold space to an S-dimensional space θ, and $f(\theta)$ can exactly represent the manifold. In the specific case of physiological signals such as ECG, the signal can be considered as a manifold that is represented as a continuous function of the diagnostic features. There has been significant research in this area which shows that a sensing matrix populated with Gaussian i.i.d. parameters and multiplied by a Radmacher matrix can provide a stable embedding of a manifold in the compressed measurements. This can be a possible method for preserving diagnostic features in the recovery of the signal. A fundamental bound on the number of measurements required to embed an

S-dimensional manifold M having condition number τ, geodesic covering regularity R, and volume V, with probability $1-\rho$ and conditioning ϵ, is given by

$$m \geq O\left(\frac{S \log(nVR\tau^{-1}\epsilon^{-1}) \log(1/\rho)}{\epsilon^2}\right). \tag{4.8}$$

Experience in the field: For ECG and EEG signals from nearly 50 subjects we observed that for using manifold-based shape preserving compressive sensing the minimum number of measurements required to accurately recover 1000 samples of the signal is nearly 400 for ECG and 675 for EEG. Hence, using manifold in the field we are promised a compression of no better than 2.5.

4.5 Advancements in compressive sensing

4.5.1 Personalized basis

The solution to the optimization problem in Eq. (4.15) depends on the functions $G(.)$ and $f(.)$. Typically $G(.)$ is highly non-linear while $f(.)$ is not smooth. Hence the non-linear basis pursuit algorithms such as [40] may not be applicable since they assume that the non-linear sparsity model can be represented fairly accurately using Taylor series expansion. A potential advancement can be to extend some techniques applied to CS for linear sparsity models for application on non-linear shape preserving sparsity models.

An idea is to use personalized bases learned from short period of signal recordings in reconstruction of compressively sensed signals. When used on ECG signals, it has been shown to achieve nearly 10 times compression for linear sparsity models. Dictionary learning of signals can successfully capture individual features and characteristics from the signal vector space to enhance signal sparsity and thereby to significantly improve under-sampling ratio when combined with compressive sensing. The primary hypothesis of personalized dictionary learning is that the sparsest domain may not represent unique morphological parameters of signals.

Considering the collected signal recording of each individual are formed into a matrix $\boldsymbol{X} \in \mathbb{R}^{n\times L}$ by evenly partitioning the samples into L segments each with a window size of n, a set of n-dimensional training data is obtained as $x_1, x_2, \ldots, x_L \in \mathbb{R}^n$. The problem of training a basis $\boldsymbol{D} \in \mathbb{R}^{n\times P}$ (also called dictionary) to sparsify the training data $\boldsymbol{X}$ (based on a linear model) is to find P elementary waveforms (also called atom) $d_1, d_2, \ldots, d_p \in \mathbb{R}^n$ and k-sparse coefficient vectors $\gamma_1, \gamma_2, \ldots, \gamma_L \in \mathbb{S}_k^P$ such that each x_i can be well approximated by a linear combination of at most k elementary waveforms as $\boldsymbol{D}\gamma_i$. To exploit the redundancy for reducing the signal sparsity k, the dictionary size P must be chosen to be greater than the signal dimension n. Therefore, the dictionary learning problem [47] can be formulated as

$$\min_{\boldsymbol{D},\, \gamma_1,\, \gamma_2,\, \ldots\, \gamma_L} \sum_{i=1}^{L} \|x_i - \boldsymbol{D}\gamma_i\|_2^2 + \lambda \|\gamma_i\|_1, \tag{4.9}$$

where λ is a regularization parameter to balance the trade-off between signal sparsity k and approximation error $\|\boldsymbol{x}_i - \mathbf{D}\gamma_i\|_2$. Increasing the value of λ forces the ℓ_1 penalty term to be small, leading to sparser representations. On the other hand, reducing the value of λ plays less importance on sparsity, thereby forcing the approximation error to be small. The problem in (4.9) is often solved by fixing one of $\mathbf{D}$ and $\boldsymbol{\Gamma} \in \mathbb{R}^{P \times L}$ to update the other in alternating and iterative fashions. Well-known algorithms include K-SVD [48], geometric multi-resolution analysis (GMRA) [19, 49,50], and Online Dictionary Learning (ODL) [51].

For a linear model of the signals we have $\boldsymbol{X} \approx \boldsymbol{D\Gamma}$, and the trained dictionary $\boldsymbol{D}$ can be seamlessly incorporated into the compressive sensing framework in (4.1) and (4.2). A direct approach is to replace the pre-determined orthogonal basis $\boldsymbol{\Phi}$ used in (4.2) with the trained basis $\mathbf{D}$. By using the trained basis in signal recovery, the individual features learned from the training data will be directly applied as additional prior knowledge to enhance reconstruction. Moreover, according to the condition on m, we can expect that the under-sampling ratio will be also improved given that a high signal sparsity can be enforced by applying a small value of λ.

A new possibility in this approach is that dictionary learning can be performed in a certain feature (instead of time) domain to offload the sparsity constraint required by compressive sensing. This characteristic is especially useful in case of reconstructing signals that are equivalent in some features (diagnostically equivalent). For such cases, the training data $\boldsymbol{X} \in \mathbb{R}^{n \times L}$ can be transformed into a certain feature domain $\boldsymbol{\Phi} \in \mathbb{R}^{n \times n}$ to get the coefficients $\boldsymbol{B} \in \mathbb{R}^{n \times L}$, expressed as $\boldsymbol{B} = \boldsymbol{\Phi X}$. Then a feature basis $\boldsymbol{D}$ can be trained through the dictionary learning problem in (4.9) by using the coefficients $b_1, b_2, \ldots, b_n$ as the training input. Note that the trained feature basis $\boldsymbol{D}$ now sparsifies the signal b_i on the feature domain $\boldsymbol{\Phi}$ rather than the signal x_i itself, expressed as $\boldsymbol{B} \approx \boldsymbol{D\Gamma}$. Given $\boldsymbol{X} = \boldsymbol{\Phi}^{-1}\boldsymbol{B}$, the signal can be modeled as $\boldsymbol{X} = \boldsymbol{\Phi}^{-1}\boldsymbol{D\Gamma}$, where $\boldsymbol{\Gamma}$ are composed of k-sparse vectors. With the new sparsity model, a sparse representation γ_i on the trained basis can be recovered by solving the basis pursuit problem:

$$\min_{\gamma_i} \|\gamma_i\|_1, \quad \text{subject to} \quad \|y - \boldsymbol{\Theta\Phi}^{-1}\boldsymbol{D}\gamma_i\|_2^2 \leq \epsilon, \tag{4.10}$$

and the signal x_i can be reconstructed as $x_i = \boldsymbol{\Phi}^{-1}\boldsymbol{D}\gamma_i$. Note that the sparsity constraint is now enforced by the dictionary learning upon $\boldsymbol{B}$, the transformation domain $\boldsymbol{\Phi}$ is no longer required to be the sparse domain ($\boldsymbol{B}$ can be dense) for successful recovery. This offloads the sparsity constraint and allows the feature exploration on an arbitrary domain to be seamless combined with compressive sensing. Therefore, this approach opens up new possibilities of performance enhancement when there exist non-sparse features that can better represent the vector space structure of signals.

Challenges

C1: Reconstruction of a signal that is sparse in non-linear domain Personalized basis basically represents a signal by the summation of a number of signal templates. Essentially templates can be used as generative models as shown in [52] on PPG generative model. For a given mathematical form of generative model, the atoms d_i

can be short term simulation of the generative model with various parameters. This approach can be further explored in order to employ the personalized basis theory for shape preserving reconstruction of signals. There are several pertinent research questions associated with this approach: 1) How much sparsity improvement can be achieved by dictionary learning as compared to pre-determined basis? 2) What is the trade-off between signal dimension, dictionary size P, and signal sparsity k? 3) What is the lower bound of signal sparsity k in theory and practice? 4) Is it necessary and feasible to perform a partial dictionary update?

To best answer these questions, the sparsity characteristic of physiological signals such as ECG, EMG, or EEG on pre-deterministic orthogonal bases, including DCT, various types of DWT, as well as joint basis that are composed of multiple orthogonal bases, have to be explored. For orthogonal bases, orthogonal transformations can be applied to the collected data and the sparsity level for achieving different reconstruction quality can be analyzed. The reconstruction quality can be measured by both signal fidelity metrics such as Reconstructed Signal-to-Noise Ratio (RSNR) and diagnosis equivalence metrics provided by our collaborators from Mayo clinic. To analyze the ECG sparsity on joint basis, OMP [13] to find the sparse approximation of the collected data can be used.

Subsequently, dictionary learning in (4.9) on the collected data with different experiment settings of n, P, and λ can also be employed to study the trade-off between signal sparsity, approximation error, and dictionary size. Note that a unique basis will be trained for each individual ECG recording for sparsity analysis. To understand the effectiveness of dictionary learning, the sparsity results can be compared to that of pre-determined bases. The experiments will be conducted using K-SVD, GMRA, and ODL algorithm, respectively. The training time, sparsity performance, and the convergence rate over different length of the training data can be compared to find the most efficient algorithm.

One advantage of the ODL algorithm is that it does not rely on the matrix factorization of the entire training data (unlike K-SVD and GMRA). This property presents an opportunity to perform partial dictionary update or dictionary extension by processing additional training data only. The solution will have to be developed for a closed-loop framework that tightly bridges signal reconstruction in compressive sensing with dictionary learning, where dictionary learning provides trained bases to reconstruction for performance enhancement, and reconstruction provides sparsity feedback to inform dictionary update.

C2: Performance evaluation of dictionary learning approach The challenges are in answering the following questions: 1) Can the trade-off between sample size m and reconstruction error in compressive sensing be projected by that of sparsity k and approximation error in dictionary learning? 2) How much improvement on undersampling ratio can be achieved by using personalized basis? 3) How important the individual variability is to compressive sensing performance? 4) What is the impact of dictionary noise?

To answer such questions, compressive sensing and reconstruction have to be performed on the collected physiological signals using both pre-determined and personalized bases to investigate in the trade-off between under-sampling ratio m/n and the reconstruction quality (using both signal fidelity and diagnosis equivalence metrics). To understand importance of individual variability, the compressive sensing performance can be evaluated by applying personalized basis learned from different individual recordings to each other and analyzing the performance degradation. Since dictionary learning finds the most elementary waveforms to compactly represent morphology rich signals, the learned dictionary will contain noise if the trained data does so. It is also essential to evaluate the impact of dictionary noise on compressive sensing performance by comparing the results between dictionary learning upon raw and de-noised training data. To accomplish this, filtering and noise shaping techniques [53] [54] [55] have to be explored for the removal of baseline wandering, high-frequency interferences, and random noise from the training data.

C3: Exploration of diagnostic feature domain dictionary learning The research questions to answer in this task include: can performing dictionary learning in diagnostic feature domain further improve compressive sensing performance?

With close relation to the RIP condition [11] [12], the distortion δ of signal representations in a feature domain $\boldsymbol{\Phi}$ can be measured as $\delta = \frac{\|\boldsymbol{\Phi} x\|_2^2 - \|x\|_2^2}{\|x\|_2^2}$. Convex approaches such as the Numax method [56] [57] provide viable solutions to pursue the feature domain that minimizes the distortion of domain transformation given the training data.

4.5.2 Non-linear model-based compressive sensing

The core idea is to show that: a) the generative model and the requirement of equivalence reconstruction are non-linear sparsifying operators, b) there exists a sensing matrix that can ensure that an equivalent generative model-based representation of the signal can be derived from under-sampled signals, and c) a theoretical estimate of the compression ratio obtained for generative model-based CS.

A shape preserving generative model requires morphological m_p and temporal parameters t_p from the quasi-periodic signal x. The algorithms used for computing m_p and t_p is highly non-linear and includes functions such as peak detection and curve fitting optimizations. Since x is quasi-periodic, then there exists a sub sequence x_m of x such that x_m has the unique shape property that gets repeated at random periods. If $G(m_p, t_p)$ is the generative model of the signal x, then according to definition of a generative model we have

$$\|x_m - G(m_p, t_p)\|_2 \leq \delta_m \|x_m\|_2, \tag{4.11}$$

where δ_m is the morphology tolerance level of the generative model.

The morphological properties of the signal x can be suppressed to derive a signal x_t, which has only the temporal properties of x, i.e., $f(x_t) = f(x)$. This can be done with the help of digital filters that can suppress certain frequencies that are affected by the specific shape criteria. Typically, such filters can be realized as a linear transformation on x, $x_t = Dx$. For the generative model, the temporal parameter is again within certain tolerance bound of the parameters derived from x as shown in Eq. (4.12). We have

$$\|f(x_t) - f(G(m_p, t_p))\|_2 \leq \delta_t \|f(x_t)\|_2, \tag{4.12}$$

Note that if x is k-sparse, then x_t is p-sparse, where $p << k$. This is because in x_t is obtained by suppressing the morphological properties in x. Hence the process of extracting the temporal parameters from x_t is a non-linear sparsifying operation. We can now define equivalence of a generative model and signal x as follows.

Definition 1. A generative model $G(m_p, t_p)$ is defined as equivalent to x if it satisfies the conditions in Eqs. (4.11) and (4.12).

We attempt to derive approximate morphological and temporal parameters $\{\tilde{m_p}, \tilde{t_p}\}$ from the randomly sampled time series y, such that $G(m_p, t_p)$ is equivalent to x. In order to have such a mechanism, we need a sensing matrix that satisfies the Restricted Isometry Property (RIP) for the generative model $G(m_p, t_p)$. Since x is a k-sparse signal, then there exists a sensing matrix $\boldsymbol{\Theta}$ that satisfies RIP for x for some conditioning parameter ϵ. We prove Theorem 4.1.

Theorem 4.1. *If $\boldsymbol{\Theta} \in \mathbb{R}^m \times \mathbb{R}^n$ satisfies RIP of order $2k$ for x then $\boldsymbol{\Theta}$ also satisfies RIP for $G(m_p, t_p)$.*

Proof. If x is k-sparse then there exists $\boldsymbol{\Theta} \in \mathbb{R}^m \times \mathbb{R}^n$, for $m = O(k \log(\frac{n}{k}))$ such that

$$(1 - \epsilon)\|x\|_2 \leq \|\boldsymbol{\Theta} x\|_2 \leq (1 + \epsilon)\|x\|_2. \tag{4.13}$$

By combining Eqs. (4.13) and (4.11) and doing some algebraic manipulations we get

$$(1 - \epsilon - \delta\|\boldsymbol{\Theta}\|_2)\|x\|_2 \leq \|\boldsymbol{\Theta} G(m_p, t_p)\| \leq (1 + \epsilon + \delta\|\boldsymbol{\Theta}\|_2)\|x\|_2. \tag{4.14}$$

Hence $\boldsymbol{\Theta}$ satisfies RIP of order at least $2k$ for $G(m_p, t_p)$ with $\epsilon' = \epsilon + \delta\|\boldsymbol{\Theta}\|_2$. □

Thus, the generative model $G(m_p, t_p)$ provides at least k-sparsity in a signal x. Note that generative model has a higher value of conditioning parameter. This means that it classifies a larger neighborhood of x as equivalent. Hence, it provides more sparsity.

With respect to the temporal parameters we prove the following theorem.

Theorem 4.2. *There exists a sensing matrix $\boldsymbol{\Theta}' \in \mathbb{R}^s \times \mathbb{R}^n$, $s < m$ that can be used to recover a signal with a temporal property equivalent to that of x.*

Proof. Let us consider a recovery algorithm Δ that solves the problem in Eq. (4.2). Hence $\|x - \Delta(\boldsymbol{\Theta} x)\|_2 \leq \zeta$, where $\zeta > 0$ is a small number. Hence, $\|f(x) - f(\Delta(\boldsymbol{\Theta} x))\|_2 \leq \delta_t \|x\|_2$. Let us consider that there exists a digital filter D such that $x_t = Dx$ and $f(x_t) = f(x)$. Thus, using the same $\boldsymbol{\Theta}$ we can recover x_t from m samples. Note that x_t is much sparser than x. There exists $\boldsymbol{\Theta}' \in \mathbb{R}^s \times \mathbb{R}^n$, $s < m$ such that $\|x_t - \Delta(\boldsymbol{\Theta}' x_t)\|_2 \leq \zeta$. Thus, using just s samples one can recover the temporal properties of x. □

Theorem 4.2 shows that the temporal property extractor along with the equivalence condition is also a non-linear sparsifying operation.

Given a matrix with RIP property for the generative model as shown in Theorem 4.1 and a temporal parameter estimation mechanism as guided by Theorem 4.1, the following problem formulation must have a feasible solution:

$$\begin{aligned} &\text{find } \tilde{m_p}, \tilde{t_p} \text{ to minimize } \|G(\tilde{m_p}, \tilde{t_p})\|_1 \\ &\text{such that } y = \boldsymbol{\Theta} G(\tilde{m_p}, \tilde{t_p})\rangle \end{aligned} \tag{4.15}$$

Ideally an optimal solution to this problem should allow us to recover a diagnostically equivalent signal.

Challenges

Several important questions need to be answered. Theorem 4.1 gives us one example sensing matrix using which we can generate a diagnostically equivalent generative model. However, it only exploits k order of sparsity. Using the new value of ϵ', how can we derive a sensing matrix that exploits more sparsity as promised by the generative model?

Theorem 4.2 shows that with much fewer samples it is possible to recover the temporal parameters. However, it does not say how.

C1: Data collection and exploration of generative models of EMG and EEG Ongoing studies in several universities are collecting different physiological signals such as EMG ECG and EEG. Creating a universal database with such signals will help researchers explore new possibilities for compressive sensing. For example in IMPACT Lab EMG data is collected using Myo wristbands, while the EEG data is collected using Neurosky headsets. Currently, there exists brain relaxation data of 15 volunteers and EMG data of 10 volunteers. In addition online databases such as MIT BIH, BrainNet, and PhysioNet EMG databases will be used to develop and test generative models of EMG and EEG signals.

Since EEG data is chaotic, modeling attempts were made under controlled environment where the user was asked to perform a targeted mental task. An initial attempt has been successful in regenerating brain data for right or left hand movement decision by an individual. The frequency domain features of an EEG signal was used to train an Adaptive Neural Fuzzy Inference System (ANFIS). The ANFIS coefficient was then used to predict frequency features of subsequent brain signals for the given mental task. An inverse transform gives a time series that gets the same classification of right or left hand movement as the original signal. The current implementation has a classification accuracy of 91%. This technique needs to be refined to provide more accurate generative models of EEG and explore similar models for EMG data.

C2: Developing a generative model-based compressive sensing Generative models of signals can be of different complexities. For example for ECG the generative model consists of the morphological parameters, which can be only extracted using non-linear model learning techniques. However, if we ignore the morphological parameters, then the signal with only temporal parameter is sparse in time domain. On the other hand, for EEG signals both the morphological and the temporal parameters have to be obtained using non-linear machine learning techniques such as neural network training. For the first kind of signals the problem discussed in Eq. (4.15) can be simplified.

The shape of a beat makes ECG signal a non-sparse signal not only in time domain but also in DWT, DFT, and DCT domains. However, the temporal parameters are only related to the R peaks. Hence a signal with only R peaks in it can be approximated much more accurately using greater sparsity. Using this on field experience, an interesting idea is to sense the signal to only recover the temporal parameters (R peaks for ECG) and suppress the morphological parameters. The morphological parameters can be learned from a signal snippet sensed at the Nyquist rate. On obtaining the temporal parameters, the entire signal can be recovered by combining the morphological and temporal parameters. Initial proof of concept implementation of this technique on ECG signals show that we can obtain 20 fold improvement in compression ratio while maintaining diagnostic equivalence of the recovered signals.

For ECG signals the shape characteristics can be suppressed using a low pass and high pass filter combination as discussed in our previous work [58]. A digital filter as represented in Eq. (4.16), at low pass cut-off frequency of 5 Hz and high pass cut-off at 12 Hz, can effectively eliminate the P, Q, S, and T waves, and only keep the R peaks in the signal.

$$\begin{aligned} y[i] &= 2y[i-1] - y[i-2] + x[i] - 2x[i-6] + x[i-12], \text{ low pass filter,} \\ z[i] &= 32x[i-16] - z[i-1] + x[i] - x[i-32], \text{ high pass filter.} \end{aligned} \tag{4.16}$$

The resulting signal z as shown in Fig. 4.4B has only the temporal parameters and the morphology is suppressed. The signal z is compressible to a sparse vector and can be recovered with very less number of samples. Fig. 4.4C, shows the CS recovered signal from the original signal. The sensing matrix Φ used for this purpose was generated using a Bernoulli distribution. The transformation matrix Θ for making the original signal sparse was obtained by first converting the band-stop filter in Eq. (4.16) in a matrix form and multiplying it with the DWT matrix. The homotopy recovery algorithm was used to recover the signal with only temporal parameters. The morphological parameters of the signal were learned previously using the curve fitting technique described in [58]. The ECG beat shape is then centered at each R peak obtained from the signal in Fig. 4.4 and temporally scaled to match the heart rate, as discussed in [58] to obtain a diagnostically equivalent signal.

This technique although is a simplification of the problem, however has several research issues. One significant assumption is that the morphology of ECG is assumed to be stable over time. With this technique the sensing system is agnostic of any changes in the morphology. Moreover, the technique still relies on recovery based on ℓ_2 norm minimization. Such an assumption may restrict the compression ratio. Some of the burning research issues in this area are:

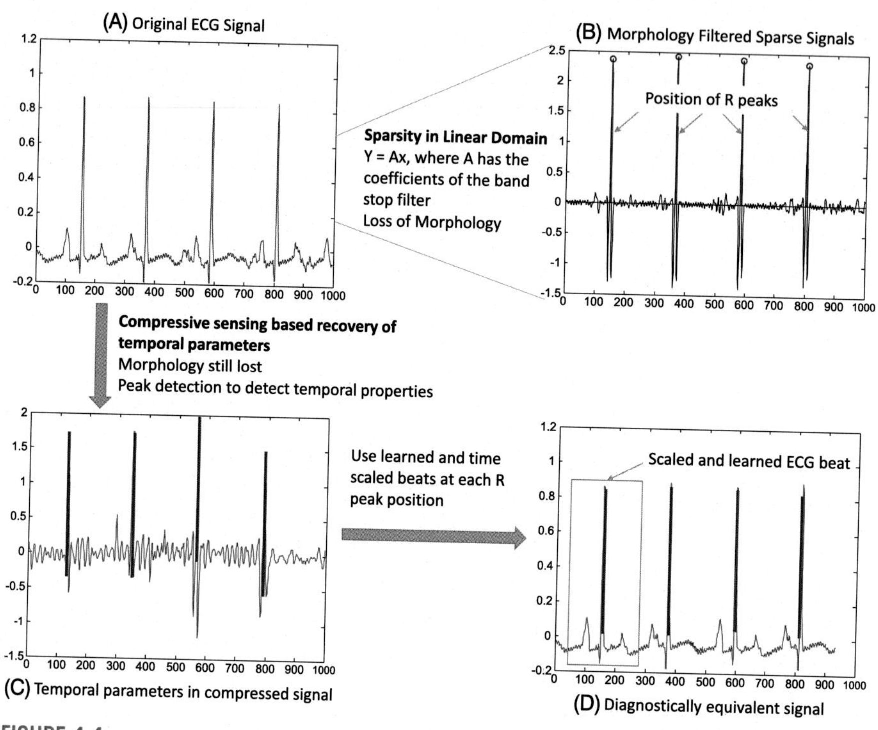

FIGURE 4.4

Proof of concept for generative model-based compressive sensing for ECG. The ECG temporal parameters can be extracted from a linear transformation which is sparse in time domain.

a) How to capture changes in morphology on the fly and re-adjust the recovery process.

b) How to solve the non-linear optimization problem in a tractable and computationally efficient manner.

4.5.3 ROI aware compressive sensing

Blocked ROI aware CS approach divides the entire image into blocks and employs different sensing matrices for each block, where every sensing matrix has a different sampling density. The blocks divide the scene x into smaller tiles. The sensing matrix $\boldsymbol{\Phi}$ is designed to distribute the total number of samples to be acquired, M, such that the blocks that overlap with the ROIs are sensed with higher number of samples. This reduces reconstruction error at the ROIs. The sensing and reconstruction steps of CS described in Section 4.2, are modified to reflect the blocked sampling and the ROI selection and update steps. The sensing step is modified as

$$y = \boldsymbol{\Phi}_t^{\boldsymbol{B}} x, \tag{4.17}$$

where $\boldsymbol{\Phi}_t^B$ is the blocked sensing matrix for acquiring observation at time t. This sensing matrix $\boldsymbol{\Phi}_t^B = [\phi_t^1, \ldots, \phi_t^B]$ is comprised of B blocks, where each block ϕ_t^b has a dimension of $r \times c$. All blocks ϕ_t^{roi} that overlap with an ROI is determined a priori based on the scene information from observations at $1, \ldots, t-1$. The sensing matrix at each block is chosen to be a Gaussian random matrix with higher number of samples M_{ROI} at each ROI overlapping block ϕ_t^{roi}. The sampling density of the sensing matrix is redistributed with the number of samples to be acquired at each non-ROI block is M_{NROI}, where $M_{NROI} < M_{ROI}$. This approach produces an overall sensing matrix with a block varying sampling density, having higher number of samples concentrated over the ROI blocks without changing the total number of scene samples to be acquired. This not only allows for reconstructing $\hat{x}$, which is more suitable for scientific analysis but also achieves the required compression. By selecting a Gaussian random sensing matrix, we ensure that every $\boldsymbol{\Phi}_t$ satisfies RIP, so that a stable reconstruction of $\hat{x}$ is possible that is not perturbed by noise. The l_1-minimization reconstruction step that exploits the sparsity of the images in the DCT domain is modified as

$$\hat{x} = \min\{\|x\|_1 \colon y = \boldsymbol{\Phi}_t^B x\}. \tag{4.18}$$

A compressive sensing step with the sensing matrix $\boldsymbol{\Phi}_t^B$ in each block is required, which is followed by block wise reconstruction of the original image. The reconstructed image $\hat{x}$ can then be analyzed using an automated approach to determining the ROIs, which in turn is used to update the sensing matrix for this scene to $\boldsymbol{\Phi}_{t+1}^B$.

Using the CS approach discussed in Section 4.2 and the blocked ROI aware CS approach, in a limited experiment we reconstruct THEMIS images of consisting of both ROI and non-ROI blocks for the same value of measurements obtained, M. For all experiments, each block in the sensing matrix is assigned the dimension 150×64. From Figs. 4.5A, B, and C the blocked approach is seen to improve PNSR at ROI significantly compared to the conventional CS approach for a given value of M for all ROI classes considered. This implies that, for each value of M, the blocked ROI aware approach retains the scientific quality of the image, making it more suitable for observations tailored to focus on specific regions in the image. By designing a sensing matrix with higher number of samples to be obtained at each ROI block, the approach achieves better reconstruction of high-frequency components of the image.

We also observe that although the ROI aware approach is formulated to improve reconstruction at ROIs at the cost of higher error at non-ROIs, this approach does not deteriorate the reconstruction error at non-ROIs as those blocks are mostly low-frequency components in the image. As non-ROIs are mostly low-frequency components in an image, comparable reconstruction is achievable by blocked ROI aware CS and conventional CS.

Challenges

C1: ROI extraction The primary challenge is to figure out region of interest automatically. If a novel area is being explored, then the definition of region of interest may not be available. In such a scenario, the primary challenge is to derive region of

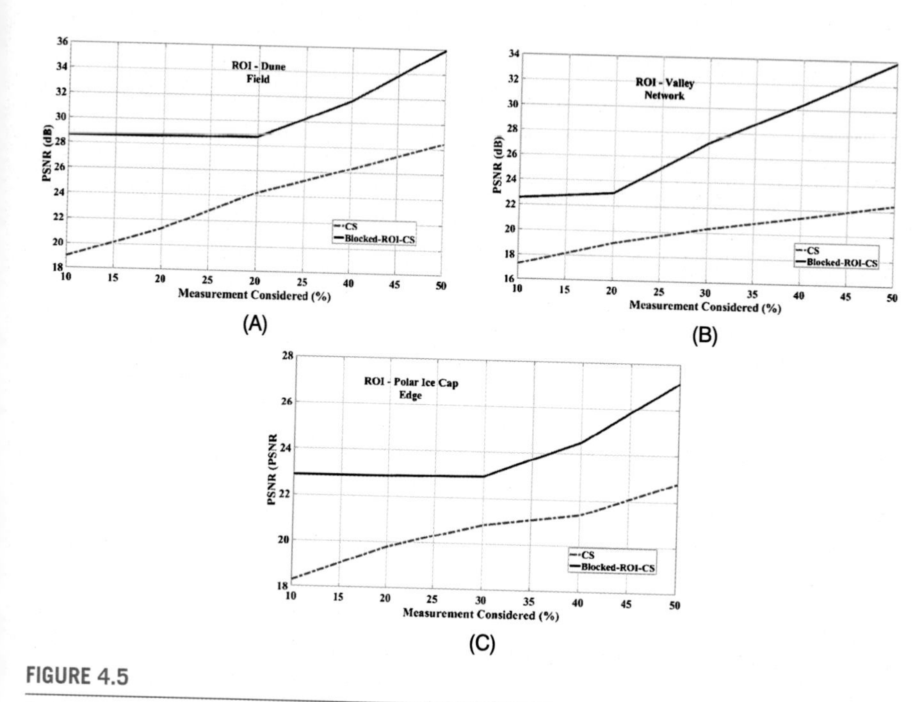

FIGURE 4.5

Comparison of PSNR of CS and Blocked-ROI-CS reconstruction of ROIs with varying measurement obtained for (A) dune field, (B) valley network, and (C) polar ice cap edge.

interest with no prior knowledge. This problem is a research topic that has regained interest in recent years. Association rule mining techniques have been initially explored by some researchers in this domain [59], however, the research area still has several paths to explore.

C2: Dynamic ROI Another challenge in this approach, is for several applications region of interest change both spatially and temporally. The ROI extraction algorithm should be able to accommodate spatio-temporal variation of ROI.

C3: Complex sensors One of the characteristics of compressive sensing is that the sensors are simple in hardware and software architecture. However, a blocked ROI-based CS will need varying sampling density for different ROIs. Building a single sensor with varying sampling density is a challenging task. As such currently there are no efforts in achieving this technology. Hence, potentially to take advantage of block ROI-based CS, multiple sensors will have to be employed each focusing on a specific ROI. This leads to increased resource requirements.

4.6 Conclusion

Compressive sensing has been a topic for extensive research in recent times. It has significant promise of high compression rates, however, in practice there are several problems. The biggest problem comes from the random sampling requirement. Reconstruction with random sampling assumption results in uniform distribution of error. Although this is not a problem as far as overall RMSE error is concerned, but it can cause high error rates on significant spatio-temporal characteristics of the data. As such several advancements are possible that perform selective sampling, blocked ROI-based compressive sensing and non-linear model driven CS. These possibilities are only tested in specific applications and have not been fully justified as consistent optimization potentials in the CS domain. Hence, a plethora of research topics are available to be explored by researchers in this domain.

Acknowledgments

The authors will like to thank Dr. Richard Heuser from St. Luke's hospital for providing with subjects to collect ECG data and also the Institutional review board for approving the study. The project is partly funded by NIH EB019202 grant.

References

[1] P. Yenduri, A. Rocca, A. Rao, S. Naraghi, M. Flynn, A. Gilbert, A low-power compressive sampling time-based analog-to-digital converter, IEEE Journal on Emerging and Selected Topics in Circuits and Systems 2 (3) (Sept 2012) 502–515.

[2] B. Shahrasbi, A. Talari, N. Rahnavard, TC-CSBP: compressive sensing for time-correlated data based on belief propagation, in: 2011 45th Annual Conference on Information Sciences and Systems (CISS), March 2011, pp. 1–6.

[3] A. Polak, M. Duarte, R. Jackson, D. Goeckel, Recovery of sparse signals from amplitude-limited sample sets, in: 2013 IEEE International Conference on Acoustics, Speech and Signal Processing (ICASSP), May 2013, pp. 4663–4667.

[4] B. Bosworth, J. Stroud, D. Tran, T. Tran, S. Chin, M. Foster, High-speed compressed sensing measurement using spectrally-encoded ultrafast laser pulses, in: 2015 49th Annual Conference on Information Sciences and Systems (CISS), March 2015, pp. 1–4.

[5] B. Bosworth, M. Foster, High-speed flow imaging utilizing spectral-encoding of ultrafast pulses and compressed sensing, in: 2014 Conference on Lasers and Electro-Optics (CLEO), June 2014, pp. 1–2.

[6] J. Wu, W. Wang, Q. Liang, X. Wu, B. Zhang, Compressive sensing-based signal compression and recovery in UWB wireless communication system, Wireless Communications and Mobile Computing 14 (13) (2014) 1266–1275, https://doi.org/10.1002/wcm.2228.

[7] M.F. Duarte, M.A. Davenport, D. Takhar, J.N. Laska, T. Sun, K.F. Kelly, R.G. Baraniuk, Single-pixel imaging via compressive sampling, IEEE Signal Processing Magazine (2008).

[8] T.W. Cabral, M. Khosravy, F.M. Dias, H.L.M. Monteiro, M.A.A. Lima, L.R.M. Silva, R. Naji, C.A Duque, Compressive sensing in medical signal processing and imaging systems, in: Sensors for Health Monitoring, Elsevier, 2019, pp. 69–92.

[9] A. Banerjee, S.K.S. Gupta, Clinical evaluation of generative model based monitoring and comparison with compressive sensing, in: Intl. Conf. on Wireless Health, ACM, October 2015, http://dx.doi.org/10.1145/2811780.2811946.

[10] A. Shukla, A. Majumdar, Row-sparse blind compressed sensing for reconstructing multi-channel EEG signals, Biomedical Signal Processing and Control 18 (2015) 174–178, https://doi.org/10.1016/j.bspc.2014.09.003.
[11] E.J. Candes, M.B. Wakin, An introduction to compressive sampling, IEEE Signal Processing Magazine 25 (2) (2008) 21–30.
[12] D.L. Donoho, Compressed sensing, IEEE Transactions on Information Theory 52 (4) (2006) 1289–1306.
[13] J. Tropp, A.C. Gilbert, et al., Signal recovery from random measurements via orthogonal matching pursuit, IEEE Transactions on Biomedical Engineering 53 (12) (2007) 4655–4666.
[14] D. Needell, R. Vershynin, Uniform uncertainty principle and signal recovery via regularized orthogonal matching pursuit, Foundations of Computational Mathematics 9 (3) (2009) 317–334.
[15] F. Ren, A scalable VLSI architecture for real-time and energy-efficient sparse approximation in compressive sensing systems, Doctor of Philosophy in Electrical Engineering, University of California, Los Angeles, Los Angeles, Jan. 2015.
[16] S.S. Chen, D.L. Donoho, M.A. Saunders, Atomic decomposition by basis pursuit, SIAM Journal on Scientific Computing 20 (1) (1998) 33–61.
[17] R. Chartrand, W. Yin, Iteratively reweighted algorithms for compressive sensing, in: IEEE International Conference on Acoustics, Speech and Signal Processing (ICASSP), IEEE, 2008, pp. 3869–3872.
[18] P. Garrigues, L. El Ghaoui, An homotopy algorithm for the lasso with online observations, in: Proc. NIPS, 2008.
[19] W.K. Allard, G. Chen, M. Maggioni, Multi-scale geometric methods for data sets ii: geometric multi-resolution analysis, Applied and Computational Harmonic Analysis 32 (3) (2012) 435–462.
[20] S. Chakraborty, A. Banerjee, S.K.S. Gupta, P.R. Christensen, Region of interest aware compressive sensing of THEMIS images and its reconstruction quality, in: 2018 IEEE Aerospace Conference, IEEE, 2018, pp. 1–11.
[21] J. Goodman, K. Forsythe, B. Miller, Efficient reconstruction of block-sparse signals, in: Statistical Signal Processing Workshop (SSP), 2011, IEEE, June 2011, pp. 629–632.
[22] Z. Zhang, T.-P. Jung, S. Makeig, B. Rao, Compressed sensing for energy-efficient wireless telemonitoring of noninvasive fetal ECG via block sparse Bayesian learning, IEEE Transactions on Biomedical Engineering 60 (2) (Feb 2013) 300–309.
[23] R. Baraniuk, V. Cevher, M. Duarte, C. Hegde, Model-based compressive sensing, IEEE Transactions on Information Theory 56 (4) (April 2010) 1982–2001.
[24] C. Hegde, P. Indyk, L. Schmidt, Nearly linear-time model-based compressive sensing, in: J. Esparza, P. Fraigniaud, T. Husfeldt, E. Koutsoupias (Eds.), Automata, Languages, and Programming, in: Lecture Notes in Computer Science, vol. 8572, Springer, Berlin Heidelberg, 2014, pp. 588–599, [online], available: http://dx.doi.org/10.1007/978-3-662-43948-7_49.
[25] F. Ren, W. Xu, D. Markovic, Scalable and parameterised VLSI architecture for efficient sparse approximation in FPGAs and SOCs, IET Electronics Letters 49 (23) (Nov. 2013) 1440–1441, [online]: http://ieeexplore.ieee.org/xpls/abs_all.jsp?arnumber=6675715.
[26] H. Mamaghanian, N. Khaled, D. Atienza, P. Vandergheynst, Compressed sensing for real-time energy-efficient ECG compression on wireless body sensor nodes, IEEE Transactions on Information Theory 58 (9) (2011) 2456–2466.
[27] B. Rajoub, et al., An efficient coding algorithm for the compression of ECG signals using the wavelet transform, IEEE Transactions on Biomedical Engineering 49 (4) (2002) 355–362.
[28] N. Ahmed, P.J. Milne, S.G. Harris, Electrocardiographic data compression via orthogonal transforms, IEEE Transactions on Biomedical Engineering 6 (1975) 484–487.
[29] F. Ren, R. Dorrace, W. Xu, D. Markovic, A single-precision compressive sensing signal reconstruction engine on FPGAs, in: Proceedings of the 23rd International Conference on Field Programmable Logic and Applications (FPL), IEEE, Sep. 2013, pp. 1–4, [online], available: http://ieeexplore.ieee.org/xpls/abs_all.jsp?arnumber=6645574.

[30] F. Ren, D. Markovic, 18.5 A configurable 12-to-237KS/s 12.8 mW sparse-approximation engine for mobile ExG data aggregation, in: Proceedings of the 2015 IEEE International Solid-State Circuits Conference (ISSCC), IEEE, Feb. 2015, pp. 1–3, [online], available: http://ieeexplore.ieee.org/xpls/abs_all.jsp?arnumber=7063062.
[31] P.S. Addison, Wavelet transforms and the ECG: a review, Physiological Measurement 26 (5) (2005) R155.
[32] B. Liu, Z. Zhang, G. Xu, H. Fan, Q. Fu, Energy efficient telemonitoring of physiological signals via compressed sensing: a fast algorithm and power consumption evaluation, Biomedical Signal Processing and Control 11 (2014) 80–88, https://doi.org/10.1016/j.bspc.2014.02.010.
[33] Z. Zhang, T.-P. Jung, S. Makeig, Z. Pi, B. Rao, Spatiotemporal sparse Bayesian learning with applications to compressed sensing of multichannel physiological signals, IEEE Transactions on Neural Systems and Rehabilitation Engineering 22 (6) (Nov 2014) 1186–1197.
[34] Z. Zhang, B. Rao, T.-P. Jung, Compressed sensing for energy-efficient wireless telemonitoring: challenges and opportunities, in: 2013 Asilomar Conference on Signals, Systems and Computers, Nov 2013, pp. 80–85.
[35] T. Ince, A. Nacaroglu, N. Watsuji, Nonconvex compressed sensing with partially known signal support, Signal Processing 93 (1) (2013) 338–344, https://doi.org/10.1016/j.sigpro.2012.07.011.
[36] A. Dremeau, C. Herzet, L. Daudet, Boltzmann machine and mean-field approximation for structured sparse decompositions, IEEE Transactions on Signal Processing 60 (7) (July 2012) 3425–3438.
[37] M. Khajehnejad, W. Xu, A. Avestimehr, B. Hassibi, Analyzing weighted l_1 minimization for sparse recovery with nonuniform sparse models, IEEE Transactions on Signal Processing 59 (5) (May 2011) 1985–2001.
[38] L. He, L. Carin, Exploiting structure in wavelet-based Bayesian compressive sensing, IEEE Transactions on Signal Processing 57 (9) (Sept 2009) 3488–3497.
[39] P. Indyk, E. Price, K-median clustering, model-based compressive sensing, and sparse recovery for earth mover distance, in: Proceedings of the Forty-Third Annual ACM Symposium on Theory of Computing, STOC'11, ACM, New York, NY, USA, 2011, pp. 627–636, [online], available: http://doi.acm.org/10.1145/1993636.1993720.
[40] H. Ohlsson, A. Yang, R. Dong, S. Sastry, Nonlinear basis pursuit, in: 2013 Asilomar Conference on Signals, Systems and Computers, Nov 2013, pp. 115–119.
[41] A. Abdulghani, A. Casson, E. Rodriguez-Villegas, Compressive sensing scalp eeg signals: implementations and practical performance, Medical & Biological Engineering & Computing 50 (11) (2012) 1137–1145, https://doi.org/10.1007/s11517-011-0832-1.
[42] M. Chen, J. Silva, J. Paisley, C. Wang, D. Dunson, L. Carin, Compressive sensing on manifolds using a nonparametric mixture of factor analyzers: algorithm and performance bounds, IEEE Transactions on Signal Processing 58 (12) (Dec 2010) 6140–6155.
[43] R.G. Baraniuk, M.B. Wakin, Random projections of smooth manifolds, Foundations of Computational Mathematics 9 (1) (Jan. 2009) 51–77, https://doi.org/10.1007/s10208-007-9011-z.
[44] H.L. Yap, M.B. Wakin, C.J. Rozell, Stable manifold embeddings with operators satisfying the restricted isometry property, in: 2011 45th Annual Conference on Information Sciences and Systems (CISS), March 2011, pp. 1–6.
[45] Y. Zigel, A. Cohen, A. Katz, The weighted diagnostic distortion (WDD) measure for ECG signals compression, IEEE Transactions on Biomedical Engineering (2000) 1422–1430.
[46] M.A. Davenport, M.F. Duarte, Y. Eldar, G. Kutyniok, Introduction to compressed sensing, in: Compressed Sensing: Theory and Applications, Cambridge University Press, 2011.
[47] K. Kreutz-Delgado, J.F. Murray, B.D. Rao, K. Engan, T.-W. Lee, T.J. Sejnowski, Dictionary learning algorithms for sparse representation, Neural Computation 15 (2) (2003) 349–396.
[48] M. Aharon, M. Elad, A. Bruckstein, K-SVD: an algorithm for designing overcomplete dictionaries for sparse representation, IEEE Transactions on Signal Processing 54 (11) (2006) 4311–4322.
[49] G. Chen, M. Maggioni, Multiscale geometric and spectral analysis of plane arrangements, in: IEEE Conference on Computer Vision and Pattern Recognition (CVPR), IEEE, 2011, pp. 2825–2832.
[50] G. Chen, M. Maggioni, Multiscale geometric wavelets for the analysis of point clouds, in: 44th Annual Conference on Information Sciences and Systems (CISS), IEEE, 2010, pp. 1–6.

[51] J. Mairal, F. Bach, J. Ponce, G. Sapiro, Online dictionary learning for sparse coding, in: Proceedings of the 26th Annual International Conference on Machine Learning, ACM, 2009, pp. 689–696.
[52] S. Nabar, A. Banerjee, S. Gupta, R. Poovendran, Resource-efficient and reliable long term wireless monitoring of the photoplethysmographic signal, in: WH '11: Proceedings of the 2nd Conference on Wireless Health, October 2011, pp. 1–6.
[53] P. Agante, J.M. De Sá, ECG noise filtering using wavelets with soft-thresholding methods, in: Computers in Cardiology, 1999, IEEE, 1999, pp. 535–538.
[54] N.V. Thakor, Y.-S. Zhu, Applications of adaptive filtering to ECG analysis: noise cancellation and arrhythmia detection, IEEE Transactions on Biomedical Engineering 38 (8) (1991) 785–794.
[55] S. Vorobyov, A. Cichocki, Blind noise reduction for multisensory signals using ICA and subspace filtering, with application to EEG analysis, Biological Cybernetics 86 (4) (2002) 293–303.
[56] C. Hegde, A.C. Sankaranarayanan, W. Yin, R.G. Baraniuk, NuMax: a convex approach for learning near-isometric linear embeddings, IEEE Transactions on Signal Processing (2015).
[57] C. Hegde, A. Sankaranarayanan, W. Yin, R. Baraniuk, A convex approach for learning near-isometric linear embeddings, in preparation, August 2012.
[58] S. Nabar, A. Banerjee, S.K.S. Gupta, R. Poovendran, GeM-REM: generative model-driven resource efficient ECG monitoring in body sensor networks, in: 2011 International Conference on Body Sensor Networks (BSN), may 2011, pp. 1–6.
[59] S. Chakraborty, S. Das, A. Banerjee, S.K.S. Gupta, P.R. Christensen, Expert guided rule based prioritization of scientifically relevant images for downlinking over limited bandwidth from planetary orbiters, in: Proceedings of the AAAI Conference on Artificial Intelligence, vol. 3, no. 01, Jul. 2019, pp. 9440–9445, [online], available: https://wvvw.aaai.org/ojs/index.php/AAAI/article/view/4995.
[60] A. Goldberger, L. Amaral, L. Glass, J.M. Hausdorff, P.Ch. Ivanov, R.G. Mark, J.E. Mietus, C.K. Peng, H.E. Stanley, PhysioBank, PhysioToolkit, and PhysioNet: components of a new research resource for complex physiologic signals, in: Circulation, 2003, pp. 9440–9445 [online], available: https://www.physionet.org/content/mitdb/1.0.0/.

CHAPTER

5 A review of deterministic sensing matrices

Daniel Ramalho[a], Katia Melo[a], Mahdi Khosravy[b], Faramarz Asharif[c], Mir Sayed Shah Danish[d], Carlos A. Duque[a]

[a]*Department of Electrical Engineering, Federal University of Juiz de Fora, Juiz de Fora, Brazil*
[b]*Media Integrated Communication Laboratory, Graduate School of Engineering, Osaka University, Suita, Osaka, Japan*
[c]*School of Earth, Energy and Environmental Engineering, Kitami Institute of Technology, Hokkaido, Japan*
[d]*Strategic Research Projects Center, University of the Ryukyus, Senbaru, Okinawa, Japan*

5.1 Introduction

The evolution of technology makes the signal generation ever and ever of larger scale. Storage has been a huge problem, once that there is not hardware with sufficient storage to archive all the data. To solve this one compresses the signal and these techniques usually discard lower coefficients.

Compressive sensing compresses the signal during this processing by getting a lower number of samples randomly, optimizing the time and getting a lower computational cost of this acquisition and this storage.

Besides, this technique allows the signal to be processed at a lower rate than the Nyquist rate. Thus, a lot of possibilities appear, because there can be used less memory to process the same amount of data, so, this hardware has a lower cost. Nowadays, compressive sensing has been used in a wide field of applications from biomedical [1] to power quality analysis [2]. Besides, there still is a great potential for application of compressive sensing to agriculture equipment [3,4], signal processing [5,6], blind signal processing [7–12], acoustic OFDM [13], ECG processing [14–20], power quality analysis [2,21,22], power line communications [23], public transportation system [24], power system planning [25,26], location based services [27], text data processing [28], smart environment [29], human motion analysis [30], texture analysis [31], image adaptation [32], image enhancement [33,34], medical image processing [35–37], human–robot interaction [38], telecommunications [39–44], sentiment mining [45], data mining [46,47], software intensive systems [48], etc.

In this work we will use deterministic matrices to compress the signal, and the matrices involved, so it makes them easier to decode.

Compressive Sensing in Healthcare. https://doi.org/10.1016/B978-0-12-821247-9.00010-X

5.2 Compressive sensing

Typically to obtain a compressed paraphrase of a signal expressed by a vector $s \in \mathbb{R}^n$ one must first take a proper basis, calculate the coefficients of s on that basis and next engage only the $k < n$ higher coefficients [49].

Thus, denoting by $x = \Psi s$ the coefficients of a signal vector s in the proper base, this indicates that we acquire an estimate for s in the set Σ_k that is the set of k-sparse vectors:

$$\Sigma_k := \{s \in \mathbb{R}^n : \#\mathrm{supp}(s) \leq k\} \tag{5.1}$$

where $supp(.)$ indicates the support set of the vector s. The support of a vector s is the set of i such that $s_i \neq 0$. # indicates the number of members of the set. One can obtain an optimal point [49] by means of the approximation process in some of the rules for $\|s\|_p$. We have

$$\sigma_k(s)_p := \inf_{z \in \Sigma_k} \|s - z\|_p. \tag{5.2}$$

This estimating process is considered as adaptive. It is because the indices of these retained coefficients deviate from a signal to another. The view expressed in [50], [51], [52], and [53] is that, since we retain some of these coefficients, it should be possible to compute some linear non-adaptive measures by keeping the information necessary to build a compact representation. This set of ideas gave rise to Compressive Sensing (CS).

CS has become a structure in the area of signal processing, gaining a lot of attention in recent years [54–56]. Mathematically it is about recovering an unknown sparse signal $x \in \mathbb{R}^n$ of a measurement $y \in \mathbb{R}^m$ such that

$$y = \Phi.x \tag{5.3}$$

where Φ is an array of size $M \times N$ called the measurement matrix or encoder [49,54]. The construction of this matrix with good properties is the main focus of CS. When the number of measurements m equals the number of samples n of the signal, the linear system is easily solved.

However, we seek to recover the signal using fewer samples than it has, so $m < n$ and the linear system (5.3) becomes indeterminate with infinite solutions.

If the coded signal y was generated from a sparse signal, where sparse means to have a few samples, but not zero, we can retrieve it as follows: find the sparse signal that forms y by minimizing the l_0:

$$\min\{\|x\|_0 : \Phi.x = y\}. \tag{5.4}$$

But Model (5.4) has a nonconvex objective and so it is very difficult to solve. Then the problem can be solved by making an approximation with convex optimization by making use of the norm l_1. This approach is called Basis Pursuit:

$$\min\{\|x\|_1 : \Phi.x = y\}. \tag{5.5}$$

To achieve this optimization a breakthrough in the literature was the determination of the condition of restricted isometry property (RIP) for the compressive sensing matrix Φ [52,57,58]. Thus, a matrix $\Phi_{m\times n}$ must satisfy the restricted isometry property of order k. That means that a constant δ_k $(0<\delta_k<1)$ should exist satisfying the following relation:

$$\forall \boldsymbol{v} \in \mathbb{R}^m, \ \|\boldsymbol{v}\|_0 \leq k \qquad (1-\delta_k).\|\boldsymbol{v}\|_2^2 \leq \|\Phi.\boldsymbol{v}\|_2^2 \leq (1+\delta_k).\|\boldsymbol{v}\|_2^2. \tag{5.6}$$

The methods of constructing Φ can be divided into two categories: random methods [59,60] and deterministic methods [61–65]. In this paper we will focus on deterministic methods.

A well-designed random array can approach optimal recovery performance with high probability. But due to the lack of structure in the matrix, its computational costs are enormous. However, deterministic measurement matrices have regular structures and can circumvent this problem to a certain point.

In this work, we will discuss some characteristics of the deterministic matrices presenting some techniques used for specific signals, such as binary signals, Reed–Muller codes and chirp.

5.3 Chirp codes compressive matrices

The problem of the signal recovery can be approached by observing that finding the signal corresponds to finding the smaller linear combinations of the columns which $\Theta=\Phi.\Psi$ from y. In particular designing for chirp signals forming the columns we remember that this signal of length K has the following form [66]:

$$v_{m,r}=\alpha.e^{\frac{j.2.\pi.m.l}{K}+\frac{j.2.\pi.r.l^2}{K}} : m,r \in \mathbb{Z}_K. \tag{5.7}$$

m is called the 'base frequency' and r is called the 'chirp rate'. Thus for a signal of length K will be required K^2 combinations, implying that Θ will have size $K\times K^2$ where Θ columns are filled by all K uni-modular chirp signals.

Consider a vector $\mathbf{y}$ made of the linear combination of chirp signals as follows:

$$\mathbf{y}(l)=s_1 e^{\frac{j.2.\pi.m_1.l}{K}+\frac{j.2.\pi.r_1.l^2}{K}}+s_2 e^{\frac{j.2.\pi.m_2.l}{K}+\frac{j.2.\pi.r_2.l^2}{K}}+\ldots \tag{5.8}$$

where the base frequencies are the m_i and the chirp rates are the r_i. The chirp rates are collected from y. It is found by searching through $f(l)=\bar{y}(l)y(l+T)$. Note that the index $l+T$ is obtained as its mod to K. This gives [66]:

$$f(l)=\sum_i |s_i|^2.e^{\frac{j.2.\pi}{K}.(m_i.T+r_i.T^2)}.e^{\frac{j.2.\pi.(2.r_i.l.T)}{K}}+CT \tag{5.9}$$

where the CT is for cross-terms as

$$s_p.\bar{s}_q.e^{\frac{j.2.\pi}{K}.(m_p.T+r_p.T^2)}.e^{\frac{j.2.\pi}{K}.l.(m_p-m_q+2.T.r_p)}.e^{\frac{j.2.\pi}{K}.l^2.(r_p-r_q)}, \tag{5.10}$$

which are chirps. Note that $f(l)$ is a sinusoidal signal with discrete frequencies $2.r_i.T \mod K$. In order to have a bijection between chirp rates and FFT bins, K must be a prime number [67]. In addition, the remainder consisting of the cross-terms is ignorable. The energy of the cross-terms is spread across all the frequencies.

Since $\boldsymbol{x}$ is sparse, $\boldsymbol{y}$ consists of sufficiently few chirps. Thus, the FFT obtained from $f(l)$ will have a spectrum with significant peaks in $2.r_i.T \mod K$ where we will remove the chirp rates

With the chirp rates r_i we can transform the signals with that rate into senoids. To do this we must multiply the signal $\boldsymbol{y}(l)$ by $e^{\frac{-j.2.\pi.r_i.l^2}{K}}$. Using the FFT in the resulting signal we get the values of m_i and s_i.

In this way choosing Θ as

$$\Theta_{k,l} = e^{\frac{j.2.\pi.m.l}{K} + \frac{j.2.\pi.r.l^2}{K}} : k = K.r + m \in \mathbb{Z}_{K^2} \tag{5.11}$$

we can then write Θ as a function of a set of matrices $U_t (t = 1, \ldots, k)$ such that

$$\Theta = [U_{r_1}, U_{r_2}, \ldots, U_{r_k}] \tag{5.12}$$

where U_{r_t} is matrix $K \times K$ whose columns are formed by chirp signal vectors of a fixed chirp rate of $r = r_t$ and base frequency of m ranging from 0 to $K - 1$. Thus, for example, for $k = 2$ and $r, m, l \in \{0, 1\}$, we have

$$\Theta = [U_{r_1}, U_{r_2}] = \begin{bmatrix} 1 & 1 & 1 & 1 \\ 1 & e^{j.\pi} & e^{j.\pi} & e^{j.\pi + j.2.\pi} \end{bmatrix}. \tag{5.13}$$

Note that $\boldsymbol{y} = \Theta.\boldsymbol{s}$ has the form (5.8). As $\boldsymbol{y}$ is given and Θ can be formed, the summary of the algorithm is as follows [66]:

Step 1: Select a $T \in \mathbb{Z}_K$, where $T \neq 0$.
Step 2: Select a stopping criterion ϵ.
Step 3: Obtain $f(l) = \bar{\boldsymbol{y}}(l)\boldsymbol{y}(l + T)$.
Step 4: Take length-K FFT from $f(l)$.
Step 5: Find the peak location in the obtained FFT. Take it as $2.r_i.T \mod K$. Keep the unique r_i matching the location.
Step 6: Multiply $\boldsymbol{y}(l)$ by $e^{\frac{-j.2.\pi.r_i.l^2}{K}}$, then take length-$K$ FFT.
Step 7: Find the peak location and keep it as m_i. The value of the peak can be used for recovery of s_i.
Step 8: Update $\boldsymbol{y}$ by subtracting $s_i.e^{\frac{j.2.\pi}{K}.(m_i.l + r_i.l^2)}$ from $\boldsymbol{y}$.
Step 9: Go through steps (3)–(8) again until $\|\boldsymbol{y}\|_2^2 < \epsilon$ or having iterated M times.

We must think through the usage of chirp codes in the case of sparsity of the Fourier transform [66].

5.3.1 Φ for sparse spectrum signals

Even with the purpose of determining the combination of columns of Θ, the measurements will be made of x using the matrix Φ ($\Phi = \Theta.\Psi^{-1}$); therefore, we are worried about Φ so it can be implemented. Thus, as Ψ is a Fourier matrix, x is a sparse superposition of sinusoids. Therefore, the structure of Φ can be found directly.

In (5.11) we define Θ for l and k, the construction groups the chirp columns into blocks of chirp rates. Since Ψ is the Fourier matrix, Φ is a matrix whose lines are formed by the Fourier transform of size K^2 of the Θ lines given by

$$\begin{aligned}
\sum_{k=0}^{K^2-1}[\Theta]_{l,k}.e^{\frac{-j.2.\pi.k.\omega}{K^2}} &= \sum_{r=0}^{K-1}\sum_{m=0}^{K-1}.e^{\frac{j.2.\pi.l}{K}.(r.l+m)}.e^{\frac{-j.2.\pi.\omega}{K^2}.(K.r+m)} \\
&= \left[\sum_{r=0}^{K-1}e^{\frac{j.2.\pi.r}{K}(l^2-\omega)}\right].\left[\sum_{m=0}^{K-1}e^{\frac{j.2.\pi.m}{K^2}(K.l-\omega)}\right] \\
&= K.\delta_{(l^2-\omega)}.e^{j.\pi.(K.l-\omega)(\frac{1}{K}-\frac{1}{K^2})}.\frac{\sin(\pi(K.l-\omega)/K)}{\sin(\pi(K.l-\omega)/K^2)}
\end{aligned} \tag{5.14}$$

where δ_i is the delta, and i is taken as its mod to K. Thus, the rows of Φ are indeed periodic trains of delta functions modulated by a sinc function. The meaning of this is that the measurements y can be made as a sparsely weighted summation of a number of samples in x. Therefore, the encoding of y has a low computational load [66].

5.4 Second order Reed–Muller compressive sensing matrix

Reed–Muller codes are really old and have a lot of applications. They were discovered by Muller and Reed, who made a decoding algorithm for it. These codes were initially given as binary but there exist q-ary code generalizations. In this case of sensing matrices, we will use the binary case. There are several ways to describe these codes and we will choose one of them [50].

As there exists a duality between time and frequency in the world, the sinusoids are the eigenfunctions of shifts in time. This is the starting point of the description of binary Hamming space wherein there exist discrete equivalents of shifts in time and frequency. Walsh functions are the analogs of sinusoids. The second-order Reed–Muller functions are the analogs of chirps.

It is important that Walsh functions are first-order Reed–Muller functions, and they are functions $\mathbb{Z}_2^m \to \mathbb{R}$ defined by

$$\phi_{0,b}(a) = \frac{1}{\sqrt{2^m}}.(-1)^{b^T.a}. \tag{5.15}$$

The Walsh functions create an orthonormal basis for $\mathbb{R}^{2m}$ and therefore H_m is unitary. They can be viewed as the sinusoids of the binary world, where b defines a binary frequency.

There exist 2^m Walsh functions in $\mathbb{Z}_m^2$. They lead to the Hadamard matrix H_m rows, and they make an orthonormal basis. So H_m is unitary. These functions are eigenfunctions of the following shifts:

$$\phi_{0,b}(a+\varepsilon) = (-1)^{b^T.\varepsilon}.\phi_{0,b}(a). \tag{5.16}$$

And these shifts can be defined as the equivalent chirps for the binary world.

A sinusoid is called linearly-chirped as its frequency changes linearly in time. We can construct functions in the binary case where the matrix P is a binary symmetric:

$$\phi_{P,b}(a) = \frac{(-1)^{w.t.a}}{\sqrt{2^m}}.i^{(2.b+P.a)^T.a} \tag{5.17}$$

These functions are second-order Reed–Muller functions, their parameters are P and b which are, respectively, binary symmetric matrices and binary vectors. The Walsh functions are associated with $P = 0$.

These second-order Reed–Muller functions have a number important properties. Like, as P is a fixed binary symmetric matrix, the set $F = \{\phi_{P,b} | b \in \mathbb{Z}_2^m\}$ makes an orthonormal basis for $\mathbb{R}^{2.m}$. Each function on this basis is an eigenfunction of a commutative group of time–frequency shift operators defined by P [68].

As focusing on real second-order Reed–Muller functions, the matrices P have zeros on the main diagonal.

Let the U_p be unitary matrix corresponding to the orthonormal basis F_p. In [69] it was proposed that $2^m \times 2^{\frac{m.(m+1)}{2}}$ compressive sensing matrices are as follows:

$$\Phi_{RM} = (U_{P_1}, U_{P_2}, \ldots, U_{P_{2^{\frac{m.(m+1)}{2}}}}). \tag{5.18}$$

Being normalized by $\frac{1}{\sqrt{2^m}}$, the elements in matrix are ± 1. The columns of Φ_{RM} form a tight frame for $\mathbb{R}^{2.m}$ with redundancy $2^{\frac{m.(m-1)}{2}}$, that is,

$$\Phi_{RM}.\Phi_{RM}^T = 2^{\frac{m.(m-1)}{2}}.I_{2^m \times 2^m}. \tag{5.19}$$

If m is even, we have the proof that is helpful to additional limit the set of P matrices to DG_{2h} sets of zero-diagonal symmetric binary $m \times m$ matrices with the property that for any distinctive couple of matrices $P, Q \in DG_{2h}$ the rank of $P + Q$ is at least $m - 2.h$. DG_{2h} are nested sets, in such a way that $DG_0 \subseteq DG_2 \subseteq \ldots$, being related with sub-codes of the second-order Reed–Muller code [69].

RIP is essential to compressive sensing. It is the situation that the eigenvalues of all Gram matrices defined by sets of k columns of the sensing matrix are in an interval $[1-\delta, 1+\delta]$, however, this characteristic is normal for a random matrix. This was examined in [69]; a matrix with m=6 is almost the same as a random Gaussian matrix.

5.4.1 Fast reconstruction algorithm

The purpose is to create a deterministic compressive sensing matrix exploiting the formation of the second-order Reed–Muller functions to build computationally practical recovery algorithm.

An essential characteristic of second-order Reed–Muller to this state is as follows. Take any $\varepsilon \in \mathbb{Z}_2^m$, then

$$\phi_{P,b}(a+\varepsilon).\bar{\phi}_{P,b}(a) = \frac{(-1)^{b^T.\varepsilon}.(-1)^{\varepsilon^T.P.a}}{2^m}. \tag{5.20}$$

This will produce a function of Walsh with frequency P_e.

Assume we were given an unknown second-order Reed–Muller function φ, a quick algorithm for recovering the parameters is given here. For each unit weight $(\varepsilon_j)_k = \delta_{j,k}$, calculate the product in (5.20) and implement the fast Hadamard transform.

The highest element of the Hadamard transform will be at $P_{\varepsilon,j}$. This concerns the column j of P. Then "dechirping" Ψ is performed by multiplying with $\phi_(P,0)$, and an extra Hadamard transform leads to b. The parameters of an unknown second-order Reed–Muller function can be obtained with complexity of $(m+1)^2.2^m$ multiplications and the search for 2^m terms in each Hadamard transform.

From Eq. (5.3), the data vector becomes

$$y = \sum_{i=1}^{k} c_i.\phi_{P_i,b_i}. \tag{5.21}$$

The quick algorithm of [69] is established on the utilization of shifting–multiplying the data with the result

$$y(a+\varepsilon).\bar{y(a)} = \frac{1}{2^m}\sum_{i=1}^{k} c_i^2.(-1)^{b_i^T.\varepsilon}.(-1)^{\varepsilon^T.P_i.a} + chirps. \tag{5.22}$$

The term on the right of (5.22) is a linear combination of Walsh functions added to the chirps consisting of cross-terms. The cross-terms are second-order Reed–Muller functions with nonzero P_s. Implementing the Hadamard transform to 5.22 ends in peaks at frequencies $P_{l\varepsilon}$ for each P_l. As second-order Reed–Muller functions with nonzero P possess Hadamard transforms distributing in frequency, it results in peaks with a general background of interference. Having concluded that P_l is one of the signal components, then a robust estimation of the corresponding b_l is possible by de-chirping the data. We have

$$y(a).\bar{\phi}_{P,0}(a) = c_i(-1)^{wt(b)}.(-1)^{b^T.a} + chirps. \tag{5.23}$$

We proceed by implementing the Hadamard transform.

Therefore, the step-by-step algorithm is:

1. Obtain the next P_j the largest energy ($|c_j|^2$) component of the compressively sensed data y'.
2. Obtain the frequency b_j.
3. Obtain the c_i minimizing $\|y\|_2$ and substitute it to obtain the compressed data

$$y' = y - \sum_{i=0}^{j} \widehat{c_i}.\phi_{P_i,b_i}.$$

4. Repeat the above-steps until $\|y'\| < \varepsilon$

A problem found with this basic method happens when two components have P_s with a common column. If this occurs two Walsh functions with the same frequency occur in (5.22). They might be reinforced or canceled. In the case of cancellation, it is problematical and may head to a missed column.

The computational load of the algorithm is not dependent on N the length of the signal targeted for recovery, but rather on the number of measurements n.

Baron et al. [70] have proposed a rule according to Donoho and Tanner study [71,72] of the probability of efficient recovery by basis pursuit. That is the number of measurements n that is essential for a reliable reconstruction; it satisfies $n > k\log_2(1+\frac{N}{k})$. Therefore, the only required change for the algorithm is to the stopping criterion.

5.5 Amini approach to RIP-satisfying matrices

The condition of RIP in compressive sensing is an essential requirement which ensures the recovery of sparse signal vectors under a noisy condition. However, there is a problem in affirmation of the RIP property for a given detection array: that of its nature of an NP-hard problem [58] which may require metaheuristic evolutionary optimization [20,73–83]. As a result, the most popular matrix fulfilling RIP is a random structure matrix. Therefore, there is an effort of designing CS matrices in the direction of fulfilling RIP.

Amini and Marvasti give an approach to make RIP-satisfying matrices as explained in the sequence [62]. Consider $A_{m\times n}$ as a real matrix composed of normalized columns in such a way that the dot product of any two columns does not surpass a certain value λ. Consider $B_{m\times k}$ as a matrix made of any of k columns in A. Then, we have the Grammian matrix $G_{k\times k} = B^T B$. Because the columns B are normal, the values on the diagonal of G are 1 and the other elements have absolute values smaller than λ. Therefore

$$\forall 1 \le i \le k: \sum_{j,\ j\neq i} |g_{i,j}| \le \lambda.(k-1) \tag{5.24}$$

where the $g_{i,j}$ are the elements of G. By means of the Gershogorin circle theorem, the eigenvalues of G are limited to the range of $[1-\lambda(k-1), 1+\lambda(k-1)]$. Therefore, having a sufficiently small k in a way that $\delta_k = \lambda(k-1) < 1$, the matrix $A_{m\times n}$ fulfills the RIP condition. The Amini–Marvasti approach to satisfy the RIP condition can is summarized as follows.

Amini–Marvasti approach to satisfy RIP: A compressive sensing matrix A composed of normalized columns $\boldsymbol{a}_i$s and the size of $m \times n$ where $\forall i, j \neq i, \langle \boldsymbol{a}_i, \boldsymbol{a}_j \rangle \leq \lambda$ has the RIP of order k if

$$\lambda < \frac{1}{k-1}. \tag{5.25}$$

5.6 Amini–Marvasti binary matrix

Amini and Marvasti introduce binary matrices which are RIP-satisfying [62]. A binary sampling matrix is one satisfying the RIP condition whose elements are 0 and 1 before its columns are normalized. A subset of these matrices has been used before as optical orthogonal codes (OOC) [84]. Consider a binary vector composed of 0 and 1, and its weight w as the number of 1. If $R(m, w, \lambda)$ represents the maximum number of binary vectors $m \times 1$ for which their weights are the w and the dot product of any two of them is not higher than λ ($\lambda \in \mathbb{Z}$), then

$$R(m, w, \lambda) \leq \left\lfloor \frac{m}{w} \cdot \left\lfloor \frac{m-1}{w-1} \cdot \left\lfloor \cdots \left\lfloor \frac{m-\lambda}{w-\lambda} \right\rfloor \cdots \right\rfloor \right\rfloor \right\rfloor \tag{5.26}$$

where $\lfloor . \rfloor$ represents the biggest integer number that is not greater than its operand. In the application of these codes in multi-access telecommunications, the small value of the dot-product of different symbols is a key to differentiate between the users when symbols are synchronized; and when they are not synchronized, the autocorrelation and cross-correlation has this important role [62]. Therefore instead of λ, two parameters λ_a and λ_c are deployed which are, respectively, the maximum autocorrelation and maximum cross-correlation.

Alternatively, a quick calculable property for the compressive matrix is the mutual coherence $\mu(\Phi)$. For a given matrix $\Phi_{m\times M}$, the mutual coherence is the maximum absolute internal product of any two normalized columns of $\Phi_{m\times M}$.

$$\mu(\Phi) = \max_{1\leq i,j\leq M, i\neq j} \left[\frac{|\phi_i^T . \phi_j|}{[\|\phi_i\|]_2 . [\|\phi_j\|]_2} \right] \tag{5.27}$$

where ϕ_i is the ith column of Φ. Proposition 1 relates δ_s and $\mu(\Phi)$ [85].

Proposition 1. *Suppose that $\phi_1, \phi_2, \ldots, \phi_M$ are columns of the matrix Φ. They are of unit norm, and Φ has coherence μ. Then, Φ fulfills the RIP of order s with constant $\delta_s = (s-1).\mu$, when $\delta_s < 1$.*

As a result of this proposition, one can approach projects of sensory matrices constructing them with low mutual coherence. In the area of the construction of binary-sensing matrices, there are known technical variants, they use algebra, graph theory, and coding theory. In this paper, three techniques will be addressed: the BCH-code [62], Extremal Set Theory [58] and Unbalanced Expander Graphs [86].

5.6.1 BCH-code

BCH codes (Bose–Chaudhuri–Hocquenghem) make two sets of algorithms for cyclic error correction. They are built utilizing polynomials on a finite field (Galois field) [87].

The main highlight of these codes is that throughout the design, there exists a high level of precision control over the code error frequency as it is correctable by the code. It is feasible to create these codes with the capability of correcting the bit errors. Also, they have the advantage of decoding by an algebraic method.

For BCH codes application in binary matrices aimed at compressing sensing one can use OOC vectors characterized by $(m, w, \lambda_a, \lambda_c)$. There is a possibility that the parameters (λ_a and λ_c) are equal ($\lambda_a = \lambda_c = \lambda$). In this case the OOC is pointed to as (m, w, λ) code.

Supposing A is an OOC vector set of length m. It has the weight of w and equal parameters ($\lambda_a = \lambda_c = \lambda$). Applying the determinations of OOC [84] it is observed that the dot product of any two vectors of A is restricted by λ. Then it is possible to build a matrix $A_{m\times n}$ by the normalized columns in A where $n = |A|$. Concerning the upper bound of the inner product of the vectors in A it is easy to verify that the matrix A fulfilling the RIP of order $k < 1 + \frac{w}{\lambda}$ [88].

Consider $q = 16^a$ where $a \in \mathbb{N}$. $\mathbb{F} = GF(q)$ as α is the primitive root. Although Ref. [62] considers the relation of $5|q-1$, indicating 5 is a factor of $q-1$ as a clear matter. We give it here as a small lemma with proof.

Lemma 1.

$$\forall a \in \mathbb{N}, \quad 5|(16^a - 1), \tag{5.28}$$

where $\mathbb{N}$ is the set of natural numbers, and $|$ indicates 'being a factor of'.

Proof. The proof of the lemma is by mathematical induction. As the statement is for a natural number a, first we show it is true for $a = 1$, second, we show that if the statement is true for $a = k$ then is true for $a = k + 1$ where k is a particular but arbitrary natural number. Mathematically we prove that

1

$$a = 1 \rightarrow 5|(16^a - 1), \tag{5.29}$$

2

$$k \in \mathbb{N},\ 5|(16^k - 1) \rightarrow 5|(16^{k+1} - 1). \tag{5.30}$$

The first one is a clear fact as we substitute $a = 1$, the result is $5|15$, which is true. To prove the second induction statement, if $5|(16^k - 1)$, then clearly $5|\big(16 \times (16^k - 1)\big)$, which can be rewritten as

$$\begin{aligned} &5|(16^{k+1} - 16)\\ &5|(16^{k+1} - 15 - 1)\\ &5|(16^{k+1} - 1). \end{aligned} \tag{5.31}$$

The lemma is proved. □

As $5|(16^a - 1)$, there exists an integer d where $q = 5d + 1$. Define

$$D_i = \{\alpha^{d+1}, \alpha^{2.d+1}, \ldots, \alpha^{5.d+1}\}, \quad 0 \le i \le d - 1. \tag{5.32}$$

Because at most of the cases the 1 number considerably less than 0, OOC vectors are commonly represented by the location of the 1. In this configuration, the code length is $q - 1$, which means $m = 16^a - 1$ and the locations of ones at each code are given by $C_i{}_{i=1}^{d-1}$ [62]:

$$C_i = \log_a(D_i - 1), 1 \le i \le d - 1. \tag{5.33}$$

A matrix with structure independent of OOC codes is given in [61]. It is made of a $p^2 \times p^{r+1}$ binary matrix. The columns have weights of p (before normalization). The dot product of any two columns is not bigger than r, and after normalization being not larger than $\frac{r}{p}$. p is a power of a prime number. The structure of the matrix is based on polynomials in $GF(p)$. These matrices are asymptotically optimal when $\frac{p}{r^2} \to \infty$ [62]:

$$\begin{aligned} \lim_{\frac{p}{r^2}\to\infty} \frac{p^{r+1}}{R(p^2, p, r)} &\ge \lim_{\frac{p}{r^2}\to\infty} \prod_{i=0}^{r} \frac{p.(p-i)}{p^2 - i}\\ &\ge \lim_{\frac{p}{r^2}\to\infty} \frac{p - r}{p}^{r+1}\\ &\ge \lim_{\frac{p}{r^2}\to\infty} e^{-\frac{r(r+1)}{p}} = e^0 = 1. \end{aligned} \tag{5.34}$$

Consider $C(\widetilde{n}, \widetilde{k}; 2)$ as a linear binary block code and $1_{\widetilde{n}\times 1}$ a unitary column vector of $\widetilde{n} \times 1$. If $1_{\widetilde{n}\times 1} \in C$ then C is symmetric. Thus for symmetric codes we can define $a_{n\times 1}$ as a code vector, thus by code linearity the complement of $a_{n\times 1}$ defined as $a_{n\times 1} \bigoplus 1_{\widetilde{n}\times 1}$ is a valid code vector too. So the vectors are made of complementary pairs.

Theorem 1 of Ref. [62] gives a binary compressive sensing matrix. It considers $C(\widetilde{n}, \widetilde{k}; 2)$ as a symmetric code of minimum distance $\widetilde{d}_{min}$. Then it composes $\widetilde{A}_{\widetilde{n}\times 2^{\widetilde{k}-}}$

matrix by the codes in columns. The matrix involves exactly one of each complement couple. Then one defines

$$A_{\tilde{n}\times 2^{\tilde{k}-1}} \triangleq \frac{1}{\sqrt{\tilde{n}}}\cdot\left[2.\tilde{A}_{\tilde{n}\times 2^{\tilde{k}-1}} - (1)_{\tilde{n}\times 2^{\tilde{k}-1}}\right]. \tag{5.35}$$

The above defined matrix fulfills RIP with the constant $\delta_k = (k-1)(1-2.\frac{\tilde{d}_{min}}{\tilde{n}})$ for $k < \frac{\tilde{n}}{\tilde{n}-2.\tilde{d}_{min}}$. k is the RIP order.

The proof of Theorem 1 can be found in [62]. This theorem implies that $\tilde{d}_{min}$ must be close to the $\frac{\tilde{n}}{2}$ value; typically this is not common in binary systems due to sending parity bits to protect the data sent. But using BCH code you can design this type of algorithm.

As previously said BCH code are a class of cyclic codes such that $\tilde{n} = 2^{\tilde{m}-1}+1$ produced by a $g(x) \in GF(2)[x]$ respecting $g(x)|x^{2^{\tilde{m}-1}}+1$ [89]. Using the theory of Galois, we have

$$x^{2^{\tilde{m}-1}}+1 = \prod_{r\in GF(2^{\tilde{m}})}^{r\neq 0} (x-r). \tag{5.36}$$

In this way we can decompose the BCH-generating polynomial into products of linear factors of $GF(2^{\tilde{m}})$. Thus, let $\alpha \in GF(2^{\tilde{m}})$ be a primitive root of the field and α^i a root of $g(x)$. Since we have $g(x) \in GF(2)[x]$ all the conjugated elements of α^i will also be roots of $g(x)$. Thus, if $\alpha^{i_1}, \alpha^{i_2}, \ldots, \alpha^{i_d}$ are different roots of $g(x)$, such that $i_1, i_2, \ldots, i_d$ forms a geometric progression, then $\tilde{d}_{min} \geq d+1$.

We must then design a polynomial $h(x)$ for parity checking. Thus, letting $l < \tilde{m}-1$ we have

$$G_{\tilde{m}}^{(l)} = \{\alpha^0, \alpha^1, \ldots, \alpha^{2^{\tilde{m}-1}+2^l-1}\}. \tag{5.37}$$

Reference [62] constructs a compressive matrix in the following steps:

1. While k is given, $i = [\log_2(k)]$ and acquire $\tilde{m} \geq i$. The length of the compressed signal vector will be $m = 2^{\tilde{m}-1}$.
2. Consider H_{seq} as the set of all $\tilde{m}$-length binary vectors having at least i zeros. Consider H_{dec} as a set of decimal numbers. The corresponding binary representations of H_{dec} are in H_{seq}.
3. Address α, a primitive roots of $GF(2^{\tilde{m}})$. Then define

$$H = \{\alpha^r | r \in H_{dec}\}.$$

4. Determine the parity check. Also set the code producing polynomials:

$$h(x) = \prod_{r\in H}(x-r)$$

and

$$g(x) = \frac{x^{2^{\tilde{m}}} - 1}{h(x)}.$$

5. Consider $\widetilde{A}_{(2^{\tilde{m}}-1)\times 2^{deg(h)-1}}$ as a binary matrix made of even parity codes in columns. While columns are taken as polynomial coefficients ($\in GF(2)[x]$), each polynomial is divisible by $(x+1).g(x)$.
6. Put -1 instead of zero elements of $\widetilde{A}$. Then normalize each column. At this stage the compressive sensing matrix $A_{(2^{\tilde{m}}-1)\times 2^{deg(h)-1}}$ is obtained.

5.6.2 Extremal Set Theory

Extremal Set Theory is a theory of sets. Consider r, k, and m as positive integers in a way that $r < k < m$. Consider X a set of m elements, i.e. $X = \{1, 2, \ldots, m\}$. Also consider these definitions:

- $2^X := \{H, H \subseteq X\}$,
- $[X]^k := H \subseteq X, |H| = k$.

A subset $F \subset [X]^k$ is called a k-uniform family. Thereof, the following definitions are given [58]:

Definition 1. A subset $F_d(r, k, m) \subset [X]^k$ is r-dense if any subset of X with r elements possesses at least a member of F_d.

Definition 2. $F_s(r, k, m) \subset [X]^k$ is r-sparse if any subset of X with r elements possesses maximally a member of F_s, i.e. $|F_i \cap F_j| \leq r - 1, \forall F_i, F_j \in F_s$.

Definition 3. A subset $F_S(r, k, m) \subset [X]^k$ is a Steiner system if each subset of X with r elements fits precisely one of the F_S members.

We should consider a divisibility condition view in [90] given by

$$\forall 0 \leq i \leq r-1 \quad \binom{k-i}{r-i} \text{ divides } \binom{m-i}{r-i}. \tag{5.38}$$

In this way a Stein $F_S(r, k, m)$ system is a subset of r-sparse set $F_s(r, k, m)$. In order to apply Extreme Set Theory, we must ascertain the maximum size of the binary matrix CS of structure $(r-1, k)$ and verify the existence of a matrix with optimal column size, for each pair of (r, k) where $r < k$. To reach the same goal the following proposition [58] relates the r sparsity of the sets and a binary detection matrix with structure $(r-1, k)$.

Proposition 2. *There is a one-to-one bijection within the set containing the r-sparse k-uniform families and the set of binary-sensing matrices of $(r-1, k)$-structure [58].*

Thus using Propositions 1 and 2 we have the following proposition.

Proposition 3. *Let F be an r-sparse k-uniform family of cardinality M on m element set X. Then $\Phi_{m\times M}$, that is, the incidence matrix of F, has maximally a coherence of $\frac{r-1}{k}$. Also, $\Phi = \frac{1}{\sqrt{k}}.\Phi$ fulfills the RIP with $\delta_s = (s-1).\left(\frac{r-1}{k}\right)$, for any $s < \frac{k}{r-1} + 1$ [58].*

The process for constructing a 2 sparse k uniform family of a set $X = \{0, 1, 2, \ldots, kp-1\}$ using infinite polynomials $F_p = \{f_1 = 0, f_2, \ldots, f_p\}$, while p is a primary power, and $2 \leq k \leq p-1$, can be summarized as follows.

Proposition 4. *Let F_p be a finite field with order p. Let $\Gamma(x)$ be the set of polynomials with maximum degree of $\frac{1}{F_p}$. Then, the set F_Γ as formed above is a 2-sparse k-uniform family $F_s(2, k, kp)$ on X. Its cardinality is p^2 [58].*

From the junction of Propositions 2 and 4 we find that the incidence matrix F_Γ has the $(2, k)$-structure and a size of $(kp \times p^2)$. Also, it can be observed that any $(r-1, k)$-structure belongs to the set of k-uniform families r-sparse in a set X with cardinality m.

Finally, using the results of the Extreme Set Theory for the r sparse case one must limit the maximum size of each possible column within these matrices. Katona and Nemetz [91] proved the following result.

Proposition 5. *Let $F_s \subseteq [X]^k$ and F_s be an r-sparse family. Then*

$$|F_s| \leq \frac{\binom{m}{r}}{\binom{k}{r}}. \tag{5.39}$$

Thus, the column size of the matrix under construction of the $(r-1, k)$-structure is limited by $\frac{\binom{m}{r}}{\binom{k}{r}}$ where m is the number of rows in the structure, and k is the number of ones in columns, and $r-1$ represents the internal product between any two columns.

The theorem for the existence of such a sensing matrix having optimal column size is given below.

Theorem 1.

$$\forall (r, k), r < k, \quad \lim_{m\to\infty} n(m, k, r).\frac{\binom{k}{r}}{\binom{m}{r}} = 1 \tag{5.40}$$

where $n(m, k, r)$ indicates the maximum possible cardinality of F_s.

Thus for m large enough, using the Rodl [92] construct, it is possible to construct an (r, k)-binary detection structure.

5.6.3 Unbalanced expander graphs

In combinatorics, an Expander Graphs is a sparse graph whose strong connectivity properties are quantified using vertices, edges, or a spectral expansion. Its expansive constructions generate research in several areas such as complexity theory, computer network design that is robust, and the theory of error correction codes [93]. Namely, similar to BCH-code, in this chapter we will show how to use Unbalanced Expander Graphs to generate a binary sampling matrix.

For this, consider that $\|x\|_0$ defines the number of nonzero entries of the x signal, and that the notation $\{n\} = \{1, \ldots, n\}$ is used to any graph G formed by the vertices V and $v \in V$. We will represent the neighbor sets of v in G by $\Gamma_G(v)$ and the set including the edges v to X by $\Gamma(v, X) = \{v \in V, X \subset V\}$ [86].

Let $A = \{n\}$ and $B = \{k\}$. We have a bipartite graph $G = (A, B, E)$. d_A is the degree of nodes in A. In the acquired graph, the ϵ-extractor [94] $\forall\epsilon > 0$ will be specified afterwards. Thus, $\forall X \subset A$, $|X| \geq k$ considering D distribution of B is obtained by random acquirement of $a \in X$ and $b \in \Gamma(a)$. So

$$\sum_{b \in B} |\Pr_D(b) - \frac{1}{|B|}| \leq \epsilon. \tag{5.41}$$

For any $\epsilon > 0$, construct a ϵ-extractor such that $d_A = 2^{O(\log \log n)^E}$, $\forall E > 1$. For our demonstrations we will use $E = 2$ [86].

Thus for $t \in \mathbb{N}$, define

$$\text{overflow}_t(X) = \{b \in B, |\Gamma(b, X)| > t\}. \tag{5.42}$$

Choosing $t = 2.d_A$, we realize that $\forall b \in \text{overflow}_t(X)$, and we have $Pr_D(b) \geq \frac{2}{|B|}$. It follows that

$$|\text{overflow}_t(X)| \leq \epsilon.|B|. \tag{5.43}$$

So let H be a graph matrix adjacent to G. Consider $X = \{i \in \{n\} : x_i \neq 0\}$ where $|X| \leq r$. Using the G properties [86] for any $i \ni \text{overflow}_t(X)$, the $(Hx)_i$ value is the sum of the maximum t elements of X other than zero. By using the indices of these entries defined as $\{j_1^i, j_2^i, \ldots, j_t^i\}$, we can retrieve the set $J_i = \{j_1^i, \ldots, j_t^i\}$, and the sequence $U_i = x_{j_1^i}, \ldots, x_{j_t^i}$, but the recovery is incomplete because, at this stage, it is clear that i belongs to $\text{overflow}_t(X)$. To improve the recovery we must increase the measures of H for a bigger measurement set H', so having $H'x$, it would be possible to recover J_i, $\forall i \ni \text{overflow}_t(X)$.

For this, one can use the notion of row tensor product [95,96].

Definition 4. Having matrices $V_{v \times n}$ and $W_{w \times n}$, then the product $V \otimes W$ is a $v.w \times n$ matrix in such a way that $(V \otimes W)_{i.v+l,j} = V_{i,j}.W_{l,j}$, where $\otimes$ is a tensor product.

Indeed, $(V \otimes W)$ is made by copying each line from V to w times and implementing a coordinate multiplication of each copy with a corresponding line of W. Thus we can define $H' = (H \otimes B)$ which guarantees that we retrieve the set J_i and the sequence U_i for all i not in $Overflow_t(X)$ [86].

So to recover x from J_i we have to perform two steps.

1. One must construct a vector y that closely approximates x so that $\|x - y\|_0 = O(\epsilon.k)$.
2. Recursively apply a reduction of sparsity parameter k to fully recover x.

The algorithms used in step 1 and 2 are called REDUCE and RECOVER, respectively, and they can be found in more detail in [86].

5.7 Sensing matrices with statistical restricted isometry property

In compressed sensing, the $N \times C$ sampling matrix Φ is expected to serve as a proximal isometry on the set of k sparse signals. This guarantees the property of restricted isometry. If ϕ provides RIP, then the Basis Pursuit or Matching Pursuit algorithms are applicable for recovery of k-sparse vectors α out of N measurements Φ_α.

Some probabilistic methods make $N \times C$ matrices fulfilling RIP. However, there is a lack of practical algorithms for checking if a matrix Φ possesses RIP. This is crucial for the workability of recovery algorithms. In contrast, Ref. [97] gives simple rules for a deterministic sampling matrix fulfilling RIP.

RIP as theorized by Candès and Tao [56]: a matrix fulfills the k-Restricted Isometry Property if it performs in isometry on all k sparse vectors.

Both Basis Pursuit (BP) and Matching Pursuit (MP) depend on the multiplication of matrix and vector. They are super linear as regards the data dimension. The recovery algorithm for Reed–Muller codes that will be shown requires just vector to vector multiplication.

We may compare compressed sensing with the role of the structured codes in the communication practice, both need a fast encoding and decoding.

When working with deterministic matrices, it makes sense to work with a weaker Statistical RIP (StRIP), so we define the following.

Definition 5 ((k, ε, δ)-StRIP matrix). An $N \times C$ matrix Φ has the (k, ε, δ)-statistical RIP if

$$\forall k\text{-sparse } \boldsymbol{\alpha} \in \mathbb{R}^C (1-\varepsilon)\|\boldsymbol{\alpha}\|^2 \leq \|\frac{\boldsymbol{\Phi.\alpha}}{\sqrt{N}}\|^2 < (1+\varepsilon)\|\boldsymbol{\alpha}\|^2. \tag{5.44}$$

Let us consider a probability higher than $1 - \delta$ (concerning a uniform distribution for vectors α k-sparse in $\mathbb{R}^C$) [97].

Definition 6. An $N \times C$ matrix Φ is a (k, ε, δ)-uniqueness-guaranteed statistical RIP if Φ has (k, ε, δ)-StRIP, and

$$\{\beta \in \mathbb{R}^C; \Phi.\alpha = \Phi.\beta\} = \{\alpha\}. \tag{5.45}$$

With a probability higher than $1 - \delta$ (concerning a uniform distribution for k-sparse vectors $\boldsymbol{\alpha}$ in $\mathbb{R}^C$ of the same norm) [97].

About this topic, there are easy configuration rules. The columns of the sensing matrix make a collection under pointwise multiplication wherein all row sums disappear. Different rows are orthogonal and one expects an upper bound on the absolute value of any column sum.

In the paper [97] there are the basic definitions, lemmas and explanations about this weaker RIP, and the application of it for:

- Discrete Chirp Sensing Matrices
- Kerdock
- Delsarte–Goethals
- Second-Order Reed–Muller Sensing Matrices
- BCH Sensing Matrices

It is easy to see that with this simple criterion (a weaker RIP) as fulfilled by the matrix, one ensures recovery. Just an exponentially small part of k-sparse signals may not be recovered. These criteria are fulfilled by multiple families of deterministic matrices. It includes the one made of subcodes of the second-order binary Reed–Muller codes. It also applies to random Fourier ensembles. For further detailed information, we refer to Ref. [97].

5.8 Deterministic compressive matrix by algebraic curves over finite fields

These compressive sensing matrices are motivated by algebraic geometry codes. It is a novel deterministic structure via algebraic curves over finite fields. A real algebraic curve is a set of points on the Euclidean plane. Their coordinates are zeros of some polynomial with two variables. They have a diversity that gives us a choice on constructing sensing matrices.

By taking proper curves, we can create a binary-sensing matrix preferred to DeVore's matrices [63].

In Ref. [61], DeVore employs polynomials in finite field $\mathbb{F}_p$ for constructing $p^2 \times p^{(r+1)}$ binary-sensing matrices. p is a prime power. The constructed matrix has coherence $\frac{r}{p}$. It satisfies the RIP of order $k < \frac{p}{r} + 1$. This approach can be utilized for any finite field $\mathbb{F}_p$. Each polynomial yields an increase to a column of matrix. Remember that every polynomial develops a Reed–Solomon code-word.

By the algebraic curves and their function fields, there is a great potential for making deterministic compressive sensing matrices [98,99].

5.9 Summary

In this chapter, it becomes clear that, to construct a compressive sensing matrix, there are many ways to do it, as using random matrices, however, with some disadvantages, or to use deterministic matrices. There is a variety of deterministic matrices introduced in the literature, that are already known and easy to manipulate. The chapter gives a brief review of deterministic compressive sensing matrices.

It is worth noticing during this chapter that techniques, such as chirp sensing, can be used for harmonic analysis. A weaker RIP used for deterministic matrices and the Reed–Muller codes can be used to create deterministic sensing matrices. It should also be noted that there is a large number of techniques focused on binary signals, which are possibly relevant for the future of CS.

References

[1] T.W. Cabral, M. Khosravy, F.M. Dias, H.L.M. Monteiro, M.A.A. Lima, L.R.M. Silva, R. Naji, C.A. Duque, Compressive sensing in medical signal processing and imaging systems, in: Sensors for Health Monitoring, Elsevier, 2019, pp. 69–92.

[2] K. Melo, M. Khosravy, C. Duque, N. Dey, Chirp code deterministic compressive sensing: analysis on power signal, in: 4th International Conference on Information Technology and Intelligent Transportation Systems, IOS Press, 2020, pp. 125–134.

[3] S. Gupta, M. Khosravy, N. Gupta, H. Darbari, N. Patel, Hydraulic system onboard monitoring and fault diagnostic in agricultural machine, Brazilian Archives of Biology and Technology 62 (2019).

[4] S. Gupta, M. Khosravy, N. Gupta, H. Darbari, In-field failure assessment of tractor hydraulic system operation via pseudospectrum of acoustic measurements, Turkish Journal of Electrical Engineering & Computer Sciences 27 (4) (2019) 2718–2729.

[5] M.H. Sedaaghi, R. Daj, M. Khosravi, Mediated morphological filters, in: 2001 International Conference on Image Processing, Proceedings, vol. 3, IEEE, 2001, pp. 692–695.

[6] M. Khosravy, N. Gupta, N. Marina, I.K. Sethi, M.R. Asharif, Morphological filters: an inspiration from natural geometrical erosion and dilation, in: Nature-Inspired Computing and Optimization, Springer, Cham, 2017, pp. 349–379.

[7] M. Khosravy, M.R. Asharif, K. Yamashita, A PDF-matched short-term linear predictability approach to blind source separation, International Journal of Innovative Computing, Information & Control (IJICIC) 5 (11) (2009) 3677–3690.

[8] M. Khosravy, M.R. Asharif, K. Yamashita, A theoretical discussion on the foundation of Stone's blind source separation, Signal, Image and Video Processing 5 (3) (2011) 379–388.

[9] M. Khosravy, M.R. Asharif, K. Yamashita, A probabilistic short-length linear predictability approach to blind source separation, in: 23rd International Technical Conference on Circuits/Systems, Computers and Communications (ITC-CSCC 2008), Yamaguchi, Japan, ITC-CSCC, 2008, pp. 381–384.

[10] M. Khosravy, M.R. Alsharif, K. Yamashita, A PDF-matched modification to Stone's measure of predictability for blind source separation, in: International Symposium on Neural Networks, Springer, Berlin, Heidelberg, 2009, pp. 219–228.

[11] M. Khosravy, Blind source separation and its application to speech, image and MIMO-OFDM communication systems, PhD thesis, University of the Ryukyus, Japan, 2010.

[12] M. Khosravy, M. Gupta, M. Marina, M.R. Asharif, F. Asharif, I. Sethi, Blind components processing a novel approach to array signal processing: a research orientation, in: 2015 International Conference on Intelligent Informatics and Biomedical Sciences, ICIIBMS, 2015, pp. 20–26.

[13] M. Khosravy, N. Punkoska, F. Asharif, M.R. Asharif, Acoustic OFDM data embedding by reversible Walsh–Hadamard transform, AIP Conference Proceedings 1618 (2014) 720–723.

[14] M.H. Sedaaghi, M. Khosravi, Morphological ECG signal preprocessing with more efficient baseline drift removal, in: 7th IASTED International Conference, ASC, 2003, pp. 205–209.
[15] N. Dey, S. Samanta, X.-S. Yang, A. Das, S.S. Chaudhuri, Optimisation of scaling factors in electrocardiogram signal watermarking using cuckoo search, International Journal of Bio-Inspired Computation 5 (5) (2013) 315–326.
[16] M. Khosravy, M.R. Asharif, M.H. Sedaaghi, Morphological adult and fetal ECG preprocessing: employing mediated morphology, IEICE Technical Report, IEICE 107 (2008) 363–369.
[17] N. Dey, A.S. Ashour, F. Shi, S.J. Fong, R.S. Sherratt, Developing residential wireless sensor networks for ECG healthcare monitoring, IEEE Transactions on Consumer Electronics 63 (4) (2017) 442–449.
[18] M. Khosravi, M.H. Sedaaghi, Impulsive noise suppression of electrocardiogram signals with mediated morphological filters, in: 11th Iranian Conference on Biomedical Engineering, ICBME, 2004, pp. 207–212.
[19] N. Dey, S. Mukhopadhyay, A. Das, S.S. Chaudhuri, Analysis of P-QRS-T components modified by blind watermarking technique within the electrocardiogram signal for authentication in wireless telecardiology using DWT, International Journal of Image, Graphics and Signal Processing 4 (7) (2012) 33.
[20] M. Khosravy, N. Gupta, N. Patel, T. Senjyu, C.A. Duque, Particle swarm optimization of morphological filters for electrocardiogram baseline drift estimation, in: Applied Nature-Inspired Computing: Algorithms and Case Studies, Springer, 2020, pp. 1–21.
[21] E. Santos, M. Khosravy, M.A. Lima, A.S. Cerqueira, C.A. Duque, A. Yona, High accuracy power quality evaluation under a colored noisy condition by filter bank ESPRIT, Electronics 8 (11) (2019) 1259.
[22] E. Santos, M. Khosravy, M.A. Lima, A.S. Cerqueira, C.A. Duque, Esprit associated with filter bank for power-line harmonics, sub-harmonics and inter-harmonics parameters estimation, International Journal of Electrical Power & Energy Systems 118 (105) (2020) 731.
[23] A.A. Picorone, T.R. de Oliveira, R. Sampaio-Neto, M. Khosravy, M.V. Ribeiro, Channel characterization of low voltage electric power distribution networks for PLC applications based on measurement campaign, International Journal of Electrical Power & Energy Systems 116 (105) (2020) 554.
[24] M. Foth, R. Schroeter, J. Ti, Opportunities of public transport experience enhancements with mobile services and urban screens, International Journal of Ambient Computing and Intelligence (IJACI) 5 (1) (2013) 1–18.
[25] N. Gupta, M. Khosravy, N. Patel, T. Senjyu, A bi-level evolutionary optimization for coordinated transmission expansion planning, IEEE Access 6 (2018) 48455–48477.
[26] N. Gupta, M. Khosravy, K. Saurav, I.K. Sethi, N. Marina, Value assessment method for expansion planning of generators and transmission networks: a non-iterative approach, Electrical Engineering 100 (3) (2018) 1405–1420.
[27] M. Yamin, A.A.A. Sen, Improving privacy and security of user data in location based services, International Journal of Ambient Computing and Intelligence (IJACI) 9 (1) (2018) 19–42.
[28] C.E. Gutierrez, P.M.R. Alsharif, M. Khosravy, P.K. Yamashita, P.H. Miyagi, R. Villa, Main large data set features detection by a linear predictor model, AIP Conference Proceedings 1618 (2014) 733–737.
[29] C. Castelfranchi, G. Pezzulo, L. Tummolini, Behavioral implicit communication (BIC): communicating with smart environments, International Journal of Ambient Computing and Intelligence (IJACI) 2 (1) (2010) 1–12.
[30] G.V. Kale, V.H. Patil, A study of vision based human motion recognition and analysis, International Journal of Ambient Computing and Intelligence (IJACI) 7 (2) (2016) 75–92.
[31] S. Hemalatha, S.M. Anouncia, Unsupervised segmentation of remote sensing images using FD based texture analysis model and isodata, International Journal of Ambient Computing and Intelligence (IJACI) 8 (3) (2017) 58–75.
[32] M. Khosravy, N. Gupta, N. Marina, I. Sethi, M. Asharifa, Perceptual adaptation of image based on Chevreul–Mach bands visual phenomenon, IEEE Signal Processing Letters 24 (5) (2017) 594–598.
[33] A.S. Ashour, S. Samanta, N. Dey, N. Kausar, W.B. Abdessalemkaraa, A.E. Hassanien, Computed tomography image enhancement using cuckoo search: a log transform based approach, Journal of Signal and Information Processing 6 (03) (2015) 244.

[34] M. Khosravy, N. Gupta, N. Marina, I. Sethi, M. Asharif, Brain action inspired morphological image enhancement, in: Nature-Inspired Computing and Optimization, Springer, Cham, 2017, pp. 381–407.
[35] M. Khosravy, M.R. Asharif, M.H. Sedaaghi, Medical image noise suppression: using mediated morphology, IEICE, Technical Report 107 (461) (2008) 265–270.
[36] N. Dey, A.S. Ashour, A.S. Ashour, A. Singh, Digital analysis of microscopic images in medicine, Journal of Advanced Microscopy Research 10 (1) (2015) 1–13.
[37] M. Khosravy, M.R. Asharif, M.H. Sedaaghi, Medical image noise suppression using mediated morphology, in: IEICE Tech. Rep., IEICE, 2008, pp. 265–270.
[38] B. Alenljung, J. Lindblom, R. Andreasson, T. Ziemke, User experience in social human–robot interaction, in: Rapid Automation: Concepts, Methodologies, Tools, and Applications, IGI Global, 2019, pp. 1468–1490.
[39] M. Khosravy, M.R. Alsharif, M. Khosravi, K. Yamashita, An optimum pre-filter for ICA based multi-input multi-output OFDM system, in: 2010 2nd International Conference on Education Technology and Computer, vol. 5, IEEE, 2010, pp. V5–129.
[40] F. Asharif, S. Tamaki, M.R. Alsharif, H. Ryu, Performance improvement of constant modulus algorithm blind equalizer for 16 QAM modulation, International Journal on Innovative Computing, Information and Control 7 (4) (2013) 1377–1384.
[41] M. Khosravy, M.R. Alsharif, K. Yamashita, An efficient ICA based approach to multiuser detection in MIMO OFDM systems, in: Multi-Carrier Systems & Solutions 2009, Springer, 2009, pp. 47–56.
[42] M. Khosravy, M.R. Alsharif, B. Guo, H. Lin, K. Yamashita, A robust and precise solution to permutation indeterminacy and complex scaling ambiguity in BSS-based blind MIMO-OFDM receiver, in: International Conference on Independent Component Analysis and Signal Separation, Springer, 2009, pp. 670–677.
[43] M. Khosravy, A blind ICA based receiver with efficient multiuser detection for multi-input multi-output OFDM systems, in: The 8th International Conference on Applications and Principles of Information Science (APIS), Okinawa, Japan, 2009, 2009, pp. 311–314.
[44] M. Khosravy, S. Kakazu, M.R. Alsharif, K. Yamashita, Multiuser data separation for short message service using ICA, SIP, IEICE Technical Report 109 (435) (2010) 113–117.
[45] M. Baumgarten, M.D. Mulvenna, N. Rooney, J. Reid, Keyword-based sentiment mining using Twitter, International Journal of Ambient Computing and Intelligence (IJACI) 5 (2) (2013) 56–69.
[46] C.E. Gutierrez, M.R. Alsharif, K. Yamashita, M. Khosravy, A tweets mining approach to detection of critical events characteristics using random forest, International Journal of Next-Generation Computing 5 (2) (2014) 167–176.
[47] N. Kausar, S. Palaniappan, B.B. Samir, A. Abdullah, N. Dey, Systematic analysis of applied data mining based optimization algorithms in clinical attribute extraction and classification for diagnosis of cardiac patients, in: Applications of Intelligent Optimization in Biology and Medicine, Springer, 2016, pp. 217–231.
[48] P. Sosnin, Precedent-oriented approach to conceptually experimental activity in designing the software intensive systems, International Journal of Ambient Computing and Intelligence (IJACI) 7 (1) (2016) 69–93.
[49] A. Cohen, W. Dahmen, R. DeVore, Compressed sensing and best k-term approximation, Journal of the American Mathematical Society 22 (1) (2009) 211–231.
[50] E.J. Candes, T. Tao, Decoding by linear programming, IEEE Transactions on Information Theory 51 (12) (2005) 4203–4215.
[51] E.J. Candès, J. Romberg, T. Tao, Robust uncertainty principles: exact signal reconstruction from highly incomplete frequency information, IEEE Transactions on Information Theory 52 (2) (2006) 489–509.
[52] E.J. Candes, J.K. Romberg, T. Tao, Stable signal recovery from incomplete and inaccurate measurements, Communications on Pure and Applied Mathematics: A Journal Issued by the Courant Institute of Mathematical Sciences 59 (8) (2006) 1207–1223.
[53] D.L. Donoho, et al., Compressed sensing, IEEE Transactions on Information Theory 52 (4) (2006) 1289–1306.

[54] H. Liu, H. Zhang, L. Ma, On the spark of binary LDPC measurement matrices from complete protographs, IEEE Signal Processing Letters 24 (11) (2017) 1616–1620.
[55] D.L. Donoho, M. Elad, Optimally sparse representation in general (nonorthogonal) dictionaries via l1 minimization, Proceedings of the National Academy of Sciences 100 (5) (2003) 2197–2202.
[56] E.J. Candes, T. Tao, Near-optimal signal recovery from random projections: universal encoding strategies?, IEEE Transactions on Information Theory 52 (12) (2006) 5406–5425.
[57] E.J. Candes, The restricted isometry property and its implications for compressed sensing, Comptes Rendus Mathématique 346 (9–10) (2008) 589–592.
[58] R.R. Naidu, C.R. Murthy, Construction of binary sensing matrices using extremal set theory, IEEE Signal Processing Letters 24 (2) (2017) 211–215.
[59] H. Rauhut, Compressive sensing and structured random matrices, in: Theoretical Foundations and Numerical Methods for Sparse Recovery, vol. 9, 2010, pp. 1–92.
[60] A. Gilbert, P. Indyk, Sparse recovery using sparse matrices, Proceedings of the IEEE 98 (6) (2010) 937–947.
[61] R.A. DeVore, Deterministic constructions of compressed sensing matrices, Journal of Complexity 23 (4) (2007) 918–925.
[62] A. Amini, F. Marvasti, Deterministic construction of binary, bipolar, and ternary compressed sensing matrices, IEEE Transactions on Information Theory 57 (4) (2011) 2360–2370.
[63] S. Li, F. Gao, G. Ge, S. Zhang, Deterministic construction of compressed sensing matrices via algebraic curves, IEEE Transactions on Information Theory 58 (8) (2012) 5035–5041.
[64] X.-J. Liu, S.-T. Xia, Constructions of quasi-cyclic measurement matrices based on array codes, in: 2013 IEEE International Symposium on Information Theory Proceedings (ISIT), IEEE, 2013, pp. 479–483.
[65] S.-T. Xia, X.-J. Liu, Y. Jiang, H.-T. Zheng, Deterministic constructions of binary measurement matrices from finite geometry, IEEE Transactions on Signal Processing 63 (4) (2015) 1017–1029.
[66] L. Applebaum, S.D. Howard, S. Searle, R. Calderbank, Chirp sensing codes: deterministic compressed sensing measurements for fast recovery, Applied and Computational Harmonic Analysis 26 (2) (2009) 283–290.
[67] W. Zhang, K. Jahoda, J. Swank, E. Morgan, A. Giles, Dead-time modifications to fast Fourier transform power spectra, The Astrophysical Journal 449 (1995) 930.
[68] S.D. Howard, A.R. Calderbank, W. Moran, The finite Heisenberg–Weyl groups in radar and communications, EURASIP Journal on Applied Signal Processing 2006 (2006) 111.
[69] S.D. Howard, A.R. Calderbank, S.J. Searle, A fast reconstruction algorithm for deterministic compressive sensing using second order Reed–Muller codes, in: 42nd Annual Conference on Information Sciences and Systems, CISS 2008, IEEE, 2008, pp. 11–15.
[70] D. Baron, M.F. Duarte, S. Sarvotham, M.B. Wakin, R.G. Baraniuk, An information-theoretic approach to distributed compressed sensing, in: Proc. 45th Conference on Communication, Control, and Computing, 2005.
[71] D.L. Donoho, J. Tanner, Neighborliness of randomly projected simplices in high dimensions, Proceedings of the National Academy of Sciences 102 (27) (2005) 9452–9457.
[72] D.L. Donoho, High-dimensional centrally symmetric polytopes with neighborliness proportional to dimension, Discrete & Computational Geometry 35 (4) (2006) 617–652.
[73] M. Khosravy, N. Gupta, N. Patel, T. Senjyu, Frontier Applications of Nature Inspired Computation, Springer, 2020.
[74] N. Gupta, M. Khosravy, O.P. Mahela, N. Patel, Plant biology-inspired genetic algorithm: superior efficiency to firefly optimizer, in: Applications of Firefly Algorithm and Its Variants, Springer, 2020, pp. 193–219.
[75] C. Moraes, E. De Oliveira, M. Khosravy, L. Oliveira, L. Honório, M. Pinto, A hybrid bat-inspired algorithm for power transmission expansion planning on a practical Brazilian network, in: Applied Nature-Inspired Computing: Algorithms and Case Studies, Springer, 2020, pp. 71–95.
[76] G. Singh, N. Gupta, M. Khosravy, New crossover operators for real coded genetic algorithm (RCGA), in: 2015 International Conference on Intelligent Informatics and Biomedical Sciences (ICIIBMS), IEEE, 2015, pp. 135–140.

[77] N. Gupta, N. Patel, B.N. Tiwari, M. Khosravy, Genetic algorithm based on enhanced selection and log-scaled mutation technique, in: Proceedings of the Future Technologies Conference, Springer, 2018, pp. 730–748.
[78] N. Gupta, M. Khosravy, N. Patel, I. Sethi, Evolutionary optimization based on biological evolution in plants, Procedia Computer Science 126 (2018) 146–155.
[79] J. Kaliannan, A. Baskaran, N. Dey, A.S. Ashour, M. Khosravy, R. Kumar, ACO based control strategy in interconnected thermal power system for regulation of frequency with HAE and UPFC unit, in: International Conference on Data Science and Application (ICDSA-2019), in: LNNS, Springer, 2019.
[80] N. Gupta, M. Khosravy, N. Patel, O. Mahela, G. Varshney, Plants genetics inspired evolutionary optimization: a descriptive tutorial, in: Frontier Applications of Nature Inspired Computation, Springer, 2020, pp. 53–77.
[81] M. Khosravy, N. Gupta, N. Patel, O. Mahela, G. Varshney, Tracing the points in search space in plants biology genetics algorithm optimization, in: Frontier Applications of Nature Inspired Computation, Springer, 2020, pp. 180–195.
[82] N. Gupta, M. Khosravy, N. Patel, S. Gupta, G. Varshney, Evolutionary artificial neural networks: comparative study on state of the art optimizers, in: Frontier Applications of Nature Inspired Computation, Springer, 2020, pp. 302–318.
[83] N. Gupta, M. Khosravy, N. Patel, S. Gupta, G. Varshney, Artificial neural network trained by plant genetics-inspired optimizer, in: Frontier Applications of Nature Inspired Computation, Springer, 2020, pp. 266–280.
[84] J.A. Salehi, Code division multiple-access techniques in optical fiber networks. I. fundamental principles, IEEE Transactions on Communications 37 (8) (1989) 824–833.
[85] J. Bourgain, S. Dilworth, K. Ford, S. Konyagin, D. Kutzarova, et al., Explicit constructions of rip matrices and related problems, Duke Mathematical Journal 159 (1) (2011) 145–185.
[86] P. Indyk, Explicit constructions for compressed sensing of sparse signals, in: Proceedings of the Nineteenth Annual ACM-SIAM Symposium on Discrete Algorithms, Society for Industrial and Applied Mathematics, 2008, pp. 30–33.
[87] R.C. Bose, D.K. Ray-Chaudhuri, On a class of error correcting binary group codes, Information and Control 3 (1) (1960) 68–79.
[88] C. Ding, C. Xing, Several classes of $(2^m - 1, w, 2)$ optical orthogonal codes, Discrete Applied Mathematics 128 (1) (2003) 103–120.
[89] S. Lin, D.J. Costello, Error Control Coding: Fundamentals and Applications, 2nd edition, Prentice Hall, Englewood Cliffs, 2004.
[90] P. Keevash, The existence of designs, preprint, arXiv:1401.3665, 2014.
[91] G. Katona, T. Nemetz, M. Simonovits, On a graph-problem of Turan, Matematikai Lapok 15 (1964) 228–238.
[92] V. Rodl, On a packing and covering problem, European Journal of Combinatorics 6 (1) (1985) 69–78.
[93] S. Hoory, S. Hoory, N. Linial, A. Wigderson, Bulletin of the American Mathematical Society 43 (2006) 439.
[94] R. Shaltiel, Recent developments in explicit constructions of extractors, in: Current Trends in Theoretical Computer Science: The Challenge of the New Century 7, 2004.
[95] Graham Cormode, S. Muthukrishnan, Combinatorial algorithms for compressed sensing, in: International Colloquium on Structural Information and Communication Complexity, 2006.
[96] A. Gilbert, M.J. Strauss, J. Tropp, R. Vershynin, Algorithmic linear dimension reduction in the ℓ_1 norm for sparse vectors, preprint, arXiv:cs/0608079, 2006.
[97] R. Calderbank, S. Howard, S. Jafarpour, Construction of a large class of deterministic sensing matrices that satisfy a statistical isometry property, IEEE Journal of Selected Topics in Signal Processing 4 (2) (2010) 358–374.
[98] H. Niederreiter, C. Xing, Rational Points on Curves Over Finite Fields: Theory and Applications, vol. 288, Cambridge University Press, 2001.
[99] H. Stichtenoth, Algebraic Function Fields and Codes, Vol. 254, Springer Science & Business Media, 2009.

CHAPTER

6 Deterministic compressive sensing by chirp codes: a descriptive tutorial

Mahdi Khosravy[a], Neeraj Gupta[b], Nilesh Patel[b], Carlos A. Duque[c], Naoko Nitta[a], Noboru Babaguchi[a]

[a]*Media Integrated Communication Laboratory, Graduate School of Engineering, Osaka University, Suita, Osaka, Japan*

[b]*Computer Science and Engineering, Oakland University, Rochester, MI, United States*

[c]*Department of Electrical Engineering, Federal University of Juiz de Fora, Juiz de Fora, Brazil*

6.1 Introduction

Compressive sensing [1–5] is a novel approach in signal and image acquisition where data sampling is carried out by much fewer sensory data than the Shannon–Nyquist theorem of sampling requirement. It is by deployment of the informative aspect of characteristics of signals where naturally they possess a sparse nature. Due to the sparseness, much fewer signal samples are picked up by the few sensors that can carry a most valuable information content of the signal/image. This brings about a considerable saving advantage in the required equipment for sampling, the required data transition bandwidth, the required memory storage, and after all the required energy. The potential application realm of compressive sensing has a wide range: data mining [6,7], medical image processing [8–10], image enhancement [11,12], text processing [13], agriculture on-board data processing [14,15], signal processing [16,17], image adaptation [18] acoustic OFDM [19], telecommunications [20–25], Electrocardiogram processing [26–32], power line communications [33], power quality analysis [34–36], etc.

This chapter presents a tutorial to deterministic compressive sensing by chirp codes [37]; we call this 'chirp code compressive sensing' (CC-CS). In CC-CS, the sensing matrix is composed of columns made of chirp sequences. A very innovative aspect of CC-CS is the deployment of Fast Fourier Transforms (FFTs) at the recovery stage. The chapter gives a tutorial on the algorithm and implementation of CC-CS apart from the theoretical aspects in noise mitigation and further improvement of the technique. CC-CS is an interesting approach due to its deterministic nature, easy implementation and after all its low computation complexity, which is $O(K \log K)$ for K measurements.

Compressive Sensing in Healthcare. https://doi.org/10.1016/B978-0-12-821247-9.00011-1

Compressive sensing is based on the sparsity nature of the signal as in signal/image compression, the sparse domain of the signal is used as a means to compress by first taking the signal/image to a sparse domain, and then keeping the location and values of the dominant samples instead of the whole signal content. Compressive sensing is even more impressive, using the same concept of sparsity, not only in that the signal can be saved in compressed format in less memory storage, but the measurement can be directly in the compressed domain. Therefore, a much smaller number of measurements is required, which brings about lots of advantages, such as less requirement to the sensory devices. This especially shows its impact when the sensory devices are expensive in applications like medical magnetic resonance imaging, or they are energy and time consuming like a variety of medical imaging systems.

While using random techniques is very common in compressive sensing, deployment of pure random codes is not practically possible due to the high computational load of channel encoding and decoding. Also in random compressive sensing, norm-1 minimization has been in use for signal recovery that has a high computational load. To realize the compressive sensing in practical systems, fast recovery is highly important. CC-CS possesses fast recovery due to its deterministic structure which is explained in detail in this chapter.

The chapter's organization is as follows. Section 6.2 briefly explains the standard compressive sensing. Section 6.3 explains how the signal can be compressively sensed by chirp codes, and Section 6.4 describes the recovery process of the compressively sensed signal by this technique. Finally, Section 6.5 presents a summary of the chapter.

6.2 The standard formulation compressive sensing

Here, briefly, the standard formulation of compressive sensing is given. Having an n-length signal vector $\boldsymbol{x}$ in vector space of $\mathbb{R}^n$, it is compressively sensed as an m-length vector $\boldsymbol{y}$ in the vector space $\mathbb{R}^m$ through m linear combinations of n samples of the signal vector. In mathematical representation:

$$y_{m\times1} = A_{m\times n}\boldsymbol{x}_{n\times1}. \tag{6.1}$$

Note that the term 'compressive sensing' originates from the fact that m is much shorter than n, $m << n$, and instead of having data from n separated data sensors, the observations are m mixtures taken from linear mixtures of n sensors. A is the compressive sensing matrix. Fig. 6.1 depicts the general concept of compressive sensing. A can be indeed composed of two transforms as

$$A = \Psi\Phi \tag{6.2}$$

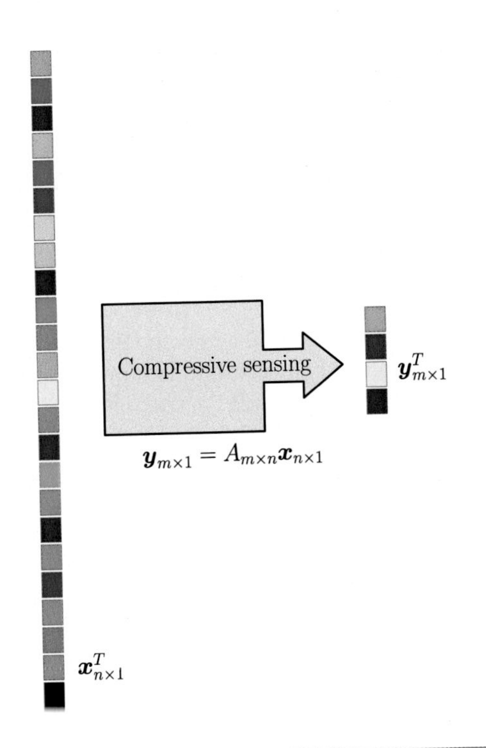

FIGURE 6.1

General schematic of compressive sensing.

where Φ transforms the signal to a sparse basis, and Ψ is the compressed sensing measurement. These two can be combined in one sensing matrix, A, which can be freely designed.

Compressive sensing is a double-sided problem where one side is the sensory side, and the other side is the recovery of the sensed data. At the recovery side where the signal vector $\boldsymbol{x}$ should be obtained from a much shorter length vector signal $\mathbf{y}$, at first glance it sounds like an ill-posed problem because one is looking for n variables as the elements of the signal vector $\boldsymbol{x}$ through a set of m linear equations while the number of variables is much smaller than the number of equations. Eq. (6.1) can be written in a set of linear equations as follows:

$$\begin{aligned} a_{11}x_1 + a_{12}x_2 + \cdots + a_{1n}x_n &= y_1, \\ a_{21}x_1 + a_{22}x_2 + \cdots + a_{2n}x_n &= y_2, \\ &\vdots \\ a_{m1}x_1 + a_{m2}x_2 + \cdots + a_{mn}x_n &= y_m. \end{aligned} \tag{6.3}$$

To have a solution for n variables, at least n independent equations must be available The problem looks like ill-posed because of the number of equations is much smalle than variables. In such an incomplete set of equations, there is an infinite number of possible solutions. So, can the true solution for $x_1, x_2, \ldots, x_n$ representing the compressive sensed vector $\boldsymbol{x}$ be found?

One answer to this question is that the signal should be sensed in a deterministic way with a corresponding deterministic recovery as is the case of this chapter. But in the case of non-deterministic sensing and recovery, the answer is quite interesting. Since compressive sensing is suggested for $\boldsymbol{x}$ with a sparsity level of k with at most k non-zero elements, if A fulfills some properties then the recovery is possible by $\boldsymbol{x}$ norm-1 minimization [38]. In this respect, the recovery solution for non-deterministic compressive sensing is comparable to blind source separation (BSS) [39–44]. In both, among infinite possible solutions, just one of them maximizes a characteristic measure of signal which is the sparseness in the case of compressive sensing and independence in the case of BSS. The defined measure of k-sparseness is by norm-0 so that its minimization has the nature of an NP-hard problem that cannot be managed by linear classic optimization techniques. It may be resolved by metaheuristic evolutionary optimization [45–55]. To overcome nonlinearity of norm-0, its replacement by norm-1 has been recommended.

In deterministic compressive sensing, the signal vector is sensed in a certain way with a corresponding certain recovery process. It is like an encoding–decoding technique. This chapter gives a descriptive tutorial to one of the effective deterministic compressive sensings that is by chirp codes.

6.3 Compressive sensing by chirp codes

Chirp Code Compressive Sensing (CC-CS) [37] is a deterministic approach by the employment of chirp codes. In this chapter, we maintain the consistency of the notation with the main reference of CC-CS [37]. Table 6.1 gives a summary of the notations.

Chirp codes are amended as the columns of the compressive sensing matrix, and at the recovery side, the characteristics of the chirp codes are deployed to extract the sparse signal vector. A chirp code is a signal vector of length K with two frequency natures, a single tone periodic part and an increasing all-frequency swiping tone that is the chirp tone. These two parts are multiplicative components. The K-length chirp code is as follows:

$$\boldsymbol{v}_{m,r}(l) = \alpha.\exp\left(j2\pi\frac{ml}{K} + j2\pi\frac{rl^2}{K}\right), \qquad l, m, r \in \mathbb{Z}_K; \tag{6.4}$$

$\exp(j2\pi\frac{m}{K}l)$ is the single tone multiplicative component, and $\exp(j2\pi\frac{r}{K}l^2)$ is the chirp tone multiplicative component. The first component contains the 'base frequency' m, and the chirp component contains 'the chirp rate' r, that is, the rate of chirping amongst the frequencies. The parameter α is a scalar and can take any value for simplification.

Table 6.1 Acronyms.

Symbol	Description
M	Sparsity level
K	The length of chirp codes
$\boldsymbol{x}$	The signal vector
$\boldsymbol{y}$	The compressive sensed measurement
m	Base frequency of a chirp code
r	The chirp rate of chirp code
$\boldsymbol{v}_{m,r}$	A chirp code with base freq. of m and chirp rate of r
l	Vector element index for chirp code
$\mathbb{Z}_K$	The space of integers with absolute value less than K

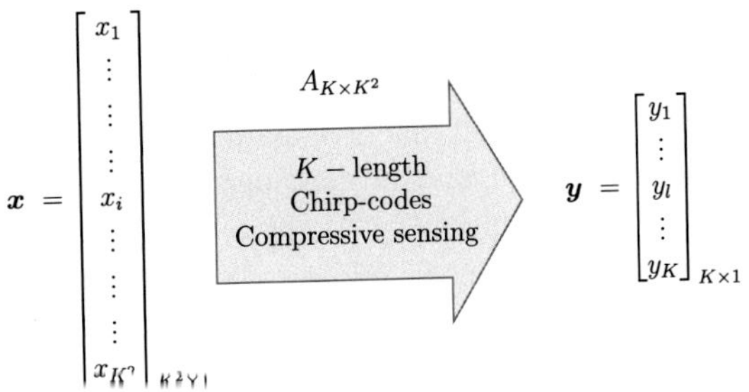

FIGURE 6.2

General schematic of compressive sensing by a matrix made of K-length chirp codes.

6.3.1 Building the compressive sensing matrix by chirp codes

Besides K that is the length of the codes, the m and r are two main parameters of a chirp code. There are K different values for each of $\frac{2\pi m}{K}$ and $\frac{2\pi r}{K}$ for all the positive integers of m and r, since they are taken from the $\mathbb{Z}_K$ space i.e. $\{0, 1, 2, \ldots, K-1\}$. Therefore, there are K^2 distinct chirp codes made by K^2 possible (m, r) tuples. The deterministic compressive sensing matrix A is formed by K^2 columns of K-lengths chirp codes as follows:

$$A_{K\times K^2} = [\boldsymbol{v}_0, \boldsymbol{v}_1, \ldots, \boldsymbol{v}_k, \ldots, \boldsymbol{v}_{K^2-1}], \qquad k = Kr + m. \tag{6.5}$$

Therefore, the elements of A are as follows:

$$[A]_{l,k} = e^{\frac{j2\pi rl^2}{K}} e^{\frac{j2\pi ml}{K}}, \qquad k = Kr + m, \tag{6.6}$$

where $r \in \mathbb{Z}_K$, $m \in \mathbb{Z}_K$. As shown in Fig. 6.2, the above compressive sensing matrix A by chirp codes has the compressive factor of K.

As the compressive sensing matrix is built, the main question is:

- Does the above $K \times K^2$ matrix have the required characteristics for compressively sensing an M-sparse signal vector?

This question should be answered by checking the Restricted Isometry Property (RIP) [38].

6.3.2 Restricted isometry property for the chirp code compressive sensing

If the compressive sensing matrix fulfills the RIP, then theoretically the recovery of the compressively sensed signals is guaranteed. The RIP for the compressive sensing matrix A indicated by Eq. (6.5) is supported as there exists a minimum value for δ_M:

$$\forall \boldsymbol{x}, |\Lambda|_1 = M, \qquad (1-\delta_M)\|\boldsymbol{x}\|^2 \le \|A_\Lambda \boldsymbol{x}\|^2 \le (1+\delta_M)\|\boldsymbol{x}\|^2, \tag{6.7}$$

where A_Λ is a sub-matrix made by the columns of A corresponding to an M sparse support vector of Λ. Indeed, although Λ is of the same length as $\boldsymbol{x}$, it has M non-zero elements. While A columns have been scaled to have a unit norm, δ_M exists and it is the same as the bounds on the eigenvalues of $A_\lambda^H A_\lambda$ which are scaled to one.

Theorem 1 of Ref. [37] indicates that in order to follow the *theorem of uniqueness* [56], in other words in order to have at most one M-sparse signal vector $\boldsymbol{x}$ for a compressively sensed vector $\boldsymbol{y}$ by A, we have the following relation between sparsity level of $\boldsymbol{x}$ and the length of chirp codes K:

$$M < \frac{\sqrt{K}+1}{2} \tag{6.8}$$

or

$$K > (2M-1)^2. \tag{6.9}$$

Since the size of the compressive matrix is $K \times K^2$, the above inequalities show the relation between the compression level and the sparsity level of the signal vectors. Therefore, the relation between the chirp code compressive sensing problem dimensions can be indicated in the two following directions:

1. The chirp code compressive sensing with K times compression is possible if M, the sparsity level, is less than $\frac{\sqrt{K}+1}{2}$.
2. If the signal vectors are of M-sparsity then the chirp codes should be longer than $(2M-1)^2$.

It gives an upper bound for sparsity level and a lower bound for the A dimensions.

6.4 Recovery of chirp code compressively sensed signal

The recovery of the compressively sensed signal vector by a chirp sensing matrix is indeed the discovery of which linear combination of the columns in A makes $\boldsymbol{y}$. Consider the M-sparse signal vector

$$\boldsymbol{x} = [x_1, x_2, \ldots, x_{K^2}]^T. \tag{6.10}$$

Having the chirp sensing matrix A, and compressive sense of $\boldsymbol{x}$ as $\boldsymbol{y} = [y_1, \ldots y_l, \ldots y_K]$, then each y_l is as follows:

$$y_l = x_1 \exp\big(j2\pi \frac{m_1}{K} l + j2\pi \frac{r_1}{K} l^2\big) + x_2 \exp\big(j2\pi \frac{m_2}{K} l + j2\pi \frac{r_2}{K} l^2\big) + \ldots \quad (6.11)$$

Note that m_i and r_i are, respectively, the base frequency and chirp rate. As it is observable, each y_l has been composed in components such that each of them is made of multiplication of a sinusoid exponential and chirp sinusoid exponential as $e^{j2\pi \frac{m_1}{K} l} \times e^{j2\pi \frac{r_1}{K} l^2}$. The chirp sinusoid exponential is explained in the appendix.

The key solution to deterministic recovery of the compressively sensed x_i values is by forming the following vector of $\boldsymbol{f}$ from $\boldsymbol{y}$:

$$\boldsymbol{f} = \begin{bmatrix} f_1 \\ \vdots \\ f_l \\ \vdots \\ f_K \end{bmatrix}, \quad (6.12)$$

where each f_l is as follows:

$$\begin{aligned} f_l &= y_l^* y_{l+T} \\ &= |x_1|^2 e^{\frac{j2\pi}{K}(m_1 T + r_1 T^2)} e^{j2\pi \frac{2r_1 T}{K} l} + |x_2|^2 e^{\frac{j2\pi}{K}(m_2 T + r_2 T^2)} e^{j2\pi \frac{2r_2 T}{K} l} + \cdots + \mathrm{CTs}, \end{aligned} \quad (6.13)$$

where y^* indicates the complex conjugate of y, and T is acquired as any non-zero value in $\mathbb{Z}_K$; also the $l + T$ in the above formula is indeed $l + T$ mod K where mod K indicates modulo K. CTs are the cross-terms as follows:

$$\mathrm{CTs} = \sum_{i, j \neq i} x_i x_j^* e^{\frac{j2\pi}{K}(m_i T + r_i T^2)} e^{\frac{j2\pi}{K} l (m_i - m_j + 2T r_i) + \frac{j2\pi}{K} l^2 (r_i - r_j)}. \quad (6.14)$$

It can be observed that f_l is composed of two types of elements; sinusoids of frequencies $\frac{2r_i T}{K}$ and chirp cross components. A chirp component energy is spread along its frequency spectrum while the energy of a sinusoid with frequency $\frac{2r_i T}{K}$ concentrates at a single frequency. This is the key to detecting the r_i and extract their corresponding components from y_l because an FFT from f_l gives us its frequency bins $\frac{2r_i T}{K}$'s as points of a spectrum showing power concentration. If for every obtained frequency bin $\frac{2r_i T}{K}$, there is one and only one corresponding chirp rate of r_i, then we can obtain the chirp rates. This means there must be a bijection between the r_i and the $\frac{2r_i T}{K}$. The following lemma gives the required condition for fulfilling this bijection as it indicates the condition as K must be a prime number. Before presenting the lemma and its proof, note that, since the r_i are taken as integer values from $\mathbb{Z}_K$, the frequency of $\frac{2r_i T}{K}$ is indeed equivalent to $\frac{2r_i T \bmod K}{K}$. Therefore the two sets $\{r_i \mid r_i \in \mathbb{Z}_K\}$, and

$\{\frac{2r_i T \bmod K}{K} \mid r_i \in \mathbb{Z}_K\}$ have K members, and the above-mentioned bijection must be fulfilled between these two sets.

Lemma 1. *If K is prime, there is a bijection between the set of chirp rates $\{r_i \mid r_i \in \mathbb{Z}_K\}$ and the set of FFT bins $\{\frac{2r_i T \bmod K}{K} \mid r_i \in \mathbb{Z}_K\}$, formally,*

$$K \in \mathbb{P} \quad \rightarrow \quad \{r_i \mid r_i \in \mathbb{Z}_K\} \twoheadrightarrow \{\frac{2r_i T \bmod K}{K} \mid r_i \in \mathbb{Z}_K\} \tag{6.15}$$

where $\twoheadrightarrow$ indicates the bijection between two sets, and $\mathbb{P}$ is the set of prime numbers.

Proof. The bijection between two sets means that for every element of each set there is one and only one corresponding element in the other set. Mathematically, for $r_1, r_2 \in \{r_i\}$ both the following conditions must be fulfilled:

1.

$$r_1 \neq r_2 \rightarrow \frac{2r_1 T \bmod K}{K} \neq \frac{2r_2 T \bmod K}{K}, \tag{6.16}$$

2.

$$\frac{2r_1 T \bmod K}{K} \neq \frac{2r_2 T \bmod K}{K} \rightarrow r_1 \neq r_2. \tag{6.17}$$

We start by an implication (6.17) that is so simple. Having $\frac{2r_1 T \bmod K}{K} \neq \frac{2r_2 T \bmod K}{K}$, let us suppose $r_1 = r_2$, then as a matter of fact $\frac{2r_1 T \bmod K}{K} = \frac{2r_2 T \bmod K}{K}$, that is, despite the antecedent of the implication, so $r_1 \neq r_2$.

Now we prove the implication (6.16). Suppose $r_1 \neq r_2$, and $\frac{2r_1 T \bmod K}{K} = \frac{2r_2 T \bmod K}{K}$. That means

$$2r_1 T \bmod K = 2r_2 T \bmod K = \mathcal{R}. \tag{6.18}$$

In other words

$$\begin{aligned} 2r_1 T &= \alpha K + \mathcal{R}, \\ 2r_2 T &= \beta K + \mathcal{R}. \end{aligned} \tag{6.19}$$

The difference of the above equations gives

$$2r_3 T = \gamma K, \tag{6.20}$$

where $r_3 = |r_1 - r_2| \neq 0, \in \mathbb{Z}_K$, $\gamma = |\alpha - \beta| \in \mathbb{Z}_K$, $\gamma \neq 0$. At this point we have the equality of (6.20), for which, in order to be fulfilled, one of the following should be true:

1. If $\gamma = 1$, then $K - 2r_3 T$. Since each of 2, r_3, and T divides K, K should be either 2 (i.e. $r_3 = 1$, $T = 1$) or not a prime number.
2. If $\gamma = 2$, then $K = r_3 T$. Since both r_3, and T belong to $\mathbb{Z}_K$, they can be dividers of K just if K is not a prime number, or it is 2.

3. $\gamma > 2$, then substituting r_3 and T by multiplication of their dividers, $2r_3T = \gamma K$ can be written as

$$2\prod_i r_i = \gamma K. \tag{6.21}$$

Taking γ to be considered as any combination of r_is, then K is the product of the remaining r_i but cannot be any of them alone since all belong to $\mathbb{Z}_k$. As well, in this case, K is not a prime number.

Therefore, since K is taken to be higher than 2, we can ensure the bijection by taking K as a prime number. □

After finding each dominant chirp rate via its dominant corresponding $2r_iT \bmod K$ in f_l, the signal vector y_l is dechirped. Dechirping the signal vector y_l from the chirp rate r_i is via its multiplication by $e^{-\frac{j2\pi r_i l^2}{K}}$. By dechirping y_l from the chirp rate r_i, the corresponding component of $x_i e^{\frac{j2\pi m_i l}{K}+\frac{j2\pi r_i l^2}{K}}$ will be the only component in the result for which its chirp rate r_i has been canceled and it is transferred to a sinusoid of the frequency $\frac{m_i}{K}$. Therefore the FFT of the dechirped signal vector of $y_l e^{-\frac{j2\pi r_i l^2}{K}}$ gives the values of x_i and m_i as, respectively, the amplitude and frequency of the dominant FFT bin. At this stage the triple of (r_i, m_i, x_i) gives a sample of signal vector $\boldsymbol{x}$ as x_i is the sample value, and $Kr_i + m_i$ is the location of the extracted sample. Having this extracted sample, its corresponding component is removed from y_l as follows:

$$y_l := y_l - x_i \exp\left(j2\pi \frac{m_i}{K} l + j2\pi \frac{r_i}{K} l^2\right). \tag{6.22}$$

The above update of y_l reduces its energy by removing a dominant component. Having the updated y_l, then at the next step f_l is also updated as in Eq. (6.13), the rest of the tuples (x_i, r_i, m_i) are obtained in the same way until all the signal vector $\boldsymbol{x}$ samples are obtained. This process can be stopped by considering a level for the remained energy in y_l after updating at each iteration.

6.4.1 Step by step in one iteration of recovery

The recovery of the compressively sensed signal vector $\boldsymbol{x}$ by the chirp-codes deterministic compressive sensing matrix as explained in the preceding section can be summarized in the following loop of steps.

Step 1: *Choose T and ϵ.*
The parameter ϵ is the ignorable energy level of the updated y_l as a stopping criterion should be chosen first. After each update of y_l, its energy goes lower because of removing the corresponding component of the last detected sample of $\boldsymbol{x}$. The parameter T is chosen from non-zero members of $\mathbb{Z}_K$.

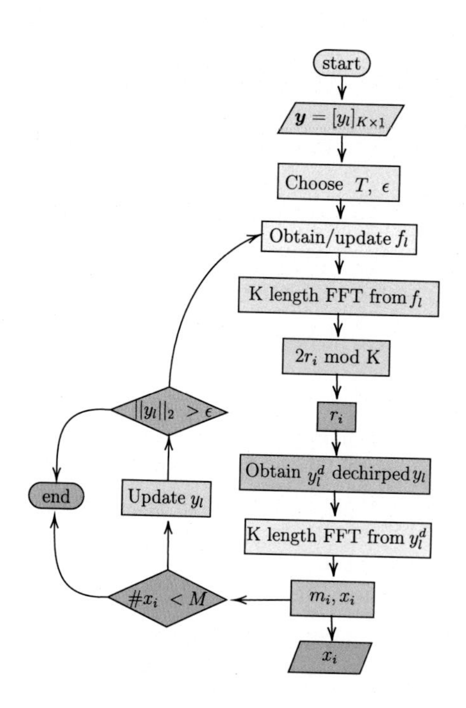

FIGURE 6.3

The block diagram of the recovery process in chirp code compressive sensing.

Step 2: *Update f_l using the last update of y_l.*
Having y_l the initial one or the updated one, then, by using Eq. (6.13), f_l can be updated.

Step 3: *Take K-length FFT from f_l.*

Step 4: *Extract $2r_i T \text{mod} K$ from the FFT sequence as its peak point, and obtain the corresponding r_i.*

Step 5: *Obtain the dechirped y_l as y_l^d using the obtained r_i.*
Dechirping is by multiplying y_l by $e^{-\frac{j2\pi r_i l^2}{K}}$.

Step 6: *Take the K-length FFT from the dechirped y_l.*

Step 7: *Obtain m_i as the peak location of the K-length FFT of the dechirped y_l.*

Step 8: *Obtain x_i as the peak amplitude of the K-length FFT of the dechirped y_l.*

Step 9: *Update y_l by removing the corresponding component to the recovered $\boldsymbol{x}$ sample.*
The update process is by using Eq. (6.22).

Step 10: *If the iteration loop has been repeated M times, or the energy of y_l is less than ϵ stop, otherwise go repeat Steps 2 to 9.*
The energy is of y_l is indeed its norm-2.

Fig. 6.3 depicts the block diagram of the recovery, and Fig. 6.4 describes the recovery loop of equations in one iteration.

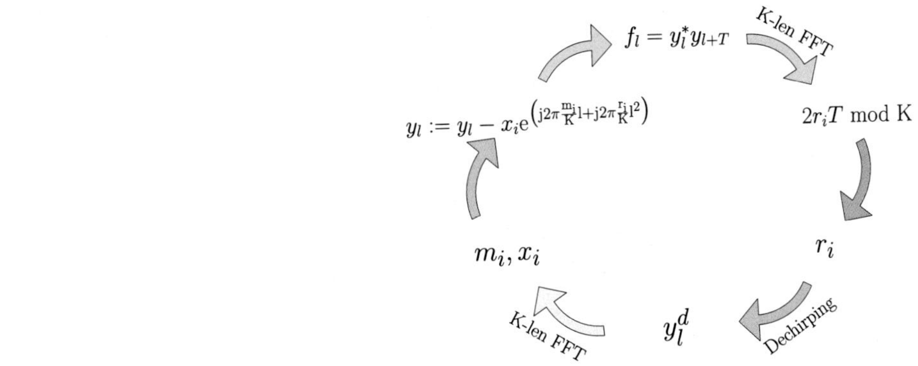

FIGURE 6.4

The recovery loop of equations in one iteration.

6.5 Summary

Deterministic compressive sensing by chirp codes as a fast and reliable technique has been described in detail in this chapter. The chapter mainly focuses on the theory and implementation of the technique and makes an effort of giving understanding of the theoretical details to the reader in a simple way. In this regard, some technical details which are not explained or only briefly mentioned in the main reference, here have been clarified. The importance of bijection fulfillment through acquiring K as a prime number has been explained and mathematically proved in a lemma. To facilitate understanding of the technique, a block diagram and iteration loop of equations besides the step-by-step process of the algorithm have been presented. The descriptive tutorial to compressive sensing by chirp codes in this chapter not only assists in understanding the approach but also may motivate the reader in practically employing the technique for different applications.

Appendix 6.A

6.A.1 Chirp sinusoid

A chirp sinusoid exponential has the following general form:

$$e^{\theta(t)} = e^{2\pi\mu t^2 + 2\pi f_0 t + \phi}, \tag{6.23}$$

where the instantaneous angular frequency is as follows:

$$\begin{aligned} \omega(t) &= \frac{d}{dt}\theta(t) \\ &= 4\pi\mu t + 2\pi f_0. \end{aligned} \tag{6.24}$$

The instantaneous frequency in hertz is

$$f(t) = 2\mu t + f_0. \tag{6.25}$$

As a matter of fact, if $\mu \neq 0$ then the frequency is not constant but it changes linearly in time, which is by nature of chirping.

References

[1] E.J. Candès, et al., Compressive sampling, in: Proceedings of the International Congress of Mathematicians, Madrid, Spain, vol. 3, 2006, pp. 1433–1452.

[2] E.J. Candès, J. Romberg, T. Tao, Robust uncertainty principles: exact signal reconstruction from highly incomplete frequency information, IEEE Transactions on Information Theory 52 (2) (2006) 489–509.

[3] E.J. Candes, J.K. Romberg, T. Tao, Stable signal recovery from incomplete and inaccurate measurements, Communications on Pure and Applied Mathematics: A Journal Issued by the Courant Institute of Mathematical Sciences 59 (8) (2006) 1207–1223.

[4] E.J. Candes, T. Tao, Near-optimal signal recovery from random projections: universal encoding strategies?, IEEE Transactions on Information Theory 52 (12) (2006) 5406–5425.

[5] D.L. Donoho, et al., Compressed sensing, IEEE Transactions on Information Theory 52 (4) (2006) 1289–1306.

[6] C.E. Gutierrez, M.R. Alsharif, K. Yamashita, M. Khosravy, A tweets mining approach to detection of critical events characteristics using random forest, International Journal of Next-Generation Computing 5 (2) (2014) 167–176.

[7] N. Kausar, S. Palaniappan, B.B. Samir, A. Abdullah, N. Dey, Systematic analysis of applied data mining based optimization algorithms in clinical attribute extraction and classification for diagnosis of cardiac patients, in: Applications of Intelligent Optimization in Biology and Medicine, Springer, 2016, pp. 217–231.

[8] M. Khosravy, M.R. Asharif, M.H. Sedaaghi, Medical image noise suppression: using mediated morphology, IEICE, Technical Report 107 (461) (2008) 265–270.

[9] N. Dey, A.S. Ashour, A.S. Ashour, A. Singh, Digital analysis of microscopic images in medicine, Journal of Advanced Microscopy Research 10 (1) (2015) 1–13.

[10] M. Khosravy, M.R. Asharif, M.H. Sedaaghi, Medical image noise suppression using mediated morphology, in: IEICE Tech. Rep., IEICE, 2008, pp. 265–270.

[11] A.S. Ashour, S. Samanta, N. Dey, N. Kausar, W.B. Abdessalemkaraa, A.E. Hassanien, Computed tomography image enhancement using cuckoo search: a log transform based approach, Journal of Signal and Information Processing 6 (03) (2015) 244.

[12] M. Khosravy, N. Gupta, N. Marina, I. Sethi, M. Asharif, Brain action inspired morphological image enhancement, in: Nature-Inspired Computing and Optimization, Springer, Cham, 2017, pp. 381–407.

[13] C.E. Gutierrez, P.M.R. Alsharif, M. Khosravy, P.K. Yamashita, P.H. Miyagi, R. Villa, Main large data set features detection by a linear predictor model, AIP Conference Proceedings 1618 (2014) 733–737.

[14] S. Gupta, M. Khosravy, N. Gupta, H. Darbari, N. Patel, Hydraulic system onboard monitoring and fault diagnostic in agricultural machine, Brazilian Archives of Biology and Technology 62 (2019).

[15] S. Gupta, M. Khosravy, N. Gupta, H. Darbari, In-field failure assessment of tractor hydraulic system operation via pseudospectrum of acoustic measurements, Turkish Journal of Electrical Engineering & Computer Sciences 27 (4) (2019) 2718–2729.

[16] M.H. Sedaaghi, R. Daj, M. Khosravi, Mediated morphological filters, in: 2001 International Conference on Image Processing, Proceedings, vol. 3, IEEE, 2001, pp. 692–695.

[17] M. Khosravy, N. Gupta, N. Marina, I.K. Sethi, M.R. Asharif, Morphological filters: an inspiration from natural geometrical erosion and dilation, in: Nature-Inspired Computing and Optimization, Springer, Cham, 2017, pp. 349–379.

[18] M. Khosravy, N. Gupta, N. Marina, I. Sethi, M. Asharifa, Perceptual adaptation of image based on Chevreul–Mach bands visual phenomenon, IEEE Signal Processing Letters 24 (5) (2017) 594–598.
[19] M. Khosravy, N. Punkoska, F. Asharif, M.R. Asharif, Acoustic OFDM data embedding by reversible Walsh–Hadamard transform, AIP Conference Proceedings 1618 (2014) 720–723.
[20] F. Asharif, S. Tamaki, M.R. Alsharif, H. Ryu, Performance improvement of constant modulus algorithm blind equalizer for 16 QAM modulation, International Journal on Innovative Computing, Information and Control 7 (4) (2013) 1377–1384.
[21] M. Khosravy, M.R. Alsharif, K. Yamashita, An efficient ICA based approach to multiuser detection in MIMO OFDM systems, in: Multi-Carrier Systems & Solutions 2009, Springer, 2009, pp. 47–56.
[22] M. Khosravy, M.R. Alsharif, B. Guo, H. Lin, K. Yamashita, A robust and precise solution to permutation indeterminacy and complex scaling ambiguity in BSS-based blind MIMO-OFDM receiver, in: International Conference on Independent Component Analysis and Signal Separation, Springer, 2009, pp. 670–677.
[23] M. Khosravy, A blind ICA based receiver with efficient multiuser detection for multi-input multi-output OFDM systems, in: The 8th International Conference on Applications and Principles of Information Science (APIS), Okinawa, Japan, 2009, 2009, pp. 311–314.
[24] M. Khosravy, S. Kakazu, M.R. Alsharif, K. Yamashita, Multiuser data separation for short message service using ICA, SIP, IEICE Technical Report 109 (435) (2010) 113–117.
[25] M. Khosravy, M.R. Alsharif, M. Khosravi, K. Yamashita, An optimum pre-filter for ICA based multi-input multi-output OFDM system, in: 2010 2nd International Conference on Education Technology and Computer, vol. 5, IEEE, 2010, pp. V5–129.
[26] M.H. Sedaaghi, M. Khosravi, Morphological ECG signal preprocessing with more efficient baseline drift removal, in: 7th IASTED International Conference, ASC, 2003, pp. 205–209.
[27] N. Dey, S. Samanta, X.-S. Yang, A. Das, S.S. Chaudhuri, Optimisation of scaling factors in electrocardiogram signal watermarking using cuckoo search, International Journal of Bio-Inspired Computation 5 (5) (2013) 315–326.
[28] M. Khosravy, M.R. Asharif, M.H. Sedaaghi, Morphological adult and fetal ECG preprocessing: employing mediated morphology, IEICE Technical Report, IEICE 107 (2008) 363–369.
[29] N. Dey, A.S. Ashour, F. Shi, S.J. Fong, R.S. Sherratt, Developing residential wireless sensor networks for ECG healthcare monitoring, IEEE Transactions on Consumer Electronics 63 (4) (2017) 442–449.
[30] M. Khosravi, M.H. Sedaaghi, Impulsive noise suppression of electrocardiogram signals with mediated morphological filters, in: 11th Iranian Conference on Biomedical Engineering, ICBME, 2004, pp. 207–212.
[31] N. Dey, S. Mukhopadhyay, A. Das, S.S. Chaudhuri, Analysis of P-QRS-T components modified by blind watermarking technique within the electrocardiogram signal for authentication in wireless telecardiology using DWT, International Journal of Image, Graphics and Signal Processing 4 (7) (2012) 33.
[32] T.W. Cabral, M. Khosravy, F.M. Dias, H.L.M. Monteiro, M.A.A. Lima, L.R.M. Silva, R. Naji, C.A. Duque, Compressive sensing in medical signal processing and imaging systems, in: Sensors for Health Monitoring, Elsevier, 2019, pp. 69–92.
[33] A.A. Picorone, T.R. de Oliveira, R. Sampaio-Neto, M. Khosravy, M.V. Ribeiro, Channel characterization of low voltage electric power distribution networks for PLC applications based on measurement campaign, International Journal of Electrical Power & Energy Systems 116 (105) (2020) 554.
[34] E. Santos, M. Khosravy, M.A. Lima, A.S. Cerqueira, C.A. Duque, A. Yona, High accuracy power quality evaluation under a colored noisy condition by filter bank ESPRIT, Electronics 8 (11) (2019) 1259.
[35] E. Santos, M. Khosravy, M.A. Lima, A.S. Cerqueira, C.A. Duque, ESPRIT associated with filter bank for power-line harmonics, sub-harmonics and inter-harmonics parameters estimation, International Journal of Electrical Power & Energy Systems 118 (105) (2020) 731.
[36] H. Zaheb, M.S.S. Danish, T. Senjyu, M. Ahmadi, A.M. Nazari, M. Wali, M. Khosravy, P. Mandal, A contemporary novel classification of voltage stability indices, Applied Sciences 10 (5) (2020) 1639.
[37] L. Applebaum, S.D. Howard, S. Searle, R. Calderbank, Chirp sensing codes: deterministic compressed sensing measurements for fast recovery, Applied and Computational Harmonic Analysis 26 (2) (2009) 283–290.

[38] E. Candes, T. Tao, Decoding by linear programming, preprint, arXiv:math/0502327, 2005.
[39] M. Khosravy, Blind source separation and its application to speech, image and MIMO-OFDM communication systems, PhD thesis, University of the Ryukyus, Japan, 2010.
[40] M. Khosravy, M. Gupta, M. Marina, M.R. Asharif, F. Asharif, I. Sethi, Blind components processing a novel approach to array signal processing: a research orientation, in: 2015 International Conference on Intelligent Informatics and Biomedical Sciences, ICIIBMS, 2015, pp. 20–26.
[41] M. Khosravy, M.R. Alsharif, K. Yamashita, A PDF-matched modification to Stone's measure of predictability for blind source separation, in: International Symposium on Neural Networks, Springer, Berlin, Heidelberg, 2009, pp. 219–228.
[42] M. Khosravy, M.R. Asharif, K. Yamashita, A probabilistic short-length linear predictability approach to blind source separation, in: 23rd International Technical Conference on Circuits/Systems, Computers and Communications (ITC-CSCC 2008), Yamaguchi, Japan, 2008, pp. 381–384.
[43] M. Khosravy, M.R. Asharif, K. Yamashita, A theoretical discussion on the foundation of Stone's blind source separation, Signal, Image and Video Processing 5 (3) (2011) 379–388.
[44] M. Khosravy, M.R. Asharif, K. Yamashita, A PDF-matched short-term linear predictability approach to blind source separation, International Journal of Innovative Computing, Information & Control 5 (11) (2009) 3677–3690.
[45] M. Khosravy, N. Gupta, N. Patel, T. Senjyu, Frontier Applications of Nature Inspired Computation, 2020.
[46] N. Gupta, M. Khosravy, N. Patel, O.P. Mahela, G. Varshney, Plant genetics-inspired evolutionary optimization: a descriptive tutorial, in: Frontier Applications of Nature Inspired Computation, Springer, 2020, pp. 53–77.
[47] C. Moraes, E. De Oliveira, M. Khosravy, L. Oliveira, L. Honório, M. Pinto, A hybrid bat-inspired algorithm for power transmission expansion planning on a practical Brazilian network, in: Applied Nature-Inspired Computing: Algorithms and Case Studies, Springer, 2020, pp. 71–95.
[48] M. Khosravy, N. Gupta, N. Patel, T. Senjyu, C.A. Duque, Particle swarm optimization of morphological filters for electrocardiogram baseline drift estimation, in: Applied Nature-Inspired Computing: Algorithms and Case Studies, Springer, 2020, pp. 1–21.
[49] G. Singh, N. Gupta, M. Khosravy, New crossover operators for real coded genetic algorithm (RCGA), in: 2015 International Conference on Intelligent Informatics and Biomedical Sciences (ICIIBMS), IEEE, 2015, pp. 135–140.
[50] N. Gupta, N. Patel, B.N. Tiwari, M. Khosravy, Genetic algorithm based on enhanced selection and log-scaled mutation technique, in: Proceedings of the Future Technologies Conference, Springer, 2018, pp. 730–748.
[51] N. Gupta, M. Khosravy, N. Patel, I. Sethi, Evolutionary optimization based on biological evolution in plants, Procedia Computer Science 126 (2018) 146–155.
[52] J. Kaliannan, A. Baskaran, N. Dey, A.S. Ashour, M. Khosravy, R. Kumar, ACO based control strategy in interconnected thermal power system for regulation of frequency with HAE and UPFC unit, in: International Conference on Data Science and Application (ICDSA-2019), in: LNNS, Springer, 2019.
[53] M. Khosravy, N. Gupta, N. Patel, O.P. Mahela, G. Varshney, Tracing the points in search space in plant biology genetics algorithm optimization, in: Frontier Applications of Nature Inspired Computation, Springer, 2020, pp. 180–195.
[54] N. Gupta, M. Khosravy, N. Patel, S. Gupta, G. Varshney, Evolutionary artificial neural networks: comparative study on state-of-the-art optimizers, in: Frontier Applications of Nature Inspired Computation, Springer, 2020, pp. 302–318.
[55] N. Gupta, M. Khosravy, N. Patel, S. Gupta, G. Varshney, Artificial neural network trained by plant genetic-inspired optimizer, in: Frontier Applications of Nature Inspired Computation, Springer, 2020, pp. 266–280.
[56] D.L. Donoho, M. Elad, Optimally sparse representation in general (nonorthogonal) dictionaries via l1 minimization, Proceedings of the National Academy of Sciences 100 (5) (2003) 2197–2202.

CHAPTER

7

Deterministic compressive sensing by chirp codes: a MATLAB® tutorial

Mahdi Khosravy[a], Naoko Nitta[a], Faramarz Asharif[b], Katia Melo[c], Carlos A. Duque[c]

[a]*Graduate School of Engineering, Osaka University, Osaka, Japan*

[b]*Kitami Institute of Technology, Kitami, Japan*

[c]*Department of Electrical Engineering, Federal University of Juiz de Fora, Juiz de Fora, Brazil*

7.1 Introduction

Compressive sensing [1–5] is a new road to signal/image data sampling. It is performed using by much less required samples than one stated by the Shannon–Nyquist theorem. It deploys the sparsity nature of the information where just some samples are associated with information, most of them not having any considerable information. Compressive sensing applies the sparsity in sensing the data, and by sensing the data in a compressed way, it needs much fewer sensors. In a wide range of applications, the compressive sensing can be applied, like data mining [6,7], text processing [8], signal processing [9,10], agriculture on-board data processing [11,12], image enhancement [13,14], acoustic OFDM [15], medical image processing [16–18], image adaptation [19] Electrocardiogram processing [20–26], telecommunications [27–32], power quality analysis [33–35], power line communications [36], etc.

This chapter is a step-by-step MATLAB® tutorial for deterministic compressive sensing by chirp codes [37].

7.2 Compressive sensing

In compressive sensing, we have an n-length signal vector $\boldsymbol{x} \in \mathbb{R}^n$. The signal vector $\boldsymbol{x}$ is sensed as a much shorter signal vector of length m as $\boldsymbol{y} \in \mathbb{R}^m$ as follows:

$$y_{m\times 1} = A_{m\times n}\boldsymbol{x}_{n\times 1}, \qquad m << n. \tag{7.1}$$

A is called the compressive sensing matrix. The compressive sensing must be such that the initial signal vector $\boldsymbol{x}$ be recoverable by using the compressively sensed signa

Compressive Sensing in Healthcare. https://doi.org/10.1016/B978-0-12-821247-9.00012-3

vector $\boldsymbol{y}$. Although the recovery stage is an issue, its possibility is clear by the theoretical considerations in the sensing stage. The recovery side of the problem wherein $\boldsymbol{x}$ is obtained from a much shorter $\boldsymbol{y}$ in length mainly is the most operative side of the problem. Although, at first view, the recovery seems impossible, it yet is possible. In the recovery problem, we are looking for n variables as samples of $\boldsymbol{x}$ having a set of m linear equations. The number of variables is much smaller than the number of equations, and that makes it sound like an ill-posed problem:

$$\begin{aligned} a_{11}x_1 + a_{12}x_2 + \cdots + a_{1n}x_n &= y_1, \\ a_{21}x_1 + a_{22}x_2 + \cdots + a_{2n}x_n &= y_2, \\ &\vdots \\ a_{m1}x_1 + a_{m2}x_2 + \cdots + a_{mn}x_n &= y_m. \end{aligned} \tag{7.2}$$

Since $m < n$, the number of equations is smaller than the number of variables, therefore, the compressive sensing seems an ill-posed problem if we do not consider the assumption of the sparsity of the signal. It is a case similar to the blind source separation (BSS) problem [38–43] where there exist infinitely many possible solutions, but since BSS works based on the assumption of independence of the sources it searches for the one with maximum independence amongst the estimate of the sources.

The common approach to compressive sensing is by random sensing matrices, as the general approach of compressive sensing for a signal vector of $\boldsymbol{x}$ of sparsity level of k. Under a certain condition for the sensing matrix A, the signal vectors can be recovered by norm-1 minimization [44]. Norm-1 is a replacement for norm-0, which is the measure of k-sparseness to overcome it NP-hard problem nature. Also, an NP-hard problem may be resolvable by metaheuristic evolutionary optimization [45–52,46,53–55]. Norm-1 is a linear optimization problem.

The focus of this chapter is on deterministic compressive sensing wherein the signal is sensed by a deterministic reversible process. The chapter presents practical MATLAB tutorial chirp code deterministic compressive sensing.

The method of the chapter is a step-by-step tutorial from the simplest relaxed condition to the advanced case.

7.3 Chirp codes compressive sensing: MATLAB tutorial

Compressive sensing by chirp codes [37] is a deterministic approach. A descriptive tutorial to this technique has been given in Chapter 6 of this book, and we invite the reader to refer to that chapter in addition to the main reference. Since for a K-length chirp code there are K^2 different possible chirp codes, the sensing matrix is composed of K-length chirp codes in K^2 columns. Briefly, as indicated in the general framework of the compressive sensing problem, an n-length signal vector $\boldsymbol{x}$ in vector space of $\mathbb{R}^n$ is compressively sensed as an m-length vector $\boldsymbol{y}$ in the vector space $\mathbb{R}^m$ through m linear combinations of n samples of the signal vector: In deterministic compressive sensing by chirp codes, A is composed of the chirp codes shaping its columns; as explained above it is of $K \times K^2$ dimensions. The performance and capability of the technique are due to chirp codes characteristics; the presented MATLAB codes in this chapter are meant to help the reader to understand the technique. At first, let us have a close view at a chirp code in MATLAB.

7.3.1 Chirp codes

A chirp code is indeed a vector of length K. It has as regards frequency two aspects. One is a single-tone periodic part. The other is an increasing swiping-all-frequencies tone. The first one is called the base frequency and the latter one is called the chirp tone. The two frequency tones are multiplied by each other. The K-length chirp code is as follows:

$$v_{m,r}(l) = \alpha.\exp\big(j2\pi \frac{ml}{K} + j2\pi \frac{rl^2}{K}\big), \qquad l, m, r \in \mathbb{Z}_K; \tag{7.3}$$

$\exp(j2\pi \frac{m}{K} l)$ being the single-tone multiplicative component. Furthermore, $\exp(j2\pi \frac{r}{K} l^2)$ is the chirp tone multiplicative component. The first component contains the 'base frequency' m, and the chirp component contains 'the chirp rate' r, which is the rate of chirping amongst the frequencies.

A short MATLAB code is given here for a short chirp code of length 17. In this code you see how to make a chirp code by Eq. (7.3), and the resultant code is plotted.

```
%% MATLAB code: A chirp code
K=17;
alpha = 1;
r=[1:K-1 0];
m=[1:K-1 0];
l=0:K-1;
%% watching the chirp components
chirpcode=alpha*exp((1i*2*pi*m(7)*l/K)+((1i*2*pi*r(5)*(l.^2)))/K);
figure(1)
plot3(1:K,real(chirpcode),imag(chirpcode))
title('A chirp code of lenght K')
zlabel('Chirp code value')
xlabel('Real part')
ylabel('Imaginary part')
figure(2)
subplot(211);plot(1:K, real(chirpcode), 'b*-');grid on
title('(a) Real part')
subplot(211);plot(1:K, imag(chirpcode), 'm*-');grid on
title('(b) Imaginary part')
```

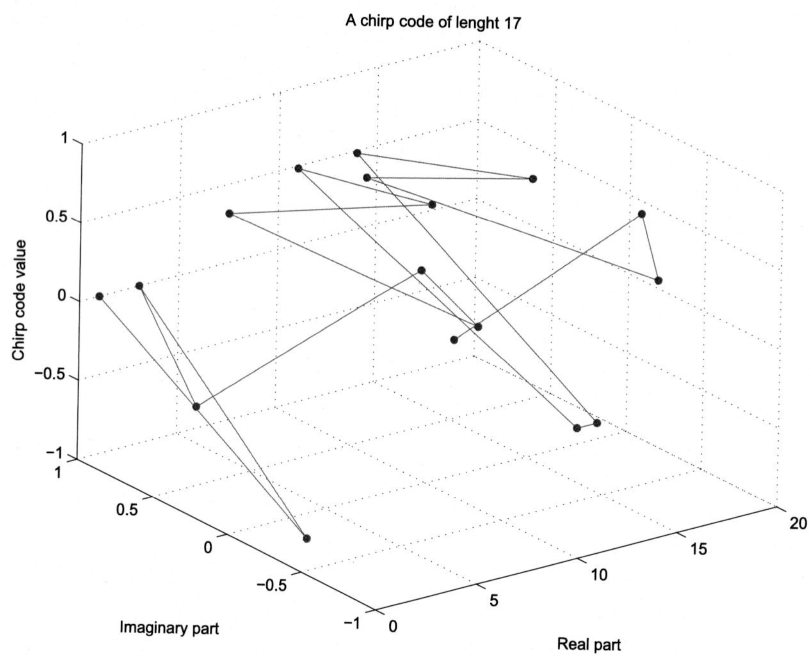

FIGURE 7.1

A chirp code as a complex number sequence.

After running the above piece of code in MATLAB, you can see how a chirp code works. As is seen in Fig. 7.1, it is a complex number sequence. Figs. 7.2A and 7.2B, respectively, depict the real and imaginary parts of the same chirp code.

Now, let us have a closer view to the multiplicative components of the same chirp code:

- $e^{j2\pi \frac{ml}{K}}$,
- $e^{j2\pi \frac{rl^2}{K}}$.

At the sequence of the former piece of the code, the following code depicts the components separately:

```
component1=alpha*exp((1i*2*pi*m(7)*1/K));
component2=alpha*exp((1i*2*pi*r(5)*(1.^ 2)/K));
figure(3);
subplot(221);plot(real(component1), 'b*-');
title('(a) Real part of component1 exp((1i*2*pi*m(7)*1/K))')
axis([1 17 -1 1]);grid on
subplot(213);plot(real(component2), 'b*-');
title('(b) Real part of component2 exp((1i*2*pi*r(5)*(1.^2)/K))')
```

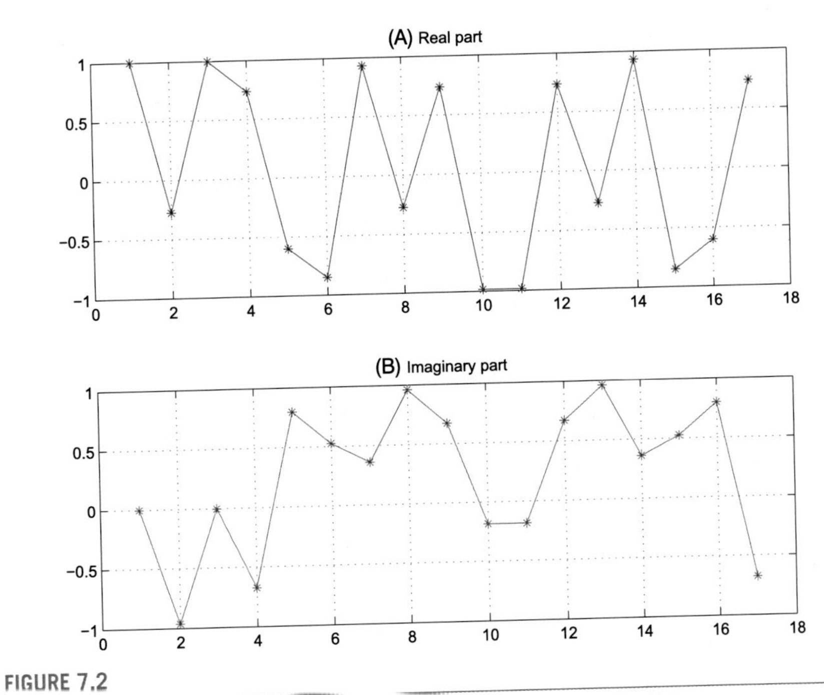

FIGURE 7.2

The real part and imaginary parts of the chirp code depicted in Fig. 7.1.

```
axis([1 17 -1 1]);grid on
subplot(223);plot(imag(component1), 'm*-');
title('(c) Imaginary part of component1 exp((1i*2*pi*m(7)*1/K))')
axis([1 17 -1 1]);grid on
subplot(214);plot(imag(component2), 'm*-');
title('(d) Imaginary part of component2 exp((1i*2*pi*r(5)*(1.^2)/K))')
axis([1 17 -1 1]);grid on
```

The plotted figures of the above code are shown in Fig. 7.3. Fig. 7.3A, Fig. 7.3B, Fig. 7.3C, and Fig. 7.3D, respectively, show the real part of the first component, the real part of the second component, the imaginary part of the first component, and the imaginary part of the second component. As can be seen between the main part made of base frequency and the part made of chirp frequency, there is no difference relevant for the detection process.

Now, let us see how the discrete Fourier transform (DFT) of the these two components work. At the sequence of the preceding MATLAB code, the following piece of code depicts the DFT of the components and the chirp code:

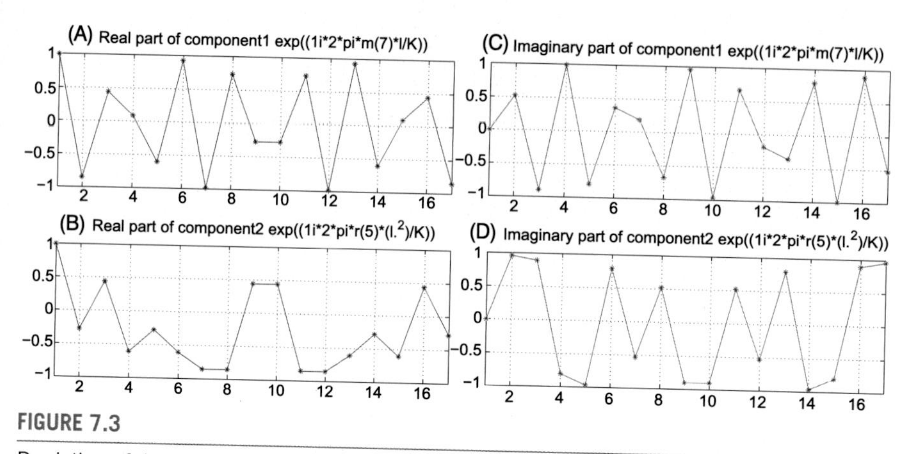

FIGURE 7.3

Depiction of the multiplicative components of the chirp code of Fig. 7.1 and Fig. 7.2, (A) real part, (B) imaginary part.

```
figure(4);
subplot(311);stem(0:K-1, abs(fft(component1)));
title('(a) DFT of exp((1i*2*pi*m(7)*l/K))'); grid on
subplot(312);stem(0:K-1, abs(fft(component2)));
title('(b) DFT of exp((1i*2*pi*r(5)*(l.^ 2)/K))'); grid on
subplot(313);stem(0:K-1, angle(fft(chirpcode)));
title('(c) DFT of exp((1i*2*pi*m(7)*l/K)+((1i*2*pi*r(5)*(l.^ 2)))/K)');
grid on
```

The plots depicted in Fig. 7.4 show how the energy of the part with chirp frequency has spread along the spectrum, but the component of the base frequency has a concentrated energy at the base frequency. However, the chirp code which is composed of both components has a spectrum with spread energy. Although the peak of the energy can be observed at m_i, keep in mind that this is for a single chirp code, and in the case of having a linear combination of chirp codes the spread of energy will be more obvious.

7.3.2 The approach for extraction of parameters

In recovery of the compressively sensed signal vector by chirp codes the vector of f has a key role. Here, we see how effectively it helps to extract the base frequency m_i

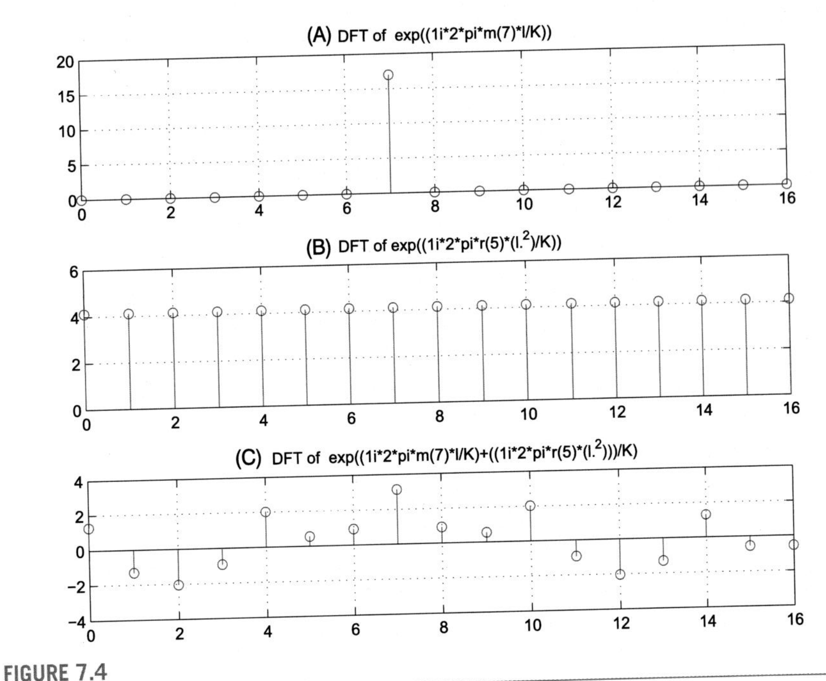

FIGURE 7.4

DFT of the chirp code component of base frequency (A), the component of chirp frequency (B), and the chirp code with both components (C).

from a chirp code. We have

$$f = \begin{bmatrix} f_1 \\ \vdots \\ f_l \\ \vdots \\ f_K \end{bmatrix}, \tag{7.4}$$

where f_l is

$$\begin{aligned} f_l &= y_l^* y_{l+T} \\ &= |x_1|^2 e^{\frac{j2\pi}{K}(m_1 T + r_1 T^2)} e^{j2\pi \frac{2r_1 T}{K} l} + |x_2|^2 e^{\frac{j2\pi}{K}(m_2 T + r_2 T^2)} e^{j2\pi \frac{2r_2 T}{K} l} + \cdots + \text{CTs}, \end{aligned} \tag{7.5}$$

where y^* is the complex-conjugate of y, and T is a nonzero value in $\mathbb{Z}_K$. The $*$ indicate a circular multiplication between y_l^* and y_l wherein the latter one shows a shift of $l + T$. So, due to the circular nature, $l + T$ can be indicated as $l + T \bmod K$

where mod K indicates modulo K. CTs are the cross-terms as follows:

$$\mathrm{CTs} = \sum_{i,j\neq i} x_i x_j^* e^{\frac{j2\pi}{K}(m_i T + r_i T^2)} e^{\frac{j2\pi}{K} l(m_i - m_j + 2Tr_i) + \frac{j2\pi}{K} l^2 (r_i - r_j)}. \tag{7.6}$$

As seen, f_l is made of sinusoids with frequency of $\frac{2r_i T}{K}$ and chirp cross components. The latter have energy spread along the spectrum, despite the sinusoid component which is associated with an energy concentration in $\frac{2r_i T}{K}$. This is used for detection of r_i via a discrete Fourier transform of y_l.

The recovery steps by using f_l is briefly given here, and in the sequence the steps are clarified though the given MATLAB practice.

Step 1: *Choosing a nonzero $T \in \mathbb{Z}_K$ and ϵ as minimum limit of energy of f_l after updating.*
Step 2: *Updating or obtaining f_l having the last update of y_l.*
Step 3: *Taking K-length FFT of f_l.*
Step 4: *Obtaining the dominant r_i from $2r_i T \bmod K$ at the peak point of the FFT of f_l.*
Step 5: *Obtain the dechirped y_l as y_l^d using the obtained r_i.*
Step 6: *Take K-length FFT from the dechirped y_l.*
Step 7: *Obtain m_i as the peak location of K-length FFT of the dechirped y_l.*
Step 8: *Obtain x_i as the peak amplitude of K-length FFT of the dechirped y_l.*
Step 9: *Update y_l by removing the corresponding component to the recovered $\boldsymbol{x}$ sample.*
The update process is by using Eq. (7.8).
Step 10: *If the iteration loop has been repeated M times, or the energy of y_l is less than ϵ stop, otherwise repeat Steps 2 to 9.*
The energy is of y_l is indeed its norm-2.

7.3.2.1 Single chirp

Here, first we see in a MATLAB code the $\boldsymbol{f}$ characteristics when $\boldsymbol{y}$ is composed of a single chirp code which means the signal vector $\boldsymbol{x}$ before being compressively sensed has only one sample which is represented by the single chirp component. Later we continue for the case of having a mixture of chirp codes corresponding to multiple signal samples.

```
K=17;
alpha = 1;
r=[1:K-1 0];
m=[1:K-1 0];
l=[0:K-1];
%% Extracting r(5) from a single chirp code
T=3;
chirpcode=alpha*exp((1i*2*pi*m(7)*l/K)+((1i*2*pi*r(5)*(l.^2)))/K);
f_l=conj(chirpcode).*circshift(chirpcode', -T)';
```

```
figure(5);
subplot(311);stem(abs(fft(chirpcode)));
title('DFT of the chirp'); grid on
r2TmodK=alpha*exp((1i*2*pi*mod(r(5)*T*2, K)*1/K));
subplot(312);stem(abs(fft(f_1)));
title('DFT of f(1)'); grid on
subplot(313);stem(abs(fft(r2TmodK)));
title('DFT of exp((1i*2*pi*mod(r(5)*TT*2, K)*1/K))'); grid on
```

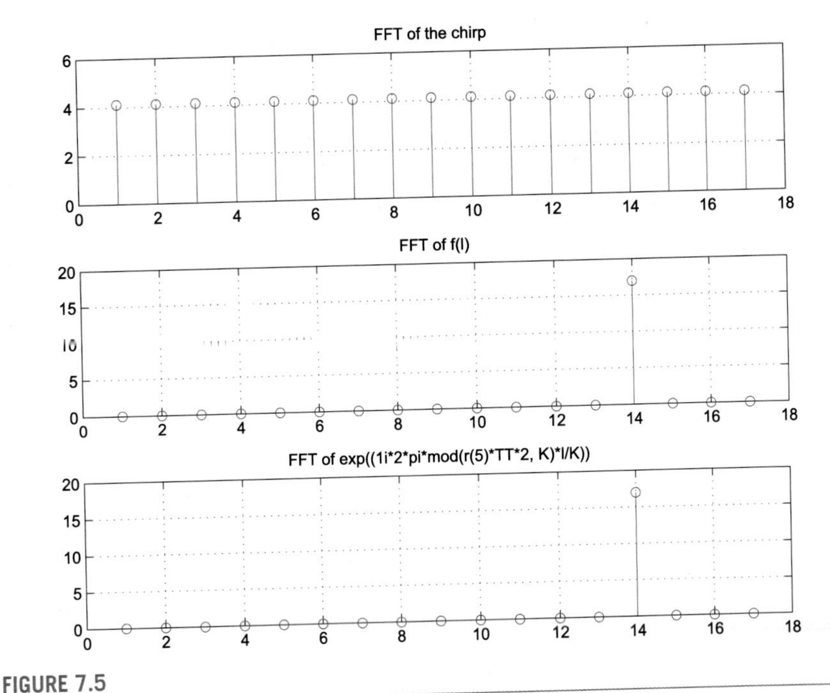

FIGURE 7.5

DFT of the chirp code (top), DFT of f (middle), and DFT of sinusoid with frequency of $r_5 2T \bmod K$ (bottom).

As is seen from Fig. 7.5, f_l for a single chirp has the FFT with energy spectrum concentration at $r_i 2T \bmod K$. Therefore, it can be used for extraction of the chirp rate r_i. This is theoretically observable in Eq. (7.4). To use this property of f_l, there must be a bijection between the set of $\{r_i\}$ and the set of $\{2r_i T \mathrm{mod} K\}$, otherwise, having a $2r_i T \mathrm{mod} K$ by DFT of f_l, we could not find the corresponding r_i.

7.3.2.2 Mixture of two chirps

In this section, the DFT of a mixture of two chirps and its resultant f vector is analyzed and depicted by using the following code:

```
clc; clear all; close all
alpha = 1;
K = 17;
r = [1:K-1 0];
m = [1:K-1 0];
l = [1:K-1 0];
r1 = r(5);
r2 = r(8);
TT = 3;
s1 = 2;
s2 = 2;
mod(r1*TT*2, K)
mod(r2*TT*2, K)
chirpcode1=alpha*exp((1i*2*pi*m(7)*l/K)+((1i*2*pi*r1*(l.^2)))/K);
chirpcode2=alpha*exp((1i*2*pi*m(9)*l/K)+((1i*2*pi*r2*(l.^2)))/K);
r2TmodK_1=alpha*exp((1i*2*pi*mod(r1*TT*2, K)*l/K));
r2TmodK_2=alpha*exp((1i*2*pi*mod(r2*TT*2, K)*l/K));
mixchirp=s1*chirpcode1+s2*chirpcode2;
fsig=conj(mixchirp).*circshift(mixchirp', -TT)';
figure(6); subplot(421);stem(abs(fft(chirpcode1)))
subplot(422);stem(abs(fft(chirpcode2)))
subplot(423);stem(abs(fft(r2TmodK_1)))
subplot(424);stem(abs(fft(r2TmodK_2)))
subplot(413);stem(abs(fft(mixchirp)))
subplot(414);stem(abs(fft(fsig)))
```

As is seen from Fig. 7.6, f_l of the mixed chirps has the FFT with energy spectrum concentration at $r_1 2T$ mod K and $r_2 2T$ mod K. So, the corresponding r_i are detectable. However, in the case of the DFT of each chirp as well as the DFT of the mixture of chirps, it can be seen that the energy of DFT is spread along all the frequency bins.

7.3.3 Bijection between the sets $\{r_i\}$ and $\{2r_i T \mathrm{mod} K\}$

The bijection between $\{r_i\}$ and $\{2r_i T \mathrm{mod} K\}$ is essential for extraction of the chirp rate which is one of the important identifying elements of a chirp code. As is indicated in main reference [37], to fulfill this necessary bijection, K must be a prime number. This issue has been proved in Chapter 6 of this book. Here, the following code illustrates this issue:

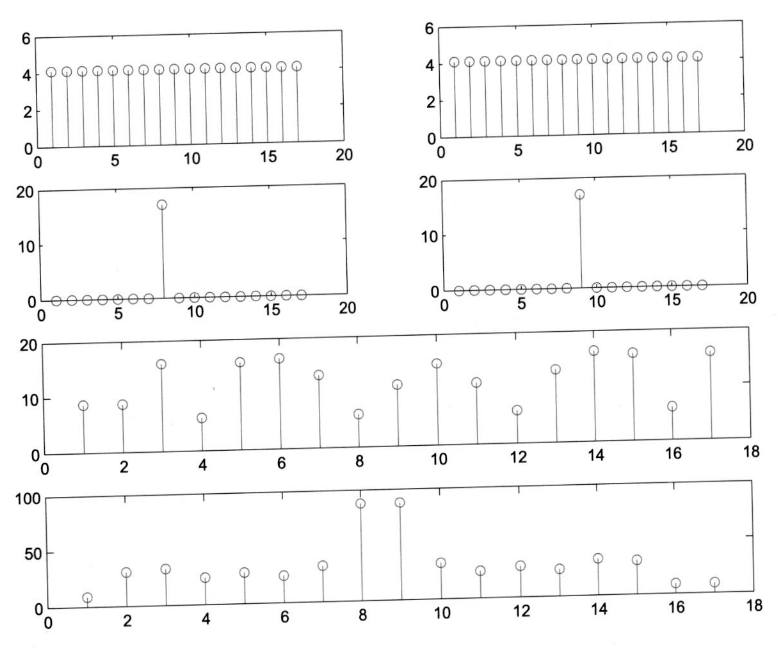

FIGURE 7.6

DFT of the chirp codes (top), DFT of corresponding sinusoids with frequency of $r_i 2T$ mod K (2nd row), DFT of the mixture of the chirp codes (3rd row), and DFT of f (bottom).

```
clc;clear all;close all
T=3;
for j=1:2
figure(6+j);
for K=6*j:6*j+5
clear fout
r=[0 1:K-1 ];
for i=1:K
fout(i)=mod(2*r(i)*T, K)
end
subplot(2,3,K-6*j+1);plot(sort(fout),'p');
title(K);xlabel('r(i)');ylabel('mod(2*r(i)*3, K)');
axis([0 K 0 K]); grid on
end
end
```

In the output figures of the above code, the correspondence between the two sets is depicted for different values of K. For simplicity, the depiction is by the order

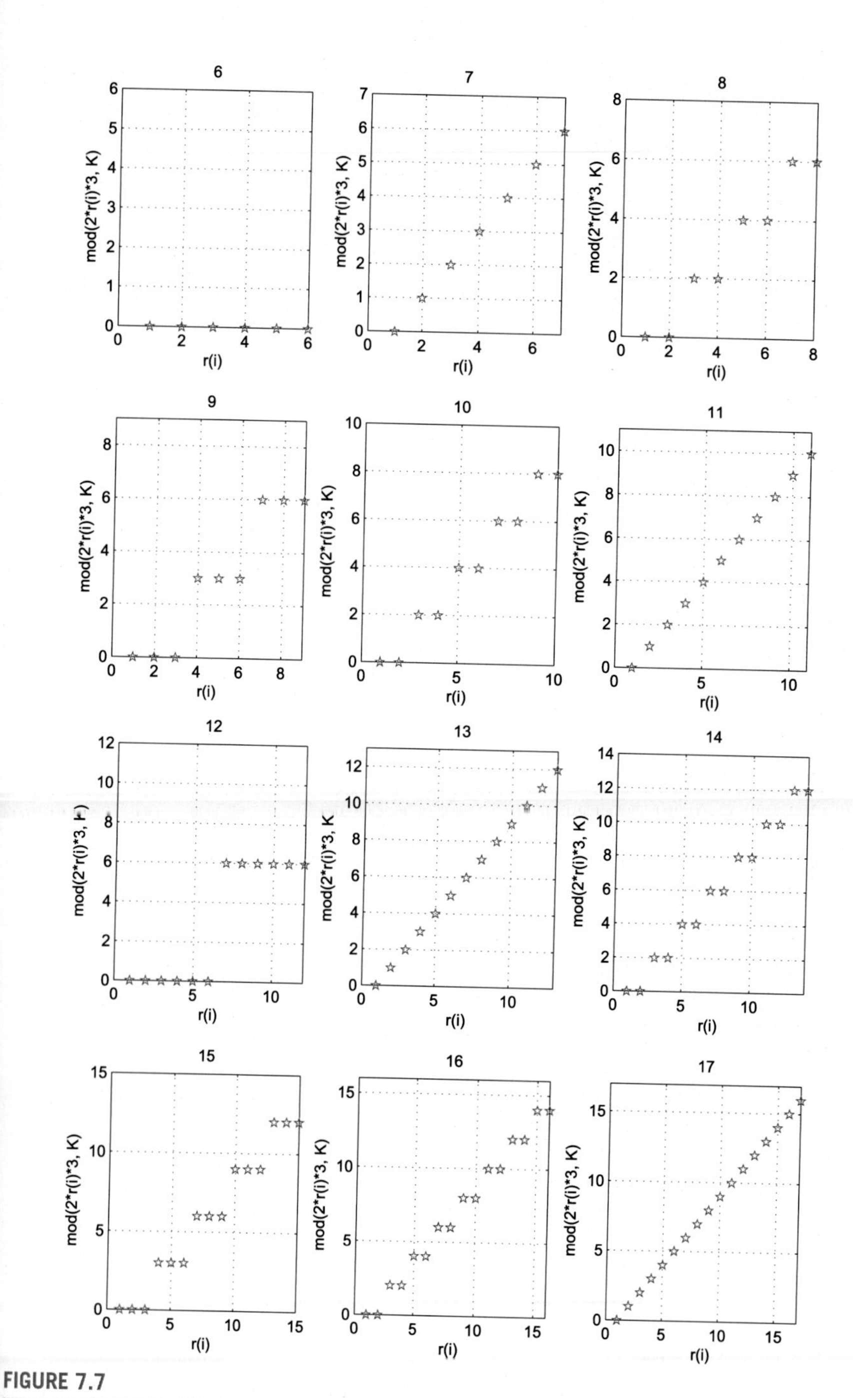

FIGURE 7.7

Illustration of the bijection between the sets $\{r_i\}$ and $\{2r_i T \bmod K\}$ for different values of K as per location of sub-figure.

of the values in the set $\{2r_i T \mathrm{mod} K\}$. As is seen in Fig. 7.7, for the K values of $6, 8, 9, 10, 12, 14, 15, 16$ there is no bijection. In the case of $K = 6$, all values of r_i correspond to the same value of 0. In the cases of $K = 8, 10, 14, 16$, each two values of r_i correspond to the same value as those in the other set. In the cases of $K = 9$ and 15, each three values of r_i correspond to the same value as in the other set. In the cases of $K = 12$, each five values of r_i correspond to the same value as in the other set. Despite all the above cases in which the required bijection does not exist, when K is a prime number, $K = 7, 11, 13$ and 17, the one-to-one correspondence is found.

7.3.4 Dechirping y_l

After detection of the r_i from the dominant frequency bins of f_l, the signal vector y_l is dechirped for the corresponding dominant r_i as in the following equation:

$$y_l^d = y_l e^{-\frac{j2\pi r_i l^2}{K}}. \tag{7.7}$$

After dechirping the signal vector y_l, since the chirp rate of r_i is removed, the corresponding energy does not spread along the spectrum; hence it is observable at m_i. The following short code illustrates the dechirping at the last three lines:

```
clc; clear all; close all
alpha = 1;
K = 17;
r = [1:K-1 0];
m = [1:K-1 0];
l = [1:K-1 0];
r1 = r(5);
r2 = r(8);
TT = 3;
s1 = 2;
s2 = 2;
mod(r1*TT*2, K)
mod(r2*TT*2, K)
chirpcode1=alpha*exp((1i*2*pi*m(7)*l/K)+((1i*2*pi*r1*(l.^2)))/K);
chirpcode2=alpha*exp((1i*2*pi*m(9)*l/K)+((1i*2*pi*r2*(l.^2)))/K);
r2TmodK_1=alpha*exp((1i*2*pi*mod(r1*TT*2, K)*l/K));
r2TmodK_2=alpha*exp((1i*2*pi*mod(r2*TT*2, K)*l/K));
mixchirp=s1*chirpcode1+s2*chirpcode2;
fsig=conj(mixchirp).*circshift(mixchirp', -TT)';
% Dechirping
dechirp1=exp(-1i*2*pi*r1*l.^2/K).*mixchirp/K;
figure(7);stem(abs(fft(dechirp1))); title('decherped mixture for r1')
```

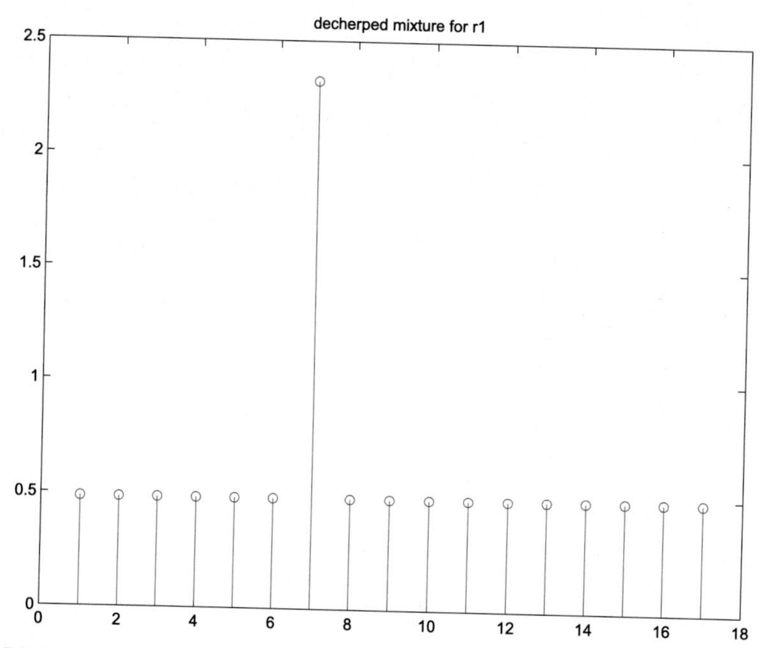

FIGURE 7.8

The DFT of dechirping the mixture of codes for the chirp rate of $r(5)$ which reveals $m(7)$ and $s_1 \approxeq 2$.

As is seen from the result of the code, the dechirped vector signal has concentration of energy around the seventh frequency bin of DFT. It indicates $m_1 = 7$ and its amplitude indicates the corresponding recovered signal sample as $s_1 \approx 2$ (Fig. 7.8).

7.3.5 Sensing and recovery

At this stage, the reader should be completely familiar with the characteristics of the chirp codes and the way they are used for compressive sensing and signal recovery. Especially, it is highly important to understand the bijection between the set of chirp rates $\{r_i \mid r_i \in \mathbb{Z}_K\}$ and the set of FFT bins $\{\frac{2r_i T \bmod K}{K} \mid r_i \in \mathbb{Z}_K\}$. In short, in chirp code compressive sensing, the sensing matrix is made of chirp codes as columns: K^2 chirp codes of length K. Then a signal of length K^2 is compressively sensed as a signal vector of length K, which is a linear combination of K length chirp codes while the samples of the initial K^2 length signal play the role of weights of the linear combination. Recovery is an iterative process where, in each loop, $\boldsymbol{f}$ is updated from the last update of $\boldsymbol{y}$. The K-length DFT of $\boldsymbol{f}$ gives the chirp rate r_i corresponding to the dominant signal sample. Dechirping $\boldsymbol{y}$ using the obtained r_i and taking the K-length DFT gives the corresponding m_i and x_i. r_i, m_i and x_i give a signal sample value and location. Therefore, having a signal sample, $\boldsymbol{y}$ can be updated by removing the corresponding chirp codes from the mixture as follows:

$$y_l := y_l - x_i \exp\left(j2\pi \frac{m_i}{K} l + j2\pi \frac{r_i}{K} l^2\right). \tag{7.8}$$

This loop repeats until the energy of the chirp mixture reaches a lower bound.

The following code gives a complete process of compressive sensing by chirp codes and recovery on a harmonic signal:

```
%% Compressive Sensing: Deterministic Chirp Codes
%% Implementation on a harmonic signal
clc; clear all; close all
K=17;
=0:K-1;
for i=1:K
bijest(i)=mod(2*r(i)*3, K);
end
%% Making Phi the CS matrix
alpha = 1;
m=0:K-1;
r=0:K-1;
l=0:K-1;
for ii=1:K
for i = 1:K
Phi(:,i+K*(ii-1))= alpha*exp((1i*2*pi*m(i)*l/K)+((1i*2*pi*r(ii)*
(l.^2)))/K);
end
end
%% The harmonic signal
N=K^2;
l2=0:N-1;
amts=[1 4 2];
frequs=[121 12 50]/N;
x=zeros(N,1);
for j=1:size(amts,2)
x=x+(amts(j)*exp(1i*2*pi*frequs(j)*l2))'; % input signal
end
%% Transferring to Fourier Sparse Domain
s=fft(x);s2=s;
figure(9);subplot(212);stem((0:N-1),abs(s))
%% Compressive Sensing
mixchirp=Phi*s2;
mixchirp=mixchirp';
%% Recovery
TT=3;
```

```
samples=[];
for jj=1:size(amts,2)
%% Updating f_1
fsig=conj(mixchirp).*circshift(mixchirp', TT)';
%% Obtaining m_i
[aa, r_max]=max(abs(fft(fsig)));
rf=r(find(bijest==r_max-1));
%% Dechirping
dechirp=exp(1i*2*pi*rf*1.^2/K).*mixchirp/K;
%% Obtaining x_i and m_i
[sampval, m_max]=max(abs(fft(dechirp)));
temp=fft(dechirp);
sampval=temp(m_max);
m_max=mod((17-m_max+2),17);
if m_max==0
m_max=17;
end
samploc=rf*17+m_max;
samples=[samples [samploc; sampval]];
%% Updating y_1
s2(samploc)=s2(samploc)-sampval;
mixchirp=Phi*s2;
mixchirp=mixchirp';
jj=jj+1;
end
s_est=zeros(size(s));
for i=1:size(samples,2)
s_est(real(samples(1,i)),1)=samples(2,i);
end
x_est=(ifft(s_est));
figure(9);subplot(211);plot(abs(s));subplot(212); plot(abs(s_est))
figure(10);plot(1:100, abs(x_est(1:100)), 1:100, abs(x(1:100)));
title('Original Signal and the recostructed on by CS')
```

Fig. 7.9 shows the result of the above listed code as the original signal and the reconstructed signal have been depicted.

7.4 Summary

The chapter gives a step-by-step MATLAB tutorial to chirp code compressive sensing. Every point used in this technique of compressive sensing has been clarified by short and simple codes with a focus on just the point under study. After explanation of

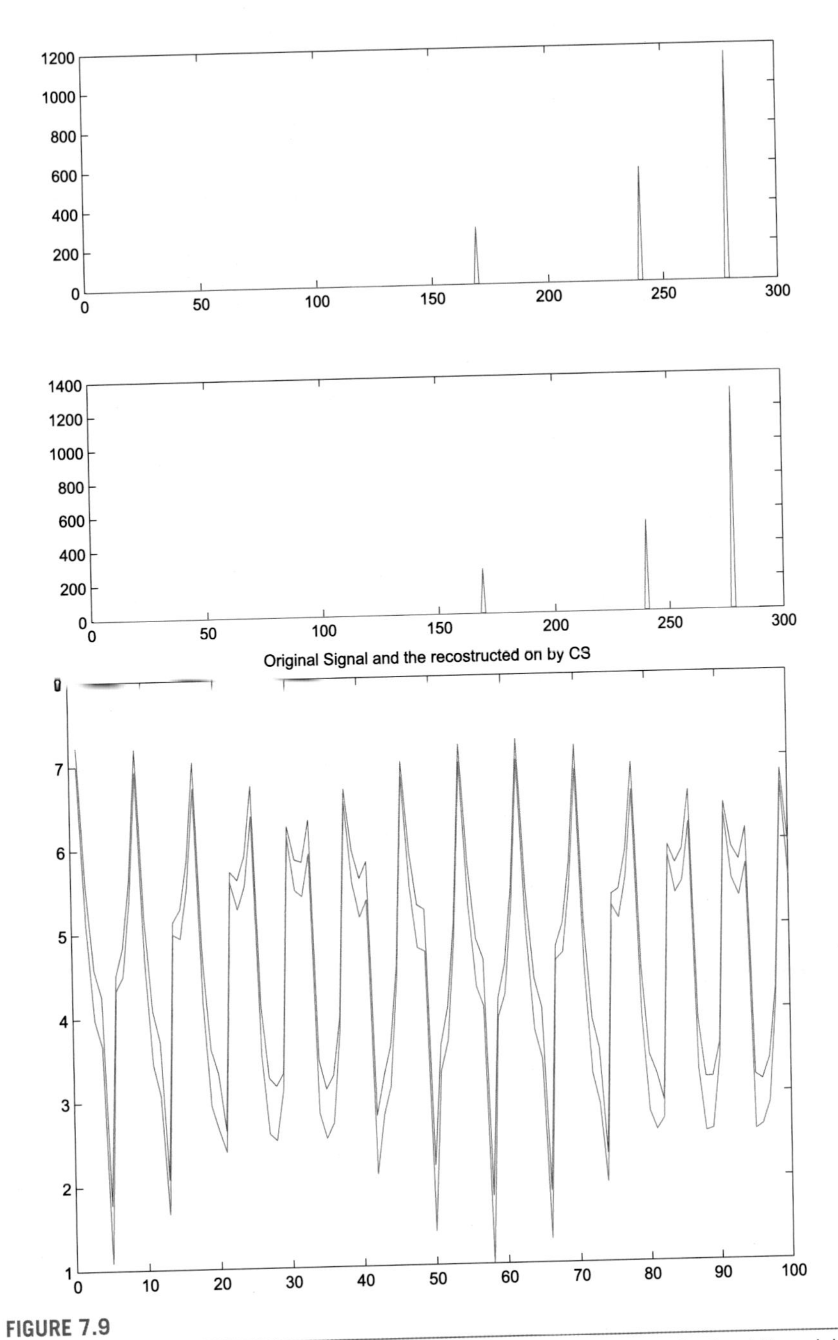

FIGURE 7.9

(Up) The spectrum of the original signal and the reconstructed signal, (bottom) the original signal (blue) and the reconstructed one from the compressed sensed data (blue).

a simple code, two codes are mixed and the process of separation of these two codes as used in chirp code compressive sensing has been clarified. At the end a complete code of chirp code compressive sensing and recovery has been presented.

References

[1] E.J. Candès, et al., Compressive sampling, in: Proceedings of the International Congress of Mathematicians, Madrid, Spain, vol. 3, 2006, pp. 1433–1452.

[2] E.J. Candès, J. Romberg, T. Tao, Robust uncertainty principles: exact signal reconstruction from highly incomplete frequency information, IEEE Transactions on Information Theory 52 (2) (2006) 489–509.

[3] E.J. Candes, J.K. Romberg, T. Tao, Stable signal recovery from incomplete and inaccurate measurements, Communications on Pure and Applied Mathematics: A Journal Issued by the Courant Institute of Mathematical Sciences 59 (8) (2006) 1207–1223.

[4] E.J. Candes, T. Tao, Near-optimal signal recovery from random projections: universal encoding strategies?, IEEE Transactions on Information Theory 52 (12) (2006) 5406–5425.

[5] D.L. Donoho, et al., Compressed sensing, IEEE Transactions on Information Theory 52 (4) (2006) 1289–1306.

[6] C.E. Gutierrez, M.R. Alsharif, K. Yamashita, M. Khosravy, A tweets mining approach to detection of critical events characteristics using random forest, International Journal of Next-Generation Computing 5 (2) (2014) 167–176.

[7] N. Kausar, S. Palaniappan, B.B. Samir, A. Abdullah, N. Dey, Systematic analysis of applied data mining based optimization algorithms in clinical attribute extraction and classification for diagnosis of cardiac patients, in: Applications of Intelligent Optimization in Biology and Medicine, Springer, 2016, pp. 217–231.

[8] C.E. Gutierrez, P.M.R. Alsharif, M. Khosravy, P.K. Yamashita, P.H. Miyagi, R. Villa, Main large data set features detection by a linear predictor model, AIP Conference Proceedings 1618 (2014) 733–737.

[9] M.H. Sedaaghi, R. Daj, M. Khosravi, Mediated morphological filters, in: 2001 International Conference on Image Processing, Proceedings, vol. 3, IEEE, 2001, pp. 692–695.

[10] M. Khosravy, N. Gupta, N. Marina, I.K. Sethi, M.R. Asharif, Morphological filters: an inspiration from natural geometrical erosion and dilation, in: Nature-Inspired Computing and Optimization, Springer, Cham, 2017, pp. 349–379.

[11] S. Gupta, M. Khosravy, N. Gupta, H. Darbari, N. Patel, Hydraulic system onboard monitoring and fault diagnostic in agricultural machine, Brazilian Archives of Biology and Technology 62 (2019).

[12] S. Gupta, M. Khosravy, N. Gupta, H. Darbari, In-field failure assessment of tractor hydraulic system operation via pseudospectrum of acoustic measurements, Turkish Journal of Electrical Engineering & Computer Sciences 27 (4) (2019) 2718–2729.

[13] A.S. Ashour, S. Samanta, N. Dey, N. Kausar, W.B. Abdessalemkaraa, A.E. Hassanien, Computed tomography image enhancement using cuckoo search: a log transform based approach, Journal of Signal and Information Processing 6 (03) (2015) 244.

[14] M. Khosravy, N. Gupta, N. Marina, I. Sethi, M. Asharif, Brain action inspired morphological image enhancement, in: Nature-Inspired Computing and Optimization, Springer, Cham, 2017, pp. 381–407.

[15] M. Khosravy, N. Punkoska, F. Asharif, M.R. Asharif, Acoustic OFDM data embedding by reversible Walsh-Hadamard transform, AIP Conference Proceedings 1618 (2014) 720–723.

[16] M. Khosravy, M.R. Asharif, M.H. Sedaaghi, Medical image noise suppression: using mediated morphology, IEICE, Technical Report 107 (461) (2008) 265–270.

[17] N. Dey, A.S. Ashour, A.S. Ashour, A. Singh, Digital analysis of microscopic images in medicine, Journal of Advanced Microscopy Research 10 (1) (2015) 1–13.

[18] M. Khosravy, M.R. Asharif, M.H. Sedaaghi, Medical image noise suppression using mediated morphology, in: IEICE Tech. Rep., IEICE, 2008, pp. 265–270.

[19] M. Khosravy, N. Gupta, N. Marina, I. Sethi, M. Asharifa, Perceptual adaptation of image based on Chevreul–Mach bands visual phenomenon, IEEE Signal Processing Letters 24 (5) (2017) 594–598.
[20] M.H. Sedaaghi, M. Khosravi, Morphological ECG signal preprocessing with more efficient baseline drift removal, in: 7th IASTED International Conference, ASC, 2003, pp. 205–209.
[21] N. Dey, S. Samanta, X.-S. Yang, A. Das, S.S. Chaudhuri, Optimisation of scaling factors in electrocardiogram signal watermarking using cuckoo search, International Journal of Bio-Inspired Computation 5 (5) (2013) 315–326.
[22] M. Khosravy, M.R. Asharif, M.H. Sedaaghi, Morphological adult and fetal ECG preprocessing: employing mediated morphology, IEICE Technical Report, IEICE 107 (2008) 363–369.
[23] N. Dey, A.S. Ashour, F. Shi, S.J. Fong, R.S. Sherratt, Developing residential wireless sensor networks for ECG healthcare monitoring, IEEE Transactions on Consumer Electronics 63 (4) (2017) 442–449.
[24] M. Khosravi, M.H. Sedaaghi, Impulsive noise suppression of electrocardiogram signals with mediated morphological filters, in: 11th Iranian Conference on Biomedical Engineering, ICBME, 2004, pp. 207–212.
[25] N. Dey, S. Mukhopadhyay, A. Das, S.S. Chaudhuri, Analysis of p-qrs-t components modified by blind watermarking technique within the electrocardiogram signal for authentication in wireless telecardiology using DWT, International Journal of Image, Graphics and Signal Processing 4 (7) (2012) 33.
[26] T.W. Cabral, M. Khosravy, F.M. Dias, H.L.M. Monteiro, M.A.A. Lima, L.R.M. Silva, R. Naji, C.A. Duque, Compressive sensing in medical signal processing and imaging systems, in: Sensors for Health Monitoring, Elsevier, 2019, pp. 69–92.
[27] F. Asharif, S. Tamaki, M.R. Alsharif, H. Ryu, Performance improvement of constant modulus algorithm blind equalizer for 16 QAM modulation, International Journal on Innovative Computing, Information and Control 7 (4) (2013) 1377–1384.
[28] M. Khosravy, M.R. Alsharif, K. Yamashita, An efficient ICA based approach to multiuser detection in MIMO OFDM systems, in: Multi-Carrier Systems & Solutions 2009, Springer, 2009, pp. 47–56.
[29] M. Khosravy, M.R. Alsharif, B. Guo, H. Lin, K. Yamashita, A robust and precise solution to permutation indeterminacy and complex scaling ambiguity in BSS-based blind MIMO-OFDM receiver, in: International Conference on Independent Component Analysis and Signal Separation, Springer, 2009, pp. 670–677.
[30] M. Khosravy, A blind ICA based receiver with efficient multiuser detection for multi-input multi-output OFDM systems, in: The 8th International Conference on Applications and Principles of Information Science (APIS), Okinawa, Japan, 2009, 2009, pp. 311–314.
[31] M. Khosravy, S. Kakazu, M.R. Alsharif, K. Yamashita, Multiuser data separation for short message service using ICA, SIP, IEICE Technical Report 109 (435) (2010) 113–117.
[32] M. Khosravy, M.R. Alsharif, M. Khosravi, K. Yamashita, An optimum pre-filter for ICA based multi-input multi-output OFDM system, 2010 2nd International Conference on Education Technology and Computer, vol. 5, IEEE, 2010, pp. V5–129.
[33] E. Santos, M. Khosravy, M.A. Lima, A.S. Cerqueira, C.A. Duque, A. Yona, High accuracy power quality evaluation under a colored noisy condition by filter bank ESPRIT, Electronics 8 (11) (2019) 1259.
[34] E. Santos, M. Khosravy, M.A. Lima, A.S. Cerqueira, C.A. Duque, ESPRIT associated with filter bank for power-line harmonics, sub-harmonics and inter-harmonics parameters estimation, International Journal of Electrical Power & Energy Systems 118 (105) (2020) 731.
[35] H. Zaheb, M.S.S. Danish, T. Senjyu, M. Ahmadi, A.M. Nazari, M. Wali, M. Khosravy, P. Mandal, A contemporary novel classification of voltage stability indices, Applied Sciences 10 (5) (2020) 1639.
[36] A.A. Picorone, T.R. de Oliveira, R. Sampaio-Neto, M. Khosravy, M.V. Ribeiro, Channel characterization of low voltage electric power distribution networks for PLC applications based on measurement campaign, International Journal of Electrical Power & Energy Systems 116 (105) (2020) 554.
[37] L. Applebaum, S.D. Howard, S. Searle, R. Calderbank, Chirp sensing codes: deterministic compressed sensing measurements for fast recovery, Applied and Computational Harmonic Analysis 26 (2) (2009) 283–290.

[38] M. Khosravy, Blind source separation and its application to speech, image and MIMO-OFDM communication systems, PhD thesis, University of the Ryukyus, Japan, 2010.
[39] M. Khosravy, M. Gupta, M. Marina, M.R. Asharif, F. Asharif, I. Sethi, Blind components processing a novel approach to array signal processing: a research orientation, in: 2015 International Conference on Intelligent Informatics and Biomedical Sciences, ICIIBMS, 2015, pp. 20–26.
[40] M. Khosravy, M.R. Alsharif, K. Yamashita, A PDF-matched modification to Stone's measure of predictability for blind source separation, in: International Symposium on Neural Networks, Springer, Berlin, Heidelberg, 2009, pp. 219–228.
[41] M. Khosravy, M.R. Asharif, K. Yamashita, A probabilistic short-length linear predictability approach to blind source separation, in: 23rd International Technical Conference on Circuits/Systems, Computers and Communications (ITC-CSCC 2008), Yamaguchi, Japan, ITC-CSCC, 2008, pp. 381–384.
[42] M. Khosravy, M.R. Asharif, K. Yamashita, A theoretical discussion on the foundation of Stone's blind source separation, Signal, Image and Video Processing 5 (3) (2011) 379–388.
[43] M. Khosravy, M.R. Asharif, K. Yamashita, A PDF-matched short-term linear predictability approach to blind source separation, International Journal of Innovative Computing, Information & Control 5 (11) (2009) 3677–3690.
[44] E. Candes, T. Tao, Decoding by linear programming, preprint, arXiv:math/0502327, 2005.
[45] M. Khosravy, N. Gupta, N. Patel, T. Senjyu, Frontier applications of nature inspired computation.
[46] N. Gupta, M. Khosravy, N. Patel, O.P. Mahela, G. Varshney, Plant genetics-inspired evolutionary optimization: a descriptive tutorial, in: Frontier Applications of Nature Inspired Computation, Springer, 2020, pp. 53–77.
[47] C. Moraes, E. De Oliveira, M. Khosravy, L. Oliveira, L. Honório, M. Pinto, A hybrid bat-inspired algorithm for power transmission expansion planning on a practical Brazilian network, in: Applied Nature-Inspired Computing: Algorithms and Case Studies, Springer, 2020, pp. 71–95.
[48] M. Khosravy, N. Gupta, N. Patel, T. Senjyu, C.A. Duque, Particle swarm optimization of morphological filters for electrocardiogram baseline drift estimation, in: Applied Nature-Inspired Computing: Algorithms and Case Studies, Springer, 2020, pp. 1–21.
[49] G. Singh, N. Gupta, M. Khosravy, New crossover operators for real coded genetic algorithm (RCGA), in: 2015 International Conference on Intelligent Informatics and Biomedical Sciences (ICIIBMS), IEEE, 2015, pp. 135–140.
[50] N. Gupta, N. Patel, B.N. Tiwari, M. Khosravy, Genetic algorithm based on enhanced selection and log-scaled mutation technique, in: Proceedings of the Future Technologies Conference, Springer, 2018, pp. 730–748.
[51] N. Gupta, M. Khosravy, N. Patel, I. Sethi, Evolutionary optimization based on biological evolution in plants, Procedia Computer Science 126 (2018) 146–155.
[52] J. Kaliannan, A. Baskaran, N. Dey, A.S. Ashour, M. Khosravy, R. Kumar, ACO based control strategy in interconnected thermal power system for regulation of frequency with HAE and UPFC unit, in: International Conference on Data Science and Application (ICDSA-2019), in: LNNS, Springer, 2019.
[53] M. Khosravy, N. Gupta, N. Patel, O.P. Mahela, G. Varshney, Tracing the points in search space in plant biology genetics algorithm optimization, in: Frontier Applications of Nature Inspired Computation, Springer, 2020, pp. 180–195.
[54] N. Gupta, M. Khosravy, N. Patel, S. Gupta, G. Varshney, Evolutionary artificial neural networks: comparative study on state-of-the-art optimizers, in: Frontier Applications of Nature Inspired Computation, Springer, 2020, pp. 302–318.
[55] N. Gupta, M. Khosravy, N. Patel, S. Gupta, G. Varshney, Artificial neural network trained by plant genetic-inspired optimizer, in: Frontier Applications of Nature Inspired Computation, Springer, 2020, pp. 266–280.

CHAPTER

8

Cyber physical systems for healthcare applications using compressive sensing

K. Keerthana, S. Aasha Nandhini, S. Radha
Department of ECE, SSN College of Engineering, Kalavakkam, Chennai, India

8.1 Introduction

A Cyber Physical System (CPS) [1] is a developing technology that has paved the way for many new academical and industrial inventions. CPS is more than software, networking and embedded computing. It is a major combination of computation, communication and control. The term physical in CPS implies to all man made and natural systems that obey all the standards and principles of physics. Here the physical and computation process are tightly combined together and with the help of various sensor networks and computers they have control on the physical process and also collect feedback from the various systems. The various characteristics of CPS are that it is networked, has a strong sensing capability, and can work on a real time environment with highly predictable behavior and has higher performance capability. CPS has so far been dependent, safe, efficient, reliable, stable and secure. It finds its applications in various sectors such as healthcare [2], military, transport, energy, infrastructure, and building systems. One fundamental component in CPS is Wireless Sensor Networks (WSNs).

WSN is a combination of various sensors, actuators and processors networked together. They collect data from a real time environment and send these data to the reviewing site or processors where the decision is made and as a result is being implemented with the help of actuators. CPS is deployed with the help of WSN technologies. In order to achieve many daily activities more efficiently, CPS acts as a bridge connecting the physical world with the virtual world. The implementation of WSN in CPS has made all real time decision making activities reliable and predictable. The wireless sensor node architecture consists of a sensor node that collects different measurement values from the physical world and converts them such that they can be understood by the user or the computer. The processing unit present in the sensor node consists of a memory unit where the data collected from the sensing unit is stored and the processor executes the information that is given to it. In addition

Compressive Sensing in Healthcare. https://doi.org/10.1016/B978-0-12-821247-9.00013-5

to it, we have a power unit to power up the nodes when they are placed in the open environment. The transceiver unit in the sensor node plays an important role as the sensors are deployed in the open space, where wired communication is not always possible. In order to overcome this we have the transceiver which can collect data from the neighbor nodes.

CPS in healthcare is one developing sector in recent trends. The prevailing healthcare systems do lack the latest technical integration. The developments in areas like WSN, cloud computing and sensors in healthcare [3] are making CPS a very strong competitor in the healthcare sector. Patients can be monitored remotely and feedback can be given to them disregarding their location. Extensive research has been carried out as regards the sensors deployed this sector. These sensors collect all the essential metrics from the patients. The data that is being collected from patients should be stored in safe and efficient ways, should be made available for the doctors at all time. Therefore, efficient database storage and management have become important.

In this chapter, real time data has been collected from patients. These data undergo a pre-processing technique through which data on the area of interest is obtained. After this the Compressing Sensing (CS) technique is applied to this data and it is further sent to the cloud for storage and is retrieved through a suitable recovery algorithm and is available for doctors for diagnosing the disease. The main aim of this chapter is to process the real time data and to reduce the amount of data that is being stored in the cloud.

8.2 Related works

A few papers related to the area of research are discussed below with their advantages and limitations. Mohammad Mehedi Hassan et al. [4] (2017) presented an efficient network system that is an integration of both WBAN and Cloud used for accurate data sharing that consists of four layers and can provide constant delivery of healthcare details. Significant data that is related to the patients, such as text, image, and voice plays an important role in the healthcare service process; easily maintaining these data from various WBAN is fundamental for many applications. Content Centric Networking is integrated (CCN) with the proposed architecture to improve the ability of the WBAN coordinator. Thus, it can support uninterrupted media healthcare content delivery and reduce packet loss through adaptive streaming technique. The practical implementation of this method is complex.

Shah J. Miah et al. [5] (2017) have proposed an e-health professional advisory system. This system consists of cloud computing that facilitates doctors to recognize and diagnose the various diseases for people residing in both urban and rural societies. This system serves as a platform for doctors in evaluating and diagnosing patient's data and medical history through intermediary clinics. The practical achievement of formerly nonexistent clinical work is thus practically accomplished by making use of this system. The limitation of this system is that Internet connectivity in rural areas may be a major concern. Yin Zhang et al. [6] (2017) provided a much more comfortable assistance and atmosphere for healthcare. Their paper proposes a patient-based healthcare applications and assistance with the aid of CPS, called Health-CPS, which is a combination of both big data and cloud technologies. This system consists of different layers each of which does its own services like storage and simultaneous computation. Using this proposed system we can improve the achievements of the healthcare system using Health-CPS. But the system has a very complex architecture.

Gao et al. [7] (2017) in their paper showed that the bonds that prevail between CPS and patients are growing strong and sturdy. The authors mainly focus on a blood glucose-insulin control system as the number of diabetic patients has increased over years. The conventional method for the treatment of diabetes is by injecting insulin manually, which is overall a very tedious process. In order to overcome these problems and to properly monitor the insulin levels, a better system is proposed. However, for the successful functioning of this system blood samples must be collected from the patients periodically. Truong Quang Vinh et al. [8] (2016) presented a system that does both capturing of real time dental images and the classification of tooth images. Here, real time dental images are captured from patients with the help of special cameras used for this purpose. In addition to this, this device has many other features, such as examining the dental images, image processing and maintaining the patient's record. Henceforth, a system which is reliable, portable and handy was developed. But the system lacks appropriate processing techniques.

Abdol Hamid Pilevar [9] (2016) proposes an innovative technique, where, based upon the content, the recovery of the dental images from the medical databases is carried out. Image attributes are made use in order to obtain standard recovery and portrayal of the dental images. A new method is used to measure the various features of the image and to classify the teeth based from the various models. Software has been developed for checking the efficiency of the system. The system has been found to be robust and provides accurate results. But a large teeth model database is required. Aosen Wang et al. [10] (2016) presented a unique Quantized Compressed Sensing (QCS) technique, which makes use of both the sampling rate and quantization in order to obtain an improved efficiency rate. In addition, to overcome the intricacy found in the computation, the authors have designed a system, furthermore, to combat the computational complexity of the configuration procedure; we propose a very quick configuration technique, called RapQCS. The proposed system is said to have achieved a 66% gain over the conventional QCS technique. It is also said to have produced 150 times more speed. The recovered signal reliability is only 2.32%.

Kuo-hui Yeh [11] (2016) has proposed a system which functions with the help of BSN architecture to provide a stable IoT-dependent healthcare system. This system aims to realize public IoT networks. Secrecy and authentication is maintained during transmission, processing and other backend processes. The practical application of the proposed healthcare system with the help of Raspberry PI domain to illustrate the efficiency and the workability and feasibility of the proposed techniques was accomplished. Surya Deekshith Gupta et al. [12] (2015) have developed a Raspberry Pi-based remote monitoring system. Here, the authors have designed a system which can investigate the electrocardiogram (EcG) and various critical parameters. The data collected from the patients are stored securely in their website which can be controlled only by those with authentication. The authors have developed a system which looks for any abnormality in the data fed to their website and if needed takes necessary action. To achieve this, the authors have made use of the MySQLdb module, Raspberry Pi and GSM. However, the proposed system may occupy a huge storage space in the website.

It is found from the above research work that the various healthcare data that has been collected from patients is very important. The efficient processing and storage of this data can be a great aid to the medical community like doctors, healthcare workers and clinicians.

8.3 Proposed work

A cyber physical system is proposed that captures real time images from patients and processes the image and efficient storage of the data to be carried out following clinical analysis. Fig. 8.1 shows the block diagram of the proposed cyber system for healthcare applications.

In the proposed system, real time medical images (for example, dental images) are captured with the aid of proper image sensors from the patients visiting the hospital. These images are raw images and appropriate pre-processing techniques are applied to these medical images for obtaining the particular region of interest. The pre-processed image is further divided into blocks and the compressive-sensing technique is being applied to these medical images. Here parallel sensing and com-

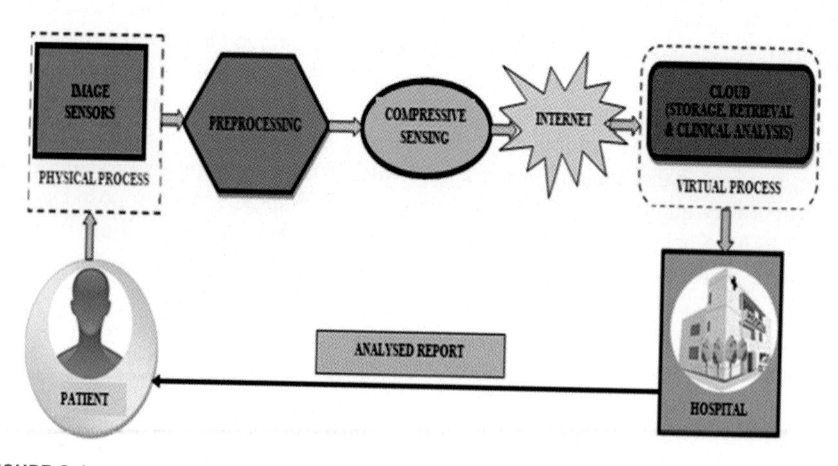

FIGURE 8.1

Overview of cyber physical system for healthcare.

pression of the medical image is done. After this the measurements are obtained. These measurements are uploaded to the cloud medical database using ThingSpeak with the help of internet. These measurements are stored in the cloud. These measurements can further be retrieved from the database using a recovery algorithm and an appropriate reconstruction algorithm is applied to these images to obtain the original images. Furthermore, the doctors can examine the patient's medical data and investigate the diseases. A detailed analysis report is sent to the patients from the doctors with suitable prescription.

8.3.1 Image acquisition

Image acquisition is always the first step in image processing. Images are collected from patients with the help of cameras and are transformed into a form that can be processed later. This process is named acquisition of image. This can be achieved with the help of image sensors.

8.3.2 Pre-processing techniques

- HSV transformation

HSV stands for Hue-Saturation-Value. It is basically a representation of color plane similar to RGB. In this color space, colors are characterized in the form of their shade (saturation or amount of gray) and their illumination rate. The *hue* (H) represents the purity of the color. The *saturation* (S) represents the whiteness of the color and the *value* (V) represents the darkness of the color. The HSV transformation of the dental image is illustrated in Fig. 8.2.

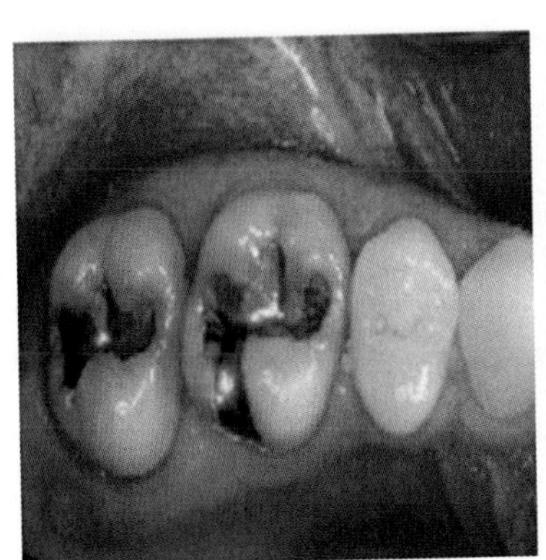
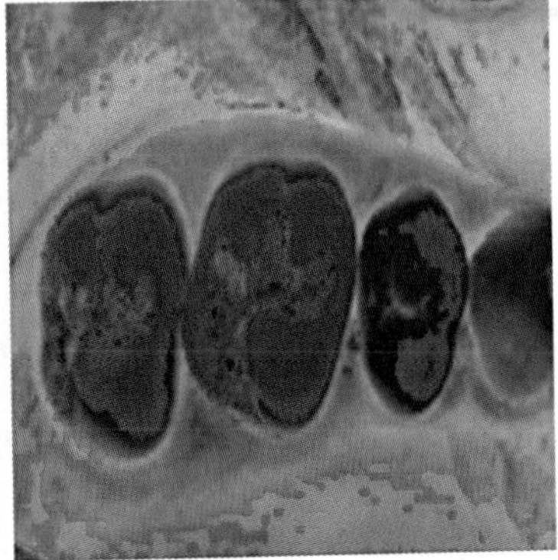

FIGURE 8.2

HSV Transformation of dental image.

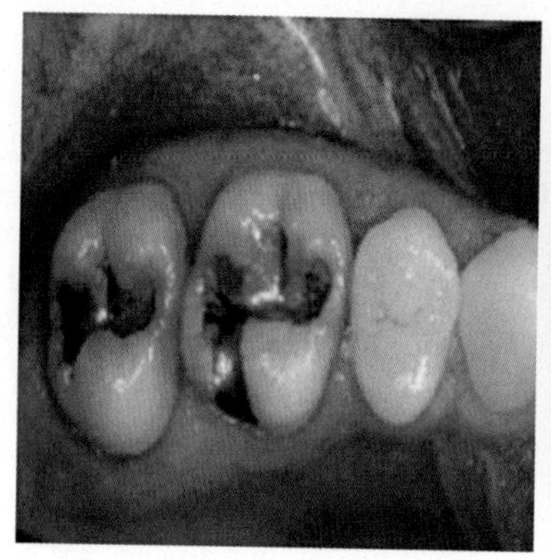
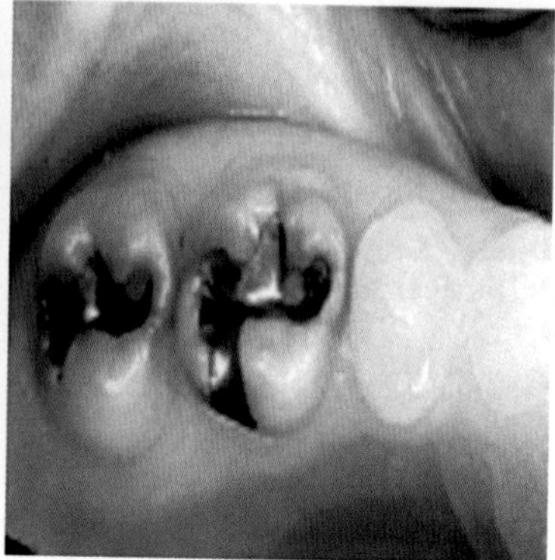

FIGURE 8.3

Contrast enhancement.

- Contrast enhancement

In order to enhance the brightness and quality of the dental images and to portray the image in an illustratable form in the perception of human eyes, contrast enhancement has been applied to the medical images. Images with high contrast have a very good quality range. Fig. 8.3 illustrates the dental images to which contrast enhancement has been applied.

- Binary image conversion

Basically there are two colors involved in the representation of the binary image, black and white. Here, two values are allocated for each pixel. A binary image is a digital image that has only two possible values for each pixel. It is a two level image in digital form. The main advantages of using this format are that it occupies very less space and also goes well with all compression techniques. Fig. 8.4 shows the binary image conversion of the medical image.

8.3.3 Block division

After the pre-processing techniques have been applied to the medical images, the images are then divided into individual blocks, which are all of the same size and do not overlap. The size of the original medical image, which is used for implementation is 256 × 256. The size into which the original image is divided is 8 × 8. Therefore, block division of the original image helps in reducing the complexity that will be faced when applying the CS process.

8.3.4 Compressive sensing

Compressive sensing [17] is an algorithm that obtains and reconstructs the original signal from using very fewer samples that may be required for the existing techniques.

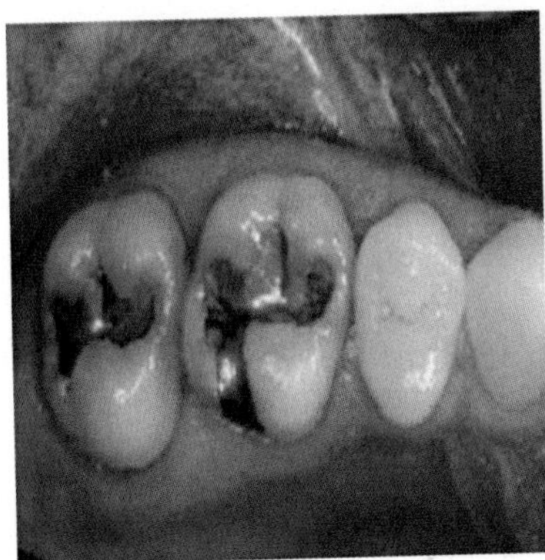

FIGURE 8.4

Binary image conversion.

This technique does compression during the sensing process. It can overcome many difficulties faced by the traditional techniques. In order to obtain exact recovery and reconstruction of the original signal, the measurement matrix and recovery algorithm must be chosen as such. Here the original signal is converted into a sparse vector and this vector is multiplied along with the measurement matrix, which is a random matrix. The random matrix is chosen here because it satisfies both the Restricted Isometric Property (RIP) and the Mutual Incoherence Property (MIP). On multiplying the two matrices, a measurement vector is obtained. At last, there is a reduction in the size of the original image since we make use of very few coefficients. On obtaining the measurements, they are uploaded to the cloud database.

8.3.5 Uploading data to cloud

In the healthcare sectors, storing the data in the cloud database has gained importance in the recent trends. Healthcare-based data broadcast being propagated using the cloud is achieving recognition. These data and pieces of information can be embellished by doctors and can be circulated in the cloud. This concept is gaining attention as of lately due to its effective management and processing of the data being stored in the cloud. ThingSpeak is one of the IoT based assistance that helps us to accumulate, visualize and to process the real time data to the cloud. The measurements that are obtained after the CS process are being uploaded to ThingSpeak. Private channels are created for this purpose and the measurements are stored here. API keys are generated for reading and writing purposes. The measurements are retrieved using MATLAB® code from the channel and the image is reconstructed from the measurements using the Orthogonal Matching Pursuit (OMP) algorithm.

8.3.6 GUI (graphical user interface)

A GUI is developed which creates an interaction between the humans and computer. The GUI gives a visual feedback to the users. A MATLAB-based and Tkinter-based GUI is developed which is used to view the patient's database.

8.4 Performance evaluation

8.4.1 Simulation results

The proposed system is implemented in simulation using MATLAB (2015a). Matrix Laboratory is commonly known as MATLAB and this is a multiple-prototype-based numerical computing medium. Medical images from patients are implemented using a simulation. A list of the parameters used for the simulation is given in Table 8.1.

In Fig. 8.5, the medical images gathered from the dental database and their reconstructed image using MATLAB are illustrated.

The medical images are obtained from the dental database. These medical images go through a process where they are divided into blocks. The original image is cast

Table 8.1 Simulation parameters.

Parameters	Value
$m_1 \times n_1$	256×256
$n \times n$	8×8
N	64
k	5
M	20/30/40

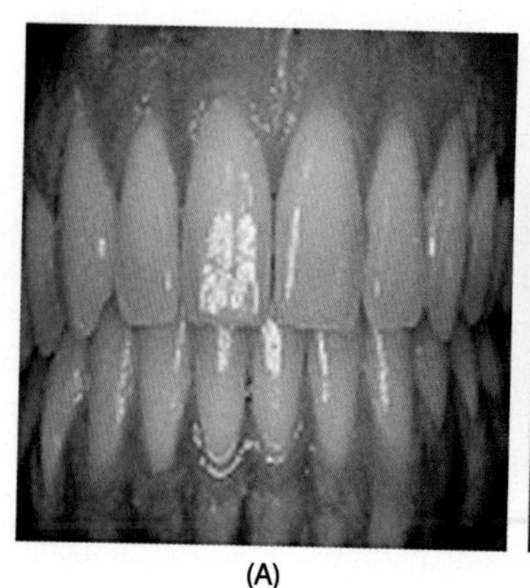

(A)

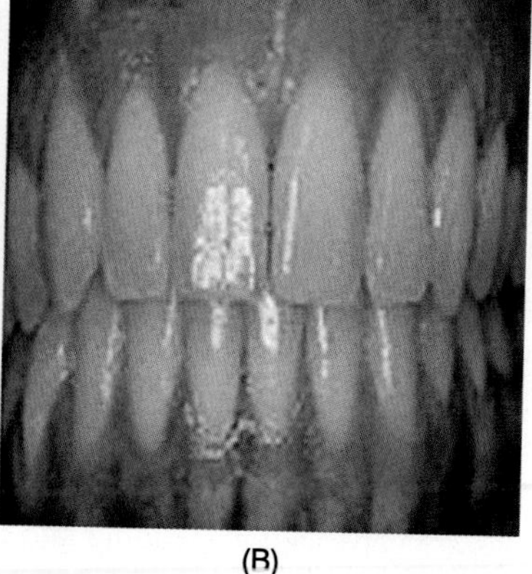

(B)

FIGURE 8.5

(A), (C), (E), (G), (I) – original image; (B), (D), (F), (H), (J) – reconstructed image with M = 30.

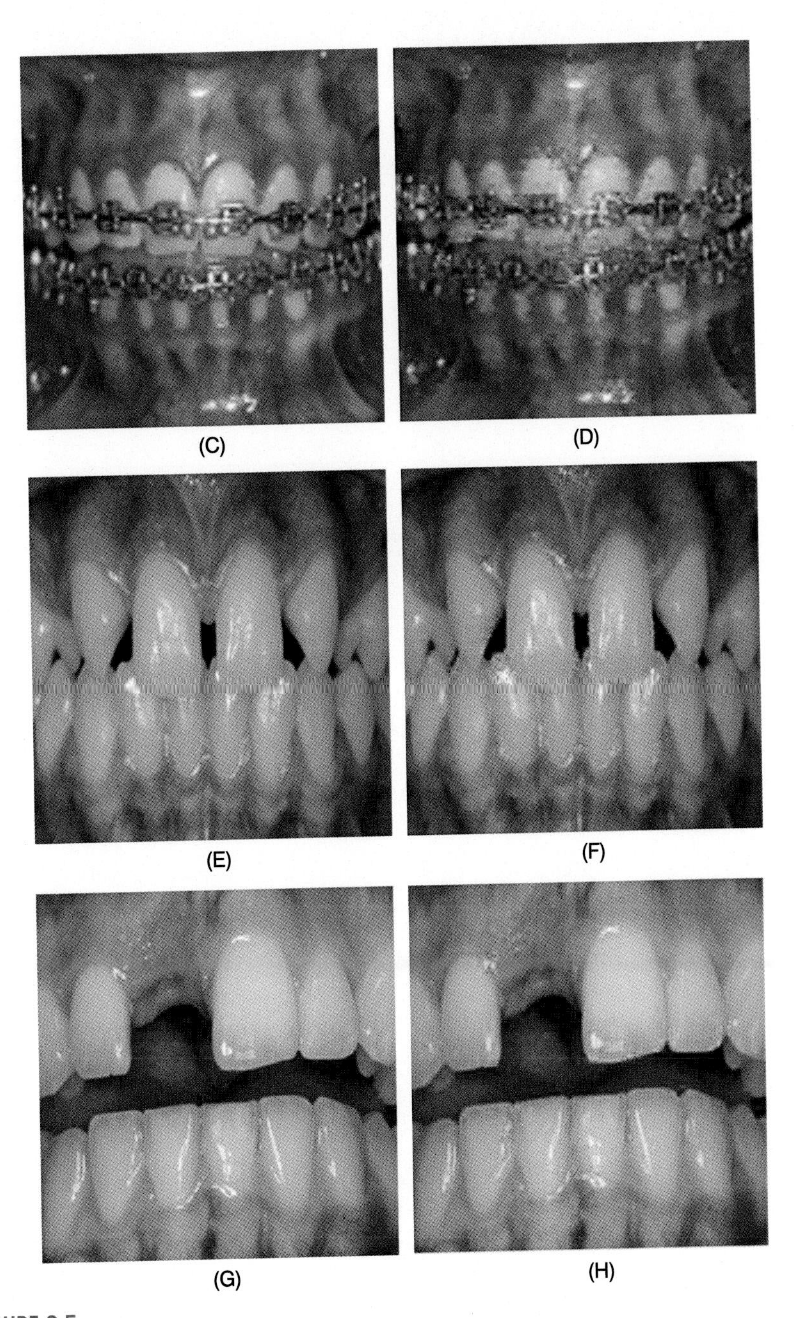

FIGURE 8.5

(*continued*)

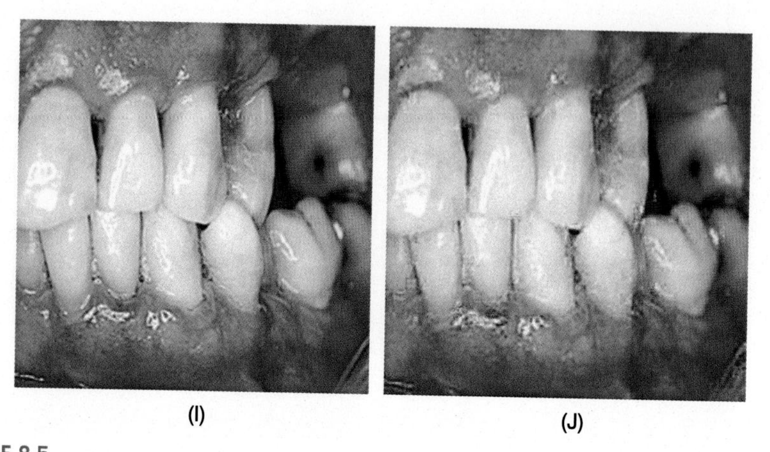

FIGURE 8.5

(*continued*)

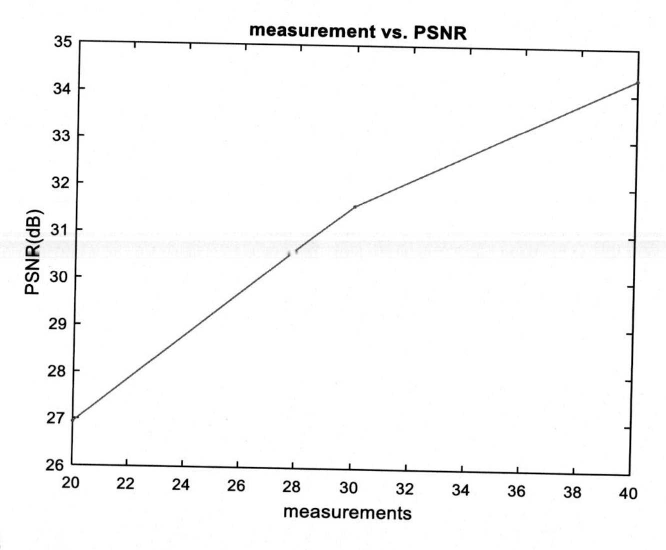

FIGURE 8.6

Graphical representation of no. of measurements vs. PSNR (dB).

into a size of 256 × 256. This image is then block divided into individual blocks of size 8 × 8. A pre-processing technique is then applied to these images. After this the compressive-sensing algorithm is applied to these individual blocks. On applying the CS algorithm, the measurement matrix is obtained. These measurements are uploaded to the cloud using the ThingSpeak software. The images can be reconstructed by using the retrieved measurements from the cloud and by making use of a proper reconstruction algorithm namely the Orthogonal Matching Pursuit Algorithm with no. of measurements M = 30. Fig. 8.6 is plotted for the number of measurements

Table 8.2 PSNR value for various dental images with M = 30.

Image	M	PSNR value
Dental image 1	30	33.3477
Dental image 2	30	28.6179
Dental image 3	30	35.7012
Dental image 4	30	34.8115
Dental image 5	30	29.5840

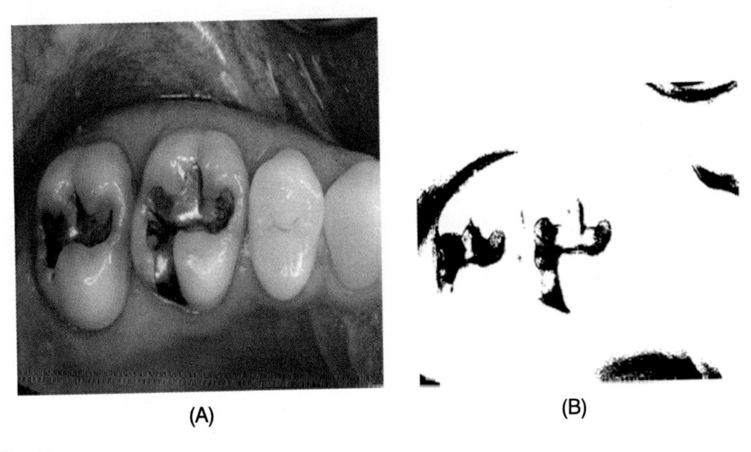

FIGURE 8.7

(A) Original Image, (B) proposed hybrid pre-processing method.

vs. PSNR. The number of measurements varies from 25 to 45. The best results were obtained at M = 30. As the PSNR value increases an increase in the measurements was observed.

Table 8.2 represents the PSNR value for various images at M=30. A hybrid pre-processing technique has been proposed. The hybrid pre-processing technique is a combination of the HSV Transformation followed by the Contrast Enhancement and then the Binary Transformation. By applying the HSV transformation to the image, the brightness value of the image is increased. On applying contrast enhancement, the quality and interpretability of the image is increased. Finally, by applying a binary transformation to the image, to each pixel are assigned two values. The hybrid pre-processing technique is applied in order to obtain the region of interest. Instead of sending the entire data, only a particular set of data is sent to the cloud. Thus, one is reducing the amount of data stored in the cloud.

Fig. 8.7 displays the original image that has been used and the corresponding pre-processed image using the proposed hybrid method.

A GUI is created using MATLAB to view the dental image after the recovery of the measurements from the cloud and the reconstruction of the dental image using the

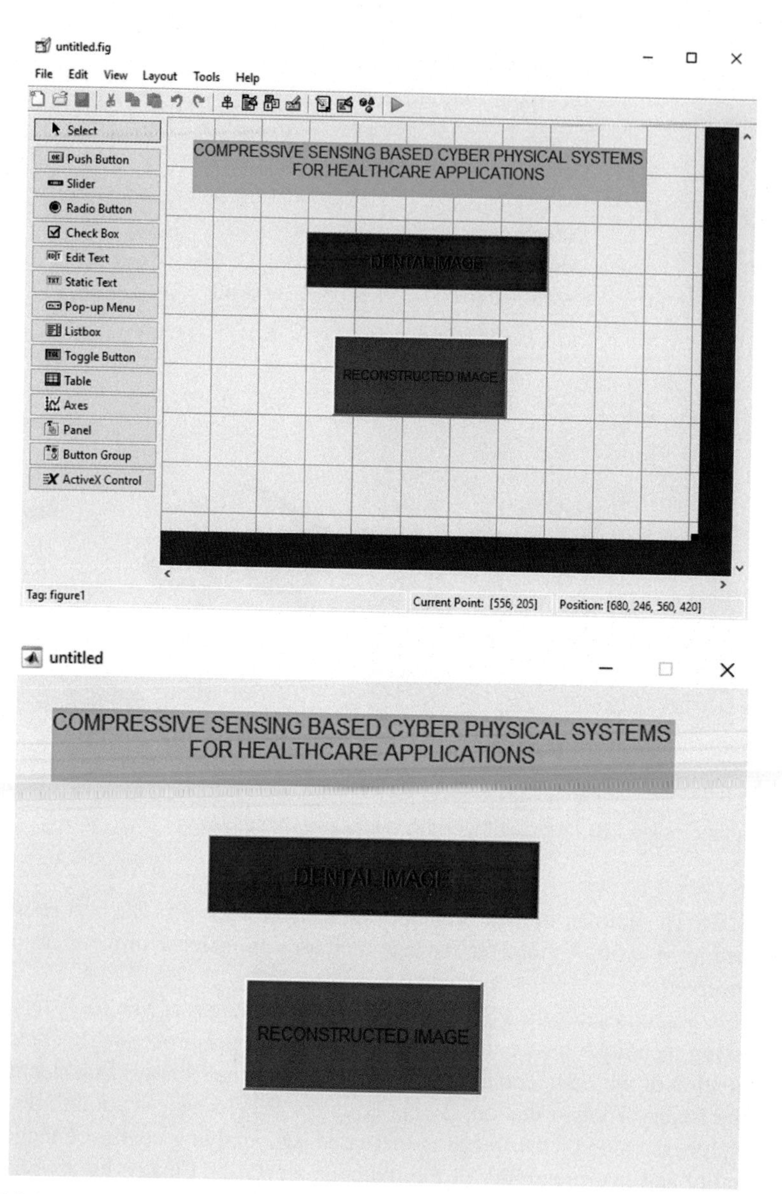

FIGURE 8.8

GUI created using MATLAB.

OMP algorithm. A MATLAB code has been written in order to retrieve the measurements from the cloud. On clicking the reconstructed image push button, the dental image that was reconstructed can be viewed. Fig. 8.8 shows the GUI that was created using the MATLAB software.

FIGURE 8.9

Experimental setup.

8.4.2 Experimental results

The proposed system is validated experimentally using Raspberry Pi hardware. The various modules involved in the hardware implementation are explained below.

(a) Raspberry Pi 3 Board B

A series of many computers are built in a single board. In this proposed system, the hardware implementation of the system is carried out using Raspberry Pi 3 Model B [14]. The Raspberry Pi 3 consists of a quad-core Cortex-A53 processor, which can perform 10 times better than a Raspberry Pi 1. It has been stated by various researchers that Pi 3 performs with 80% more speed than Pi 2.

(b) OpenCV-Python

Open Source Computer Vision Library [15] is also known as OpenCV. It is a combination of C++, C, Python and Java and is compatible with Windows, Linux, Mac OS and Android. By using OpenCV, real time data acquisition is achieved. It consists of built-in libraries for performing the real time processing of data. By using OpenCV, the system complexity is reduced. The Python [16] language has been used for coding. Fig. 8.9 displays the experimental setup for the proposed healthcare system.

The experimental setup consists of a camera module. This camera module is used to capture the real time dental images from the patients. This camera module is attached to the Raspberry Pi 3 Board. The images captured are sent to the Pi board. The Raspberry Pi has a provision for hard drive which has a SD card inserted in the slot available in the board. The Raspbian Jessie OS is installed in the board along with OpenCV software. An external power supply or an USB can be used to power

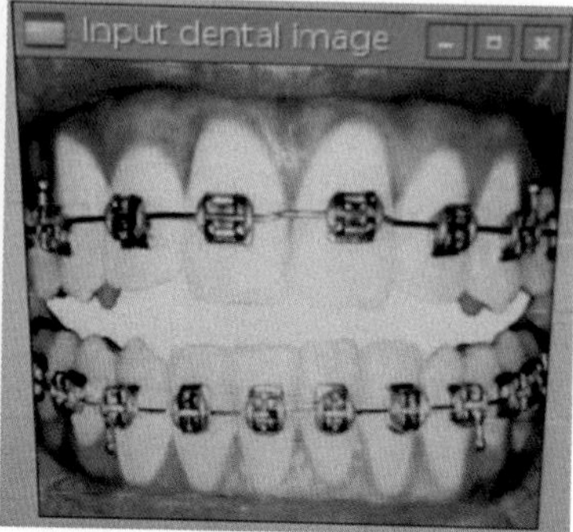

FIGURE 8.10

Screenshots of input image.

the Raspberry Pi board. The images captured can be viewed by using various display sources like a desktop monitor. The image that is acquired through the camera module and sent to the Raspberry Pi 3 Board undergoes pre-processing, and a compressive-sensing algorithm is applied and measurements are obtained. The final measurements are uploaded to the cloud using ThingSpeak.

The input dental image is taken and is resized to 256×256. The dental image is further divided into blocks of size 8×8 and the hybrid pre-processing technique is applied to it. After this the compressive-sensing algorithm is applied to the blocks. The measurement matrix was obtained for various values of M, with $M = 20, 30, 40$. Fig. 8.10 illustrates the screenshot of the input dental image that was obtained after reading the input through Raspberry Pi 3 board.

A hybrid pre-processing method which is a combination of HSV Transformation, Contrast Enhancement and Binary Transformation is applied to the dental image. OpenCV functions are used to apply these pre-processing techniques. Using the hybrid pre-processing technique, the region of interest is obtained. The advantage of this technique is that the entire data set need not be sent to the cloud. Only the required region is sent, which helps in reducing the data storage in the cloud. Fig. 8.11 shows screenshots of the pre-processing technique applied to the image, and the final output after the entire pre-processing is shown.

Fig. 8.12 illustrates the screenshot of measurements that are obtained after the pre-processing technique and the image had been divided into individual blocks and compressive sensing been applied.

ThingSpeak [13] is used to send the measurements that are obtained after the CS algorithm. A mathworks account is created for using ThingSpeak. A channel is created by selecting a new channel option in the channels page. A name is created for the channel and a description is given of the channel. Fields are created for storing the data. Each channel can hold up to eight fields. Finally, the channel settings are saved. A private channel is created so that no one can peep through the data stored in the channel. Further, API keys are generated automatically on creating a private

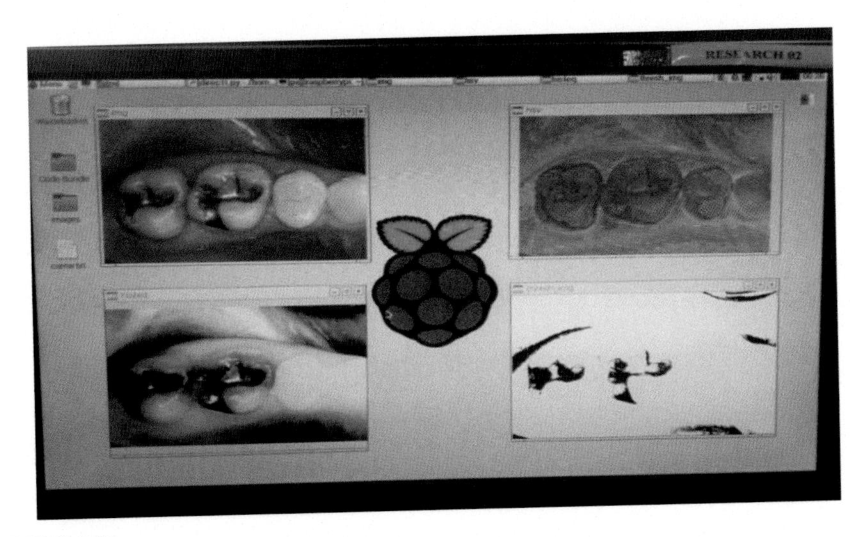

FIGURE 8.11

Screenshots of the pre-processing technique applied to the image.

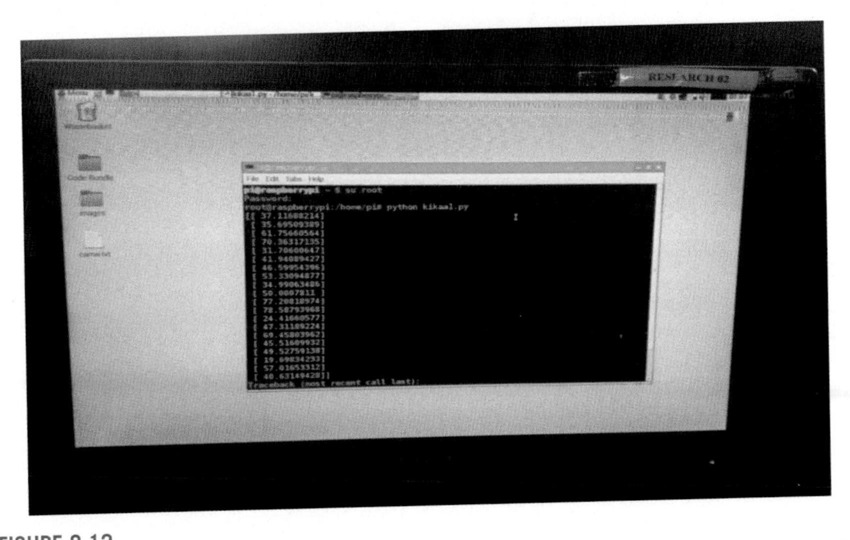

FIGURE 8.12

Screenshots of measurements obtained.

channel. Both read and write keys are generated for reading and writing data in the channel. The screenshots of the output obtained after uploading the measurements to the cloud are shown in Fig. 8.13.

In order to retrieve the data from the ThingSpeak cloud database, a MATLAB analysis code was written. Fig. 8.14 depicts the screenshot of the MATLAB analysis code in ThingSpeak.

FIGURE 8.13

Screenshots of the measurements upload to cloud.

Using the MATLAB code, the measurements are retrieved from the cloud. Data stored in each field can be retrieved using the code. Fig. 8.15 displays the measurements that were retrieved from the ThingSpeak cloud database.

A GUI is developed to view the patient's database. The GUI has options to view the patient's medical images and other details. Tkinter is used to develop a Python-based GUI. Tkinter is an easy and fast way to develop a GUI. A patient- and

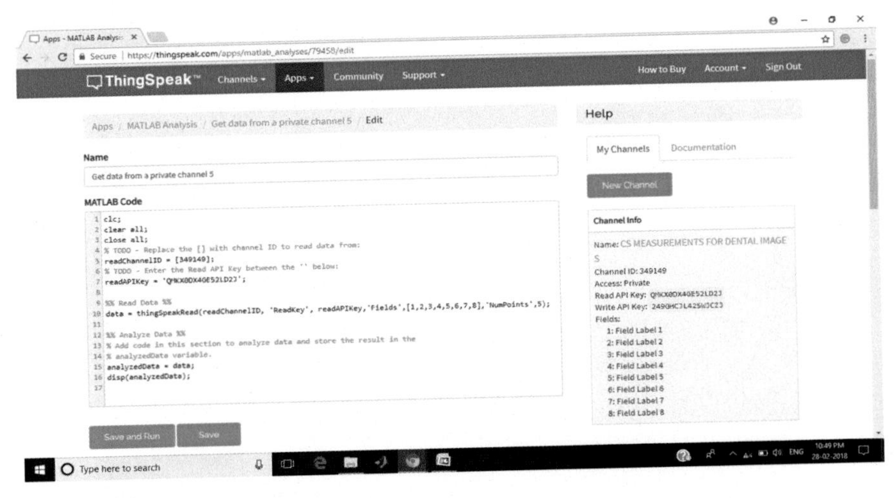

FIGURE 8.14

Screenshot of the MATLAB analysis code in ThingSpeak.

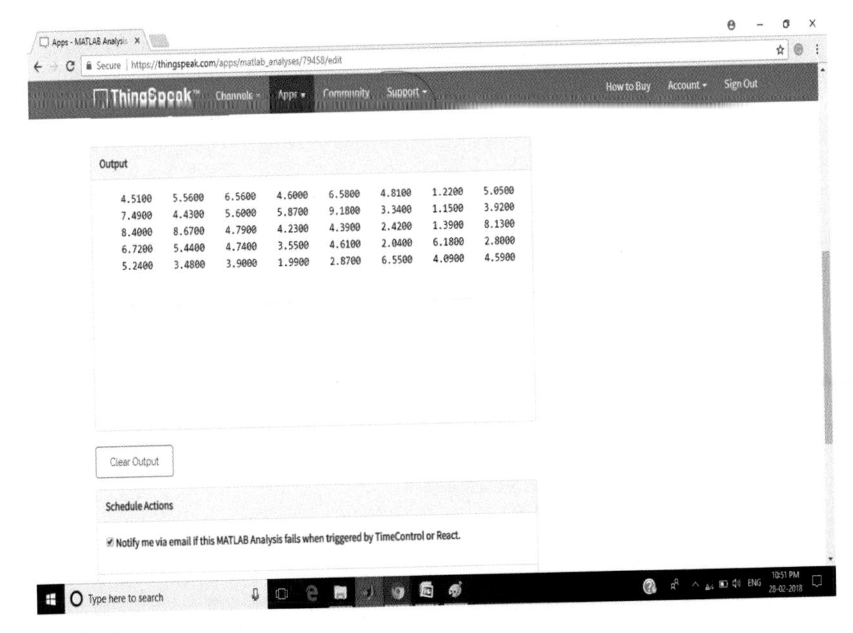

FIGURE 8.15

Screenshot of the measurements retrieved from cloud.

doctor-friendly GUI is developed, where on clicking the patient's name their details and their database can be viewed. Fig. 8.16 shows the GUI created using Tkinter.

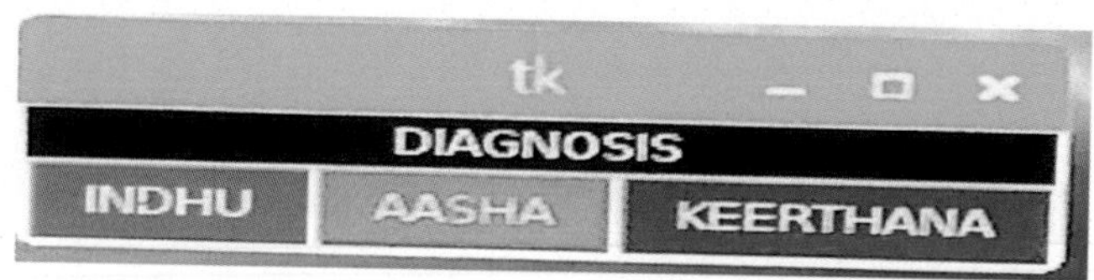

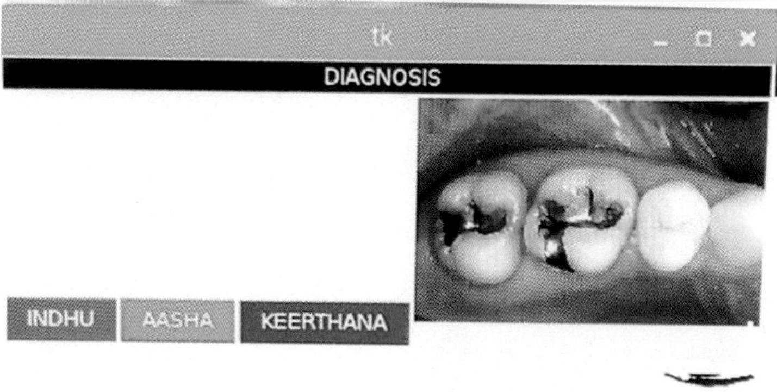

FIGURE 8.16

GUI created using Tkinter.

The percentage of reduction in the samples is calculated as follows:

$$\text{Percentage of reduction in samples} = \frac{\text{original image} - \text{compressed image}}{\text{original image}}$$
$$= \frac{65536 - 30720}{65536}$$
$$= 53\%.$$

Therefore, 53% of reduction in the samples when compared with the original image was achieved. Thereby one is reducing the amount of data stored in the cloud.

8.5 Conclusion and scope for future work

A compressive-sensing-based cyber physical system for healthcare applications was developed. The image was obtained from a1 dental healthcare database. The proposed hybrid pre-processing technique was applied to the images and the image was resized and was further divided into blocks of size 8*8. Compressive sensing was applied to these blocks. Measurements were obtained for these images. These measurements were uploaded to the cloud. The compressed image was recovered from the cloud using a suitable recovery algorithm. An effective healthcare-based system with GUI where detailed analysis report will be given to the patients from healthcare experts was proposed. The simulation results were obtained using MATLAB and hardware implementation was done in Raspberry Pi and their results were discussed in detail. The PSNR and the percentage of reduction in samples were considered. 53% of reduction in the samples was achieved. Real time implementation of the proposed project can be carried out in hospitals.

References

[1] Chih-Yu Lin, Sherali Zeadally, Tzung-Shi Chen, Chih-Yung Chang, Enabling cyber physical systems with wireless sensor networking technologies, International Journal of Distributed Sensor Networks 12 (2012) 489794.

[2] Delia Ioana Dogaru, Ioan Dumitrache, Cyber-physical systems in healthcare networks, in: IEEE International Conference on E-Health and Bioengineering – EHB, 2017.

[3] T.W. Cabral, M. Khosravy, F.M. Dias, H.L.M. Monteiro, M.A.A. Lima, L.R.M. Silva, R. Naji, C.A. Duque, Compressive sensing in medical signal processing and imaging systems, in: Sensors for Health Monitoring, Academic Press, 2019, pp. 69–92.

[4] Mohammad Mehedi Hassan, Kai Lin, Xuejun Yue, Jiafu Wand, A multimedia healthcare data sharing approach through cloud-based body area network, Future Generations Computer Systems 66 (2017) 48–58.

[5] Shah J. Miah, Jahidul Hasan, John G. Gammack, On Cloud Healthcare Clinic: an e-health consultancy approach for remote communities in a developing country, Telematics and Informatics 34 (1) (2017) 311–322.

[6] Yin Zhang, Meikang Qiu, Chun-Wei Tsai, Mohammad Mehedi Hassan, Atif Alamri, Health-CPS: healthcare cyber-physical system assisted by cloud and big data, IEEE Systems Journal 11 (1) (2017) 88–95.

[7] Jialin Gao, A smart medical system for dynamic closed-loop blood glucose-insulin control, Smart Health (2017) 18–33.

[8] Truong Quang Vinh, Bui Minh Thanh, Nguyen Ngoc Tai, Dental intraoral system supporting tooth segmentation, IEEE Access (2016) 326–329.

[9] Abdol Hamid Pilevar, DISR: dental image segmentation and retrieval, Journal of Medical Signals & Sensors 2 (1) (2016) 42.

[10] Aosen Wang, Feng Lin, Zhanpeng Jin, Wenyao Xu, A configurable energy-efficient compressed sensing architecture with its application on body sensor networks, IEEE Transactions on Industrial Informatics 12 (1) (2016) 15–27.

[11] Kuo-Hui Yeh, A secure IoT – based healthcare system with BodySensor networks, IEEE Access 4 (2016) 10288–10299.

[12] Surya Deekshith Gupta M., Vamsikrishna Patchava, Virginia Menezes, Healthcare based on iot using Raspberry Pi, in: International Conference on Green Computing and Internet of Things (ICGCLOT), 2015, pp. 796–799.

[13] https://thingspeak.com.

[14] http://docseurope.electrocomponents.com/webdocs/14ba/0900766b814ba5fd.pdf.

[15] https://opencv.org.

[16] https://docs.opencv.org/3.0beta/doc/py_tutorials/py_tutorials.html.

[17] http://inviewcorp.com/applications/compressive-sensing.

CHAPTER

9 Compressive sensing of electrocardiogram

Felipe Meneguitti Dias[a], Mahdi Khosravy[b], Thales Wulfert Cabral[a], Henrique Luis Moreira Monteiro[a], Luciano Manhaes de Andrade Filho[a], Leonardo de Mello Honório[a], Rayen Naji[c], Carlos A. Duque[a]

[a]*Department of Electrical Engineering, Federal University of Juiz de Fora, Juiz de Fora, Brazil*
[b]*Graduate School of Engineering, Osaka University, Osaka, Japan*
[c]*Medical School, Federal University of Juiz de Fora, Juiz de Fora, Brazil*

9.1 Introduction

The heart is an electromechanical pump that is activated by an electrical signal generated spontaneously in a region called the sinoatrial node. This signal propagates inside the heart through conductive tissues and sequentially stimulates different parts of the heart to contract. Correct sequential contraction of the heart allows it to pump pro-oxygenated blood into the body and deoxygenated blood to the lungs. This whole process occurs cyclically and allows the human body to function properly. Additionally, the electrical activity of the heart can be captured through electrodes placed on the skin. This electrical signal is known as an electrocardiogram (ECG). Fig. 9.1 shows an example of a 12-lead ECG of a 26-year-old man [37].

An ECG is typically used in an examination of people at risk for heart disease due to its standardized waveform for healthy people. Deviations from the standard waveform are indicative of potential cardiovascular complications. The diagnosis of some diseases, such as atrial arrhythmia, may require continuous monitoring of the ECG. Since data acquisition on these devices is constant, a large amount of data is generated. However, the data is stored on a device that has limited battery and storage capacities. A possible solution to this problem is a technique called Compressive Sensing (CS), which efficiently helps to lower the number of samples for the same length of the signal while the data content is preserved. The theory of CS was introduced by Candes, Romberg, Tao, and Donoho [1–3]. CS challenges the Nyquist–Shannon sampling theory that requires a sampling frequency twice as large as the maximum frequency of a signal. The theory of CS says that one can sample the signal at much lower frequencies and still maintain good signal quality.

Nowadays, CS has been effectively implemented in a wide range of applications from power quality analysis [4] to medical applications [5]. Besides, CS has a great potential to be applied to acoustic OFDM [6], power line communications [7], power quality analysis [8,9], public transportation system [10], power system plan-

Compressive Sensing in Healthcare. https://doi.org/10.1016/B978-0-12-821247-9.00014-7

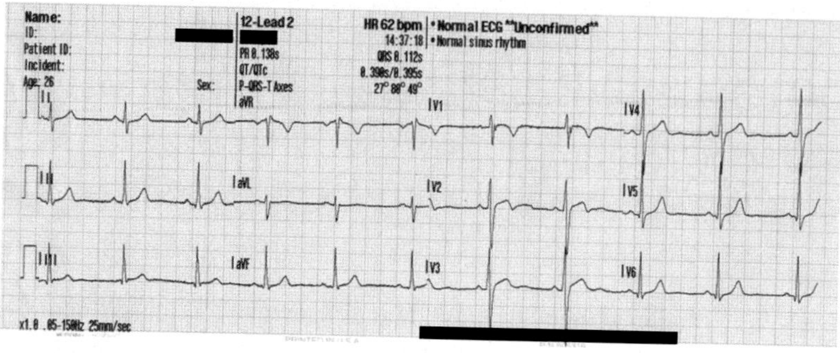

FIGURE 9.1

12-lead ECG capture of a 26-year-old man.

ning [11,12], location-based services [13], smart environment [14], texture analysis [15], text data processing [16], image enhancement [17], medical image processing [18,19], human motion analysis [20], human-robot interaction [21], electrocardiogram processing [22–24], sentiment mining [25], software-intensive systems [26], telecommunications [27–32], agriculture machinery [33,34], and data mining [35, 36].

This chapter provides a comprehensive analysis of CS applied to ECG signals, while also reviewing ECG functionality and basic heart electrophysiology.

9.2 Electrocardiogram

This section aims to provide a historical review of the ECG, introduce the anatomy of the heart, review heart electrophysiology and present an introduction to ECG acquisition mechanisms.

9.2.1 Historical background

The use of technology by doctors to aid in the diagnosis of heart disease began in the late 19th century. The advent of the thoracic x-ray in 1895 and the ECG in 1902 provided objective information on the structure and function of the heart [38]. Advances in medicine, both in new treatments and diagnostics, have been fueled by technological development ever since. Consequently, technology has allowed the increase of life expectancy of the general population [39].

The history of the onset of the ECG began in 1781 when Galvani observed that the muscles of a frog contracted when an electrical spark stimulated them. This discovery inaugurated a new research field: bioelectricity. It was later discovered that muscle and nerve fibers could propagate electricity. Since these electrical voltages have small amplitude, there was a need for a piece of equipment capable of measuring those voltages. In 1887, Augustus Waller captured the first human ECG. Waller used a sensitive electricity detector (Gabriel Lippmann capillary electrometer), but the quality of measurements was not satisfactory. In 1901, Willem Einthoven developed the string galvanometer, an instrument that had higher sensitivity and frequency response than the instrument used by Waller. Using this invention, Einthoven could capture ECGs and relate them to the functioning of the heart. Einthoven won the Nobel Prize for Physiology or Medicine in 1924 for this work [40].

In addition, to capture an ECG signal, it is necessary to place electrodes at specific points in the body. Einthoven suggested three measuring points that would form an equilateral triangle with the heart as the centre of it. This configuration became known as the Einthoven triangle. A few years later, Norman Holter invented a portable ECG device called the Holter ECG. This device's purpose was to continuously monitor the electrical activity of the heart for long periods of time.

9.2.2 Heart anatomy

The heart is a muscle pump with two main functions: i) collect oxygen-poor blood from the body and pump it to the lungs and ii) collect oxygen-rich blood from the lungs and pump it to the body [41]. It is located on the left side of the thoracic cavity between the two lungs. It is the approximate size of a man's closed fist. Also, the heart's wall has three layers: epicardium, myocardium, and endocardium. Internally, the heart has four distinct chambers: left and right atria and left and right ventricles. The atria are blood reservoirs that will be sent to the ventricles. The left atrium receives oxygenated blood from the lungs through four pulmonary veins. The right atrium receives deoxygenated blood returning from the body through the superior and inferior vena cava and from the heart through the coronary sinus. The left and right ventricles are blood-pumping chambers. The left ventricle receives oxygenated blood from the left atrium and pumps to the body through the aorta. The right ventricle receives blood from the right atrium and pumps to the lung through the pulmonary arteries.

The thickness of the chamber walls is directly related to the pumping pressure exerted by them. The atria pump blood over short distances. Therefore it has the thinnest walls. The left ventricle's wall is thicker than the right ventricle's wall because the former one pumps blood to regions of higher pressures (the arteries of the human body), and the right ventricle pumps blood to the lungs where the pressure levels are lower. In order to ensure that the blood circulates in only one direction, the heart has four valves: two atrioventricular (tricuspid and mitral) and two semilunar valves (aortic and pulmonary).

9.2.3 Conduction system

The heart's electrical impulse conduction system has as its primary function the coordination of the cardiac cycle. In this context, the sinoatrial node (SA node) is a small region located in the right atrium. The cells in this region have spontaneous stimulation, thus generating nerve impulses at a specific frequency (between 70 and 80 times per minute in a healthy adult). The sinoatrial node is known as the natural pacemaker of the human body. The signals generated by it pass through the atrium and allow its contraction. This impulse will then be sent to another particular region called the atrioventricular node (AV node). In this region, the impulses are delayed allowing the atria to have sufficient time to contract and pump blood to the ventricles. Immediately afterward, the impulses are led through a region called the Bundle of His, which branches into another region, called the area of Purkinje fibers. The electrical stimulation of the latter region allows the ventricles to contract and pump blood to their respective regions.

9.2.4 Electrophysiology

The cells of the human body have a characteristic response to electrical impulses, and those characteristic waveforms form the ECG signal measured in the human body. Electrophysiology studies the behavior of cells from an electrical point of view.

The membranes of resting cells are positively charged on their outer surface due to cation distribution. A transport mechanism called the sodium–potassium pump maintains the electrical potential of the cell. When the cell is stimulated, sodium channels of rapid conduction open up, allowing for a rapid inflow of sodium, reversing the polarity of the cell. However, these channels close quickly. This step is called rapid depolarization (phase 0). Then potassium channels open, allowing for the potassium to exit the cell, making the potential decrease slightly. This stage is called early repolarization (phase 1). Then the calcium channels allow for the calcium to enter the cell, balancing with the outward potassium flow. This stage is called plateau (phase 2). In the following, the calcium channels close, and the rapid potassium channels take the potential of the cell to approximately −90 mV. This phase is called rapid repolarization (phase 3). Finally, the cell is in the resting phase (phase 4), where the potassium outward flow maintains the potential of the cell at −90 mV. This set of events together is called the action potential. These variations of the electrical potential propagate through the heart cells. The ECG is capable of capturing the electrical potential variations of the muscle cells of the atria and the ventricles [42].

In this context, the ECG has a standard format. In the resting state, the ECG has an isoelectric shape because the electrical potentials of the cells are constant. Deflections from this point are named using letters in alphabetical order: P, Q, R, S, and T. The first deflection is the P wave. It is related to the depolarization of the atrial muscle cells. The QRS complex represents the depolarization of the muscle cells of the ventricle. Finally, the T wave represents the repolarization of the ventricle muscle cells [55]. An ECG of a healthy person has a waveform like that of Fig. 9.2 [56]. Table 9.1 shows the relationship of the physiological events and their evidence on an ECG signal [55].

9.2.5 Electrocardiogram acquisition

The ECG has a waveform that corresponds to the depolarization and repolarization cycles that occur in the heart. To capture the electrical activity of the heart, electrodes should be attached to the skin. Since the ECG has a spatial distribution along the

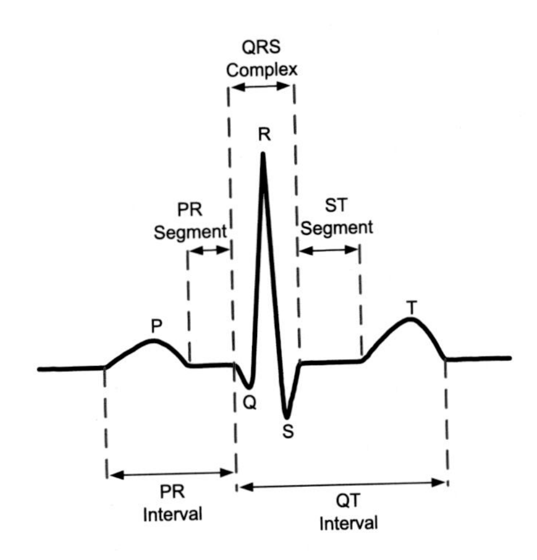

FIGURE 9.2

Electrocardiogram standard waveform.

Table 9.1 Summary of the physiological events and its correspondent ECG evidence.

Physiological event	**ECG evidence**
SA node initiates impulse	Not visible
Depolarization of atrial muscle	P Wave
Atrial Contraction	Not visible
Depolarization of AV node and Common Bundle	Not visible
Repolarization of atrial muscle	Not visible
Depolarization of ventricular muscle	QRS complex
Contraction of ventricular muscle	Not visible
Repolarization of ventricular muscle	T Wave

human body, i.e., the ECG is a vector, the electrode placed on the human body only captures a projection of the ECG vector at that point. In order to obtain more information as regards the electrical activity of the heart, it is necessary to place a higher number of electrodes.

Besides, since ECG records are very sensitive to noise, ECG preprocessing [43–45] must be employed to guarantee high-quality ECG for computer-aided processing and clinical investigation. There are several methods in the literature for such an end. A nonlinear method for ECG preprocessing is morphological filtering [46–48]. In addition, blind source separation [49–53], and blind component processing [54] as multiarray techniques can be employed for noise detection and cancellation.

9.3 Compressive sensing

Compressive Sensing (CS) was introduced by Donoho, Candès, Romberg and Tao in 2004 [1–3]. The CS theory is based on the assumption that most signals are sparse, i.e., most of the signal is zero in a certain domain, and it states that these signals can be reconstructed using far fewer samples than Nyquist–Shannon sampling theorem demands. The CS process can be segmented into two parts: signal acquisition and signal reconstruction. These two are going to be presented in this section.

9.3.1 Compressive sensing signal acquisition

The acquisition using the CS theory is mathematically defined as measuring a n-sampled signal vector $\boldsymbol{x}$ using a compressive sensing matrix Φ of the size $m \times n$, resulting in a much shorter vector signal $\mathbf{y}$ with m samples:

$$\mathbf{y}_{m\times 1} = \Phi_{m\times n}\boldsymbol{x}_{n\times 1} \qquad m << n \tag{9.1}$$

Usually, a random matrix is employed as the measurement matrix Φ. To implement this matrix in hardware, different approaches can be taken. Fig. 9.3 shows a signal acquisition model called Random Demodulator (RD) [57]. This model has three parts: i) mixer—where the input signal x is going to be multiplied by a pseudo-random sequence, ii) integrator—a low pass filter, and iii) ADC—an analog to digital converter. Additionally, there are other acquisition models, such as Modulated Wide-band Converter (MWC), Random Modulation Pre-Integrator (RMPI), and Random Equivalent Sampling (RES) [57]

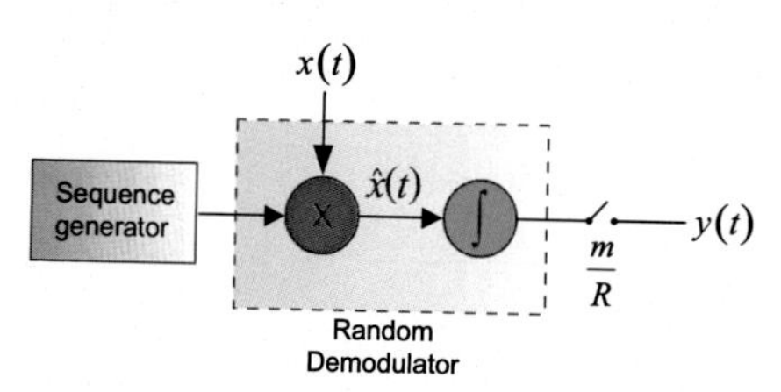

FIGURE 9.3

Random demodulator.

9.3.2 Compressive sensed signal reconstruction

Most of the signals are naturally sparse or have a sparse representation in some other domain. One can write this mapping to a sparse domain mathematically.

$$\boldsymbol{x} = \Psi\boldsymbol{s}. \tag{9.2}$$

Replacing Eq. (9.2) for Eq. (9.1) results in Eq. (9.3):

$$\mathbf{y} = \Phi\Psi\boldsymbol{s}. \tag{9.3}$$

Given $\mathbf{y}$, Φ, and Ψ, it is possible to recover $\boldsymbol{s}$ by solving an L_0-optimization problem, shown in Eq. (9.4).

$$\hat{\boldsymbol{s}} = \min_{\boldsymbol{s}} \|\boldsymbol{s}\|_0 \qquad \text{subject to } \mathbf{y} = \Phi\Psi\boldsymbol{s}. \tag{9.4}$$

The $L0$-norm is equivalent to counting all non-zero elements of the signal. Eq. (9.4) is an NP-hard problem requiring metastatic optimization [58–69] or an approximate solution. A very common linear approximation is to replace the $L0$-norm for the $L1$-norm. This is shown in Eq. (9.5):

$$\hat{\boldsymbol{s}} = \min_{\boldsymbol{s}} \|\boldsymbol{s}\|_1 \qquad \text{subject to } \mathbf{y} = \Phi\Psi\boldsymbol{s}. \tag{9.5}$$

After solving this optimization problem, the estimated signal can be obtained using Eq. (9.2). In order to guarantee the recovery of a k-sparse signal vector $\boldsymbol{x}$ from the measured signal vector $\mathbf{y}$, the condition defined in Eq. (9.6), known as the restricted isometry property (RIP), must be satisfied.

$$1 - \delta \leq \frac{\|\varphi\psi u\|_2}{\|u\|_2} \leq 1 + \delta; \tag{9.6}$$

u is a vector with the same k nonzero entries as x and $\delta > 0$ is known as the restricted isomeric constant. The RIP condition is not easy to verify. An alternative to the RIP is to use the incoherence condition. This condition requires that the measurement matrix φ and the sparse mapping matrix ψ to show incoherence. The coherence of these matrices can be calculated using Eq. (9.7):

$$\mu(\Phi, \Psi) = \sqrt{n} \max_{1 \leq i, j \leq N} |\langle \varphi_i, \psi_j \rangle|. \tag{9.7}$$

For all the processing techniques, there are always numerical indices to validate the processing technique efficiency [70].

9.3.2.1 Metrics

To evaluate the quality of the compression and how much signal was compressed, many metrics are available, such as compression ratio (CR) and percent root-mean-squared difference (PRD). The CR measures the reduction of samples after using CS. The PRD measures the reconstruction error in percentage. They are defined in Eqs. (9.8) and (9.9):

$$\mathrm{CR} = \frac{N}{M}, \tag{9.8}$$

$$\mathrm{PRD}\ (\%) = \frac{\|\boldsymbol{x} - \hat{\boldsymbol{x}}\|}{\|\boldsymbol{x}\|} \cdot 100. \tag{9.9}$$

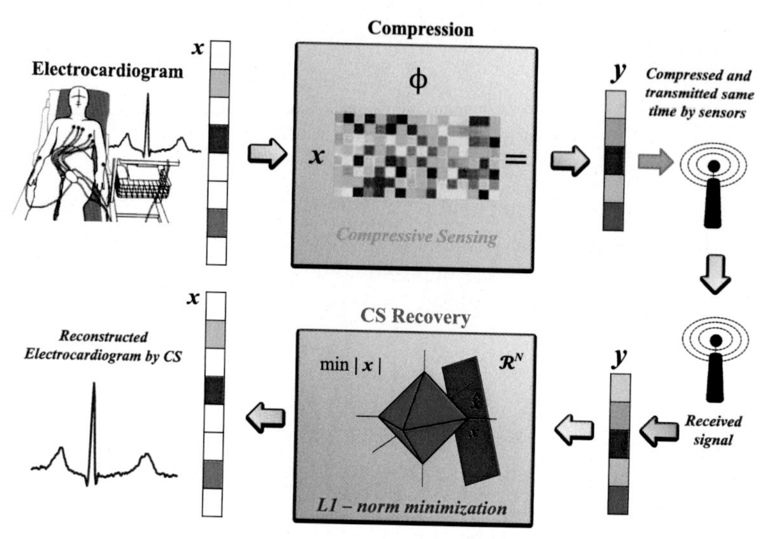

FIGURE 9.4

General process of implementation of CS in health monitoring.

9.3.3 CS importance in tele-medicine

The essence of online monitoring of vital signals of the patient besides the growing application of U-health monitoring requires better medical data management. Due to its strong theoretical ability for signal reconstruction using much fewer samples than required by the Nyquist sampling theorem, CS has many technical and clinical benefits. It can decrease the side effects of excessive exposure of the patient to the medical measurements. For example, in the case of MRI and CT scan, the imaging process becomes multiple times faster, where apart from the economic advantage and power saving, the patient receives less radiation. Also, among other issues, medical data transmission becomes feasible at a much lower rate, as well as the required processing for noise cancellation and telecommunication channel estimation becomes more efficient and faster in the case of low rate data. However, reconstructed medical signals and images by CS may need a level of enhancement that can be achieved by the available preprocessing and enhancement methods [43,71,72]. Fig. 9.4 illustrates the application of CS for health monitoring.

9.4 Compressive sensing approach to ECG

Technology and medicine have gone hand in hand for many years. This partnership was one of the main reasons for the increase in life expectancy worldwide. With increasing numbers of elderly people, the number of people suffering from heart problems has also risen, and it is the number one cause of death in the world [73].

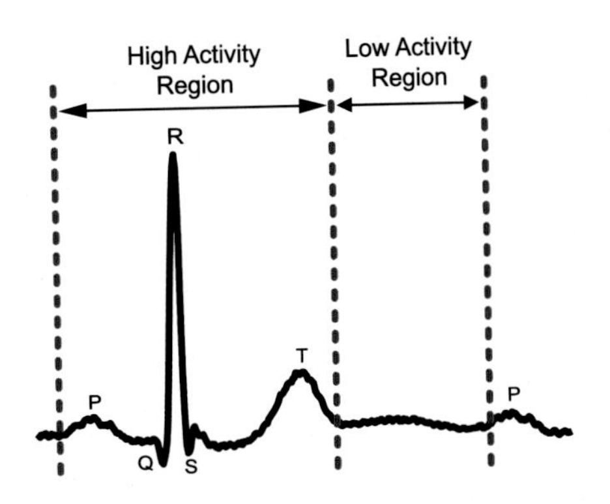

FIGURE 9.5

High and low information regions of an ECG signal.

These diseases require more medical attention and better hospital infrastructure, significantly increasing the healthcare costs [73].

The advent of body sensor technologies enables the decentralization of hospital structures, allowing each person to have a set of health meters attached to the body that can send data to health centers. This kind of technology reduces the demand for hospital services, saving both patients and hospital time and money.

Biosignal measurement equipment still faces some difficulties to be more popular, and battery life is one of the biggest obstacles to be faced [74]. The energy consumption of these devices can be split into three areas: i) data sampling, ii) wireless communication, and iii) data processing. The first one is the biggest energy consumer [74,75].

CS theory challenges the Nyquist–Shannon sampling theory, which states that the sampling frequency of a given signal should be at least twice its bandwidth to ensure reliable reconstruction of the signal. However, it was observed that sparse signals have much redundancy. In ECG signals, for example, the isoelectric region (low activity region) has no information, while the PQRST region (high activity region) has a large amount of information (see Fig. 9.5). The application of CS on ECG signals intends to decrease the amount of data to be sampled in order to reduce the energy consumption in wearable devices by decreasing the amount of data to be transmitted.

CS has a variety of applications in ECG. Several approaches, using CS theory, can positively contribute to the treatment of data extracted from the ECG signal. In this part of the chapter, we will discuss these approaches comprehensively. Nine different subsections were proposed:

- ECG signal quality (9.4.1)
- Body sensor networks and monitoring systems (9.4.2)
- Dictionary learning (9.4.3)
- ECG signal structure (9.4.4)
- Wavelets and compressive sensing (9.4.5)
- Signal compression (9.4.6)
- Reconstruction algorithms (9.4.7)

- Discrete wavelet transform (DWT) vs compressive sensing (CS) (9.4.8)
- Hardware implementation (9.4.9)

9.4.1 ECG signal quality

The goal of data compression techniques is to obtain the highest compression rates possible with the least reconstruction errors. For ECG signals, it is essential to determine what the maximum allowable reconstruction error rate is that it would provide an ECG signal of sufficient quality for later use in diagnostics. In [76], the authors present a discussion of this topic. In this study, reconstructed ECG signals are provided to cardiologists who should rate the signal as "very good", "good", "not good", and "bad". With this methodology, it was possible to correlate the reconstruction error with the diagnostic quality of the signal. The results obtained in [76] are summarized in Table 9.2 for the PRD metric [76].

Table 9.2 Diagnostic quality according to the PRD value.

PRD (%)	Signal diagnostic quality
0–2	Very good
2–9	Good
9–19	No good
19–60	Bad

By observing Table 9.2, it is seen that the maximum PRD for a reconstructed ECG signal is 9%. If they exceed this value, the signal can lose its diagnostic value.

9.4.2 Body sensor networks and monitoring systems

Currently, the demand for small devices that performs some specific function is increasing. Mobile technologies are becoming more popular in society. Nowadays, these devices can capture, display, and process different kinds of data, including biological signals. In order to perform such tasks that involve intensive processes and still saving energy, these devices require sophisticated techniques to optimize power consumption. A clever way to achieve this goal is to reduce the amount of data to be sampled, and the CS techniques are a perfect fit for this task. This is demonstrated in [77]. In this paper, the authors evaluated the impact of CS on wireless sensor networks (WSNs), especially in cardiac monitoring, using some assessments regarding reliability and improvements in the feasibility of the application. According to [77], the telecardiology sensor network is implemented using WSN with individual sensors, and the cables are replaced by radio transmission. The sampling frequency usually used for ECG acquisition is in hundreds of hertz, but that order of magnitude may be considered high for embedded devices, and thus CS can be used to reduce the sampling rate.

Body sensor networks, especially wireless, require real-time applicability and user-friendliness. In [78], a real-time ECG monitoring system using CS is proposed for a wireless network. The goal was to achieve energy efficiency using CS in a real-time application. This work used a computationally light ECG encoder in the sensor, and the decoder part was implemented in a smartphone. There was a 12.9% increase of the node life in comparison to the transmission of uncompressed data. The node life increases, demonstrating that the implemented system is more energy-efficient. The authors state that the receiver receives information accurately and can retrieve the ECG signal online using 17.7% of the mobile phone's CPU at a 50% compression rate. In [79], there is also an implementation of a real-wireless ECG monitoring focusing on energy efficiency. The implemented system uses an ultra-low-power analog front-end and bluetooth communication. The paper was able to detect a 77.37% energy consumption reduction in radio transmission, maintaining the signal undistorted (PRD < 6%).

Artifacts due to packet loss can severely distort the signal content in wireless systems. In order to solve this problem, [80] used compressive sensing techniques to mitigate these kinds of artifacts. The authors suggest that the same techniques can be used to mitigate ECG artifacts as well, e.g., power line interference and baseline wander.

9.4.3 Dictionary learning

There are sparsifying matrices available in compressive sensing, e.g., wavelets. However, these matrices may not be ideal for a particular kind of signal. Dictionary learning is a way to find a better sparse mapping matrix by the use of training data. Reference [81] proposes to adaptively learn a dictionary that exploits the multiscale sparse representation of ECG signals. The method calculates sub-dictionaries at different data scales and then exploit the correlation within each wavelet subband.

9.4.4 ECG signal structure

CS techniques usually do not take into account the structure of the signal in order to apply compression. ECG signals have a particular structure. They usually have three components of waves: P wave, QRS complex and T wave. This structure can be used to get better compression results. In [82], the sensing matrix is designed in a way to sample the signal where most of the information is located, i.e., the QRS complex. This approach improves the CR values for a given PRD. In [83], there is a preprocessing stage where the ECG QRS complex is estimated. Then the estimated QRS complex is subtracted from the original ECG signal, and the resulting signal is compressed using a DWT sparsifying matrix. In [84], the authors used compressive sensing techniques exploring the redundancy between consecutive heartbeats. Initially, there is a QRS complex detection stage followed by a sliding window to select the number of heartbeats that will be compressed.

9.4.5 Wavelets and compressive sensing

Wavelets are usually a good sparse representation of a signal. However, there are multiple wavelet families available. Therefore, a comparison between them should be made in order to use the best wavelet family. In [85,86], there is a comparison between different wavelet families, e.g., Biorthogonal, Coiflets, Daubechies, Haar, Reverse Biorthogonal, and Symlets. The authors used different metrics for the comparison: Mean Square Error (MSE), Peak Signal To Noise Ratio (PSNR), Percentage Root Mean Square Difference (PRD), and Correlation Coefficient (CoC); and different values of the compression ratio were used on different ECG signals. The authors of [86] concluded that rbio 3.9 is the best wavelet family when the compression ratio is 2:1 and 6:1, and rbio3.7 is the best wavelet family for a 4:1 compression ratio.

9.4.6 Signal compression

CS involves different steps in the signal compression pipeline. It smartly samples the signal using a simple encoder and transfers all the mathematical complexity to the encoder. Besides, CS is not the state-of-the-art compression technique; therefore, it is usually not used just for compression. This subsection will show some applications where the main focus is signal compression using CS.

In [87], a compression pipeline is developed in a multi-channel electrocardiogram signal (MECG). First, a technique called Principal Component Analysis (PCA) is applied to the MECG signal. In order to reduce even more the information dimensionality, the components with maximum variance from PCA are projected over a sparse binary sensing matrix. In [88], compressive sensing techniques are used as a lossy digital signal compression. The compression capabilities are compared to a conventional signal compression technique: Set Partitioning In Hierarchical Trees (SPIHT).

In order to estimate parameters from a compressed signal, e.g., QRS location, one has first to reconstruct the signal and then apply the appropriate technique to calculate the parameter. However, this approach is not feasible in a real-time application because signal reconstruction is a computationally complex task. The authors of [89] propose a framework to get information from a compressed ECG signal. They were able to detect the QRS complexes and estimate the heart rate in ECG signals that were compressed using CS.

Reference [90] proposes three methods of compression and compares them. It used CR, PRD, and the accuracy of the QRS detection as evaluation metrics. Initially, an Adaptive Linear Prediction method was proposed. A linear prediction was used to estimate the current sample of the signal $x[n]$ of the ECG signal from its previous samples. We write

$$\hat{x}[x] = \sum_{k=1}^{m} h^k x[n-k] \tag{9.10}$$

where $\hat{x}[n]$ is the estimate of $x[n]$ and $x[n]$ are the predictor coefficients.

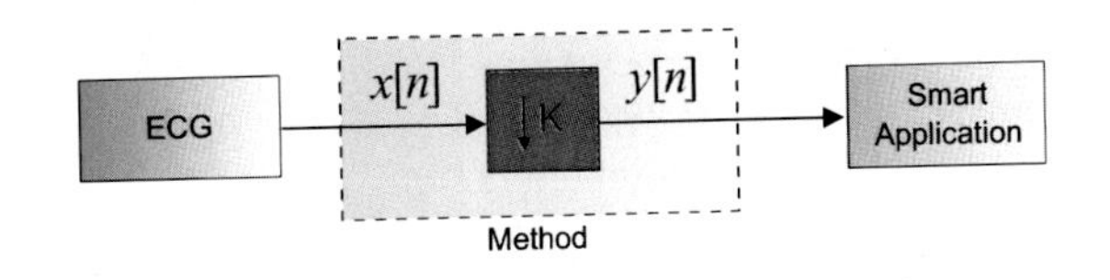

FIGURE 9.6

Compressive Sampling Matching Pursuit method.

Later, a technique known as Compressive Sampling Matching Pursuit was proposed. This method has four stages: sampling, redundancy removal, quantization, and Huffman coding. Fig. 9.6 schematically summarizes the method [90].

Table 9.3 Important information of algorithms.

Algorithm	Algorithm objective
Basis Pursuit Convex Optimization	Search the space for a solution with the minimum L_1-norm
Basis Pursuit Denoising (BPDN)	Search the space for a solution with minimum error and min L_1-norm together
Orthogonal Matching Pursuit (OMP)	Find the columns of [Φ] with the maximum correlation to the residual
Compressive Sampling Matching Pursuit (CoSaMP)	Find the top $2K$ columns of [Φ] with the maximum correlation to the residual
Normalized Iterative Hard Thresholding (NIHT)	Find the top K values of the sum of previous best guess and the signal proxy of the residual

Finally, a third method based on decimation and characterized by losses is proposed. The authors concluded that the third method was the best one.

9.4.7 Reconstruction algorithms

In the field of reconstruction methods, some algorithms are employed, aiming at the highest possible fidelity to the original signal. According to [91,92], the original CS reconstruction algorithm is the convex optimization whose reconstruction is based on the convex optimization of the $L1$-norm; such an approach is called Basis Pursuit. The authors highlight four more reconstruction algorithms: Basis Pursuit Denoising (BPDN), Orthogonal Matching Pursuit (OMP), Compressive Matching Pursuit (CoSaMP) and Normalized Interactive Hard Thresholding (NIHT). Table 9.3 summarizes key features of these algorithms [92]. [93] also deals with iterative algorithms that aid in the process of signal reconstruction. Three modified algorithms based on the known particle support (PKS) were used in an ECG application: OMP, CoSaMP, and RWLS-SL0. The modified OMP is a greedy interactive algorithm for sparse signal recovery. The reference points out that the algorithm terminates when the residual $L2$-norm falls below an approximate error threshold. The iteration steps are the same as in the OMP. The adapted CoSaMP works according to initially known information, pruning the signal to be s-sparse, and the remainder of the algorithm continues the same as the CoSaMP. The RWLS-SL0 is an iterative approach of least square reweighting based on the smooth approximation of the $L0$-norm. It is a suitable method for reconstructing scattered signals. The reference demonstrates that the modification in these algorithms results in performance improvement. Although [93] has a focus on the algorithms, the results lead to evidence of the applicability in the reconstruction of ECG signals.

9.4.8 Discrete wavelet transform (DWT) vs compressive sensing (CS)

CS and DWT are two compression techniques. Each of them has advantages over the other. These differences will be shown in this subsection.

In [94], a comparison of compression capacity was made between DWT and a CS technique using a sparse binary matrix as the sensing matrix and Basis Pursuit (BP) as the reconstruction algorithm. DWT obtained better compression results, but the techniques based on CS used less CPU and, consequently, obtained larger energy savings. In [95,96], CS techniques were applied to noisy ECG signals. These articles demonstrate a limitation of CS when it is applied to noisy signals. The DWT-TH presents much better results (CR = 5) than CS (CR = 1.67) for the same PRD value (9%).

9.4.9 Hardware implementation

From the engineering point of view, it is always a challenge to build hardware, especially delicate and small components. The development of wearable devices towards health measurement presents an even more challenging task because several concerns should be addressed. Ergonomics and safety are the main ones. The device must be comfortable and easy to use in order to improve its market acceptance. Besides, the device should not present any harm to the user. Since the ECG is a non-invasive procedure, it is a safe exam. The design of a wearable ECG device has to address two different fronts: hardware and software. On the hardware side, a set of requirements must be satisfied. It should be cheap, light, energy-efficient, and reduce the amount of processing power required by the software. A hardware implementation of the encoder part of CS would be a useful feature. It decreases the amount of sampled data and therefore decreases the amount of data to be stored and transmitted. These features make the hardware to use less energy. The software complexity should be low so that the processing power could be as low as possible.

In [78], the main focus was to improve the energy efficiency in WBSN Electrocardiogram devices. It was demonstrated that CS techniques could be used to save energy for real-time applications.

In [97], the authors presented an FPGA application. The authors have developed a CS reconstruction using the first single-precision floating-point CS reconstruction engine in a Kintex-7 FPGA. They also used OMP to achieve a good performance and hardware utilization ratio.

In [98], the authors presented the first fully-integrated compressed sensing analog-domain front-end (CS-AFE). Their goal was to build energy-efficient hardware that samples in sub-Nyquist sampling frequencies.

9.5 Conclusion

The major challenge of embedded ECG equipment is energy consumption. These devices perform continuous sampling, preprocessing, processing, and data transmission, which induces high battery utilization. The battery must also be small so that it does not bring discomfort to people using the device. In this context, the rise of CS techniques has challenged established knowledge in the area of signal processing, such as the famous Nyquist–Shannon sampling theorem, by stating that it is possible to use sampling frequencies smaller than the Nyquist frequency. This result is of great importance in several areas. Specifically, in the area of embedded ECG equipment, it is essential to emphasize that a reduction in the number of data to be sampled decreases the amount of data to be transmitted. This, in turn, decreases battery consumption and increases energy efficiency.

The ECG exam is an excellent diagnostic tool available to physicians. It is a quick, inexpensive, painless, and risk-free exam that offers a great deal of information to the doctor. Advances in embedded electronics technologies and the development of energy-efficient equipment allow for the creation of health monitoring equipment that can be coupled to the human body. These devices are of great importance in the current global scenario, where the number of elderly people is increasing. Continuous health monitoring of this segment of the population would bring significant benefits and reduced mortality. This is crucial because the earlier the diagnosis is made, the higher the chances of survival.

This review highlighted several areas and developments related to CS in ECG to inform and clarify the importance of this kind of examination in association with data processing. Thus we conclude our overview of the current state-of-the-art in CS applications to ECG signals.

References

[1] E.J. Candès, J. Romberg, T. Tao, Robust uncertainty principles: exact signal reconstruction from highly incomplete frequency information, IEEE Transactions on Information Theory 52 (2) (2006) 489–509.

[2] D.L. Donoho, Compressed sensing, IEEE Transactions on Information Theory 52 (4) (2006) 1289–1306.

[3] E.J. Candes, T. Tao, Near-optimal signal recovery from random projections: universal encoding strategies?, IEEE Transactions on Information Theory 52 (12) (2006) 5406–5425.

[4] K. Melo, M. Khosravy, C. Duque, N. Dey, Chirp code deterministic compressive sensing: analysis on power signal, in: 4th International Conference on Information Technology and Intelligent Transportation Systems, IOS Press, 2020, pp. 125–134.

[5] T.W. Cabral, M. Khosravy, F.M. Dias, H.L.M. Monteiro, M.A.A. Lima, L.R.M. Silva, R. Naji, C.A. Duque, Compressive sensing in medical signal processing and imaging systems, in: Sensors for Health Monitoring, Elsevier, 2019, pp. 69–92.
[6] M. Khosravy, N. Punkoska, F. Asharif, M.R. Asharif, Acoustic OFDM data embedding by reversible Walsh–Hadamard transform, AIP Conference Proceedings 1618 (2014) 720–723.
[7] A.A. Picorone, T.R. de Oliveira, R. Sampaio-Neto, M. Khosravy, M.V. Ribeiro, Channel characterization of low voltage electric power distribution networks for PLC applications based on measurement campaign, International Journal of Electrical Power & Energy Systems 116 (105) (2020) 554.
[8] E. Santos, M. Khosravy, M.A. Lima, A.S. Cerqueira, C.A. Duque, A. Yona, High accuracy power quality evaluation under a colored noisy condition by filter bank ESPRIT, Electronics 8 (11) (2019) 1259.
[9] E. Santos, M. Khosravy, M.A. Lima, A.S. Cerqueira, C.A. Duque, ESPRIT associated with filter bank for power-line harmonics, sub-harmonics and inter-harmonics parameters estimation, International Journal of Electrical Power & Energy Systems 118 (105) (2020) 731.
[10] M. Foth, R. Schroeter, J. Ti, Opportunities of public transport experience enhancements with mobile services and urban screens, International Journal of Ambient Computing and Intelligence (IJACI) 5 (1) (2013) 1–18.
[11] N. Gupta, M. Khosravy, N. Patel, T. Senjyu, A bi-level evolutionary optimization for coordinated transmission expansion planning, IEEE Access 6 (2018) 48455–48477.
[12] N. Gupta, M. Khosravy, K. Saurav, I.K. Sethi, N. Marina, Value assessment method for expansion planning of generators and transmission networks: a non-iterative approach, Electrical Engineering 100 (3) (2018) 1405–1420.
[13] M. Yamin, A.A.A. Sen, Improving privacy and security of user data in location based services, International Journal of Ambient Computing and Intelligence (IJACI) 9 (1) (2018) 19–42.
[14] C. Castelfranchi, G. Pezzulo, L. Tummolini, Behavioral implicit communication (BIC): communicating with smart environments, International Journal of Ambient Computing and Intelligence (IJACI) 2 (1) (2010) 1–12.
[15] S. Hemalatha, S.M. Anouncia, Unsupervised segmentation of remote sensing images using FD based texture analysis model and isodata, International Journal of Ambient Computing and Intelligence (IJACI) 8 (3) (2017) 58–75.
[16] C.E. Gutierrez, P.M.R. Alsharif, M. Khosravy, P.K. Yamashita, P.H. Miyagi, R. Villa, Main large data set features detection by a linear predictor model, AIP Conference Proceedings 1618 (2014) 733–737.
[17] A.S. Ashour, S. Samanta, N. Dey, N. Kausar, W.B. Abdessalemkaraa, A.E. Hassanien, Computed tomography image enhancement using cuckoo search: a log transform based approach, Journal of Signal and Information Processing 6 (03) (2015) 244.
[18] M. Khosravy, M.R. Asharif, M.H. Sedaaghi, Medical image noise suppression: using mediated morphology, IEICE, Technical Report 107 (461) (2008) 265–270.
[19] N. Dey, A.S. Ashour, A.S. Ashour, A. Singh, Digital analysis of microscopic images in medicine, Journal of Advanced Microscopy Research 10 (1) (2015) 1–13.
[20] G.V. Kale, V.H. Patil, A study of vision based human motion recognition and analysis, International Journal of Ambient Computing and Intelligence (IJACI) 7 (2) (2016) 75–92.
[21] B. Alenljung, J. Lindblom, R. Andreasson, T. Ziemke, User experience in social human–robot interaction, in: Rapid Automation: Concepts, Methodologies, Tools, and Applications, IGI Global, 2019, pp. 1468–1490.
[22] N. Dey, S. Samanta, X.-S. Yang, A. Das, S.S. Chaudhuri, Optimisation of scaling factors in electrocardiogram signal watermarking using cuckoo search, International Journal of Bio-Inspired Computation 5 (5) (2013) 315–326.
[23] N. Dey, A.S. Ashour, F. Shi, S.J. Fong, R.S. Sherratt, Developing residential wireless sensor networks for ECG healthcare monitoring, IEEE Transactions on Consumer Electronics 63 (4) (2017) 442–449.
[24] N. Dey, S. Mukhopadhyay, A. Das, S.S. Chaudhuri, Analysis of P-QRS-T components modified by blind watermarking technique within the electrocardiogram signal for authentication in wireless telecardiology using DWT, International Journal of Image, Graphics and Signal Processing 4 (7) (2012) 33.

[25] M. Baumgarten, M.D. Mulvenna, N. Rooney, J. Reid, Keyword-based sentiment mining using Twitter, International Journal of Ambient Computing and Intelligence (IJACI) 5 (2) (2013) 56–69.
[26] P. Sosnin, Precedent-oriented approach to conceptually experimental activity in designing the software intensive systems, International Journal of Ambient Computing and Intelligence (IJACI) 7 (1) (2016) 69–93.
[27] M. Khosravy, M.R. Alsharif, M. Khosravi, K. Yamashita, An optimum pre-filter for ICA based multi-input multi-output OFDM system, in: 2010 2nd International Conference on Education Technology and Computer, vol. 5, IEEE, 2010, pp. V5–129.
[28] F. Asharif, S. Tamaki, M.R. Alsharif, H. Ryu, Performance improvement of constant modulus algorithm blind equalizer for 16 QAM modulation, International Journal on Innovative Computing, Information and Control 7 (4) (2013) 1377–1384.
[29] M. Khosravy, M.R. Alsharif, K. Yamashita, An efficient ICA based approach to multiuser detection in MIMO OFDM systems, in: Multi-Carrier Systems & Solutions 2009, Springer, 2009, pp. 47–56.
[30] M. Khosravy, M.R. Alsharif, B. Guo, H. Lin, K. Yamashita, A robust and precise solution to permutation indeterminacy and complex scaling ambiguity in BSS-based blind MIMO-OFDM receiver, in: International Conference on Independent Component Analysis and Signal Separation, Springer, 2009, pp. 670–677.
[31] M. Khosravy, A blind ICA based receiver with efficient multiuser detection for multi-input multi-output OFDM systems, in: The 8th International Conference on Applications and Principles of Information Science (APIS), Okinawa, Japan, 2009, 2009, pp. 311–314.
[32] M. Khosravy, S. Kakazu, M.R. Alsharif, K. Yamashita, Multiuser data separation for short message service using ICA, SIP, IEICE Technical Report 109 (435) (2010) 113–117.
[33] S. Gupta, M. Khosravy, N. Gupta, H. Darbari, N. Patel, Hydraulic system onboard monitoring and fault diagnostic in agricultural machine, Brazilian Archives of Biology and Technology 62 (2019).
[34] S. Gupta, M. Khosravy, N. Gupta, H. Darbari, In-field failure assessment of tractor hydraulic system operation via pseudospectrum of acoustic measurements, Turkish Journal of Electrical Engineering & Computer Sciences 27 (4) (2019) 2718–2729.
[35] C.E. Gutierrez, M.R. Alsharif, K. Yamashita, M. Khosravy, A tweets mining approach to detection of critical events characteristics using random forest, International Journal of Next-Generation Computing 5 (2) (2014) 167–176.
[36] N. Kausar, S. Palaniappan, B.B. Samir, A. Abdullah, N. Dey, Systematic analysis of applied data mining based optimization algorithms in clinical attribute extraction and classification for diagnosis of cardiac patients, in: Applications of Intelligent Optimization in Biology and Medicine, Springer, 2016, pp. 217–231.
[37] MoodyGroove, 12 lead ECG of a 26-year-old male, [online], available: https://upload.wikimedia.org/wikipedia/commons/b/bd/12leadECG.jpg, 2007.
[38] M. AlGhatrif, J. Lindsay, A brief review: history to understand fundamentals of electrocardiography, Journal of Community Hospital Internal Medicine Perspectives 2 (1) (2012) 14383.
[39] F.R. Lichtenberg, The impact of biomedical innovation on longevity and health, Nordic Journal of Health Economics 5 (1) (2015) 45.
[40] MRI scan – overview, https://www.nhs.uk/conditions/mri-scan/.
[41] A.J. Weinhaus, K.P. Roberts, Anatomy of the human heart, in: Handbook of Cardiac Anatomy, Physiology, and Devices, Springer, 2005, pp. 51–79.
[42] T.S. Diehl, D.M. Allen, ECG Interpretation Made Incredibly Easy!, W. Kluwer/L. Williams & Wilkins, 2011.
[43] M.H. Sedaaghi, M. Khosravi, Morphological ECG signal preprocessing with more efficient baseline drift removal, in: 7th IASTED International Conference, ASC, 2003, pp. 205–209.
[44] M. Khosravy, M.R. Asharif, M.H. Sedaaghi, Morphological adult and fetal ECG preprocessing: employing mediated morphology, IEICE Technical Report, IEICE 107 (2008) 363–369.
[45] M. Khosravi, M.H. Sedaaghi, Impulsive noise suppression of electrocardiogram signals with mediated morphological filters, in: 11th Iranian Conference on Biomedical Engineering, ICBME, 2004, pp. 207–212.

[46] M.H. Sedaaghi, R. Daj, M. Khosravi, Mediated morphological filters, in: 2001 International Conference on Image Processing, 2001, Proceedings, vol. 3, IEEE, 2001, pp. 692–695.
[47] M. Khosravy, N. Gupta, N. Marina, I.K. Sethi, M.R. Asharif, Morphological filters: an inspiration from natural geometrical erosion and dilation, in: Nature-Inspired Computing and Optimization, Springer, Cham, 2017, pp. 349–379.
[48] M. Khosravy, M.R. Asharif, M.H. Sedaaghi, Medical image noise suppression using mediated morphology, in: IEICE Tech. Rep., IEICE, 2008, pp. 265–270.
[49] M. Khosravy, M.R. Asharif, K. Yamashita, A PDF-matched short-term linear predictability approach to blind source separation, International Journal of Innovative Computing, Information & Control 5 (11) (2009) 3677–3690.
[50] M. Khosravy, M.R. Asharif, K. Yamashita, A theoretical discussion on the foundation of stone's blind source separation, Signal, Image and Video Processing 5 (3) (2011) 379–388.
[51] M. Khosravy, M.R. Asharif, K. Yamashita, A probabilistic short-length linear predictability approach to blind source separation, in: 23rd International Technical Conference on Circuits/Systems, Computers and Communications (ITC-CSCC 2008), Yamaguchi, Japan, ITC-CSCC, 2008, pp. 381–384.
[52] M. Khosravy, M.R. Alsharif, K. Yamashita, A PDF-matched modification to Stone's measure of predictability for blind source separation, in: International Symposium on Neural Networks, Springer, Berlin, Heidelberg, 2009, pp. 219–222.
[53] M. Khosravy, Blind source separation and its application to speech, image and MIMO-OFDM communication systems, PhD thesis, University of the Ryukyus, Japan, 2010.
[54] M. Khosravy, M. Gupta, M. Marina, M.R. Asharif, F. Asharif, I. Sethi, Blind components processing a novel approach to array signal processing: a research orientation, in: 2015 International Conference on Intelligent Informatics and Biomedical Sciences, ICIIBMS, 2015, pp. 20–26.
[55] D.E. Becker, Fundamentals of electrocardiography interpretation, Anesthesia Progress 53 (2) (2006) 53–64.
[56] Agateller, Schematic diagram of normal sinus rhythm for a human heart as seen on ECG, [online], available: https://commons.wikimedia.org/wiki/File:SinusRhythmLabels.svg, 2007.
[57] M. Rani, S. Dhok, R. Deshmukh, A systematic review of compressive sensing: concepts, implementations and applications, IEEE Access 6 (2018) 4875–4891.
[58] J. Kalıannan, A. Baskaran, N. Dey, A.S. Ashour, M. Khosravy, R. Kumar, ACO based control strategy in interconnected thermal power system for regulation of frequency with HAE and UPFC unit, in: International Conference on Data Science and Application (ICDSA-2019), in: LNNS, Springer, 2019.
[59] M. Khosravy, N. Gupta, N. Patel, T. Senjyu, Frontier Applications of Nature Inspired Computation, Springer, 2020.
[60] N. Gupta, M. Khosravy, O.P. Mahela, N. Patel, Plant biology-inspired genetic algorithm: superior efficiency to firefly optimizer, in: Applications of Firefly Algorithm and Its Variants, Springer, 2020, pp. 193–219.
[61] C. Moraes, E. De Oliveira, M. Khosravy, L. Oliveira, L. Honório, M. Pinto, A hybrid bat-inspired algorithm for power transmission expansion planning on a practical Brazilian network, in: Applied Nature-Inspired Computing: Algorithms and Case Studies, Springer, 2020, pp. 71–95.
[62] M. Khosravy, N. Gupta, N. Patel, T. Senjyu, C.A. Duque, Particle swarm optimization of morphological filters for electrocardiogram baseline drift estimation, in: Applied Nature-Inspired Computing: Algorithms and Case Studies, Springer, 2020, pp. 1–21.
[63] G. Singh, N. Gupta, M. Khosravy, New crossover operators for real coded genetic algorithm (RCGA), in: 2015 International Conference on Intelligent Informatics and Biomedical Sciences (ICIIBMS), IEEE, 2015, pp. 135–140.
[64] N. Gupta, N. Patel, B.N. Tiwari, M. Khosravy, Genetic algorithm based on enhanced selection and log-scaled mutation technique, in: Proceedings of the Future Technologies Conference, Springer, 2018, pp. 730–748.
[65] N. Gupta, M. Khosravy, N. Patel, I. Sethi, Evolutionary optimization based on biological evolution in plants, Procedia Computer Science 126 (2018) 146–155.
[66] N. Gupta, M. Khosravy, N. Patel, O. Mahela, G. Varshney, Plants genetics inspired evolutionary optimization: a descriptive tutorial, in: Frontier Applications of Nature Inspired Computation, Springer, 2020, pp. 53–77.

[67] M. Khosravy, N. Gupta, N. Patel, O. Mahela, G. Varshney, Tracing the points in search space in plants biology genetics algorithm optimization, in: Frontier Applications of Nature Inspired Computation, Springer, 2020, pp. 180–195.

[68] N. Gupta, M. Khosravy, N. Patel, S. Gupta, G. Varshney, Artificial neural network trained by plant genetics-inspired optimizer, in: Frontier Applications of Nature Inspired Computation, Springer, 2020, pp. 266–280.

[69] N. Gupta, M. Khosravy, N. Patel, S. Gupta, G. Varshney, Evolutionary artificial neural networks: comparative study on state of the art optimizers, in: Frontier Applications of Nature Inspired Computation, Springer, 2020, pp. 302–318.

[70] M. Khosravy, N. Patel, N. Gupta, I. Sethi, Image quality assessment: a review to full reference indexes, in: Recent Trends in Communication, Computing, and Electronics, Springer, 2019, pp. 279–288.

[71] M. Khosravy, N. Gupta, N. Marina, I. Sethi, M. Asharif, Brain action inspired morphological image enhancement, in: Nature-Inspired Computing and Optimization, Springer, Cham, 2017, pp. 381–407.

[72] M. Khosravy, N. Gupta, N. Marina, I. Sethi, M. Asharifa, Perceptual adaptation of image based on Chevreul–Mach bands visual phenomenon, IEEE Signal Processing Letters 24 (5) (2017) 594–598.

[73] Cardiovascular diseases (CVDS), https://www.who.int/news-room/fact-sheets/detail/cardiovascular-diseases-(cvds).

[74] A. Milenković, C. Otto, E. Jovanov, Wireless sensor networks for personal health monitoring: issues and an implementation, Computer Communications 29 (13–14) (2006) 2521–2533.

[75] H. Cao, V. Leung, C. Chow, H. Chan, Enabling technologies for wireless body area networks: a survey and outlook, IEEE Communications Magazine 47 (12) (2009).

[76] Y. Zigel, A. Cohen, A. Katz, The weighted diagnostic distortion (WDD) measure for ECG signal compression, IEEE Transactions on Biomedical Engineering 47 (11) (2000) 1422–1430.

[77] E.C. Pinheiro, O.A. Postolache, P.S. Girao, Implementation of compressed sensing in telecardiology sensor networks, International Journal of Telemedicine and Applications 2010 (2010) 7.

[78] K. Kanoun, H. Mamaghanian, N. Khaled, D. Atienza, A real-time compressed sensing-based personal electrocardiogram monitoring system, in: Design, Automation & Test in Europe Conference & Exhibition (DATE), 2011, IEEE, 2011, pp. 1–6.

[79] K. Luo, Z. Cai, K. Du, F. Zou, X. Zhang, J. Li, A digital compressed sensing-based energy-efficient single-spot bluetooth ECG node, Journal of Healthcare Engineering 2018 (2018).

[80] H. Garudadri, P.K. Baheti, S. Majumdar, C. Lauer, F. Massé, J. van de Molengraft, J. Penders, Artifacts mitigation in ambulatory ECG telemetry, in: 2010 12th IEEE International Conference on e-Health Networking Applications and Services (Healthcom), IEEE, 2010, pp. 338–344.

[81] L.F. Polania, K.E. Barner, Multi-scale dictionary learning for compressive sensing ECG, in: 2013 IEEE Digital Signal Processing and Signal Processing Education Meeting (DSP/SPE), IEEE, 2013, pp. 36–41.

[82] F. Ansari-Ram, S. Hosseini-Khayat, ECG signal compression using compressed sensing with nonuniform binary matrices, in: 2012 16th CSI International Symposium on Artificial Intelligence and Signal Processing (AISP), IEEE, 2012, pp. 305–309.

[83] M.M. Abo-Zahhad, A.I. Hussein, A.M. Mohamed, Compression of ECG signal based on compressive sensing and the extraction of significant features, International Journal of Communications, Network and System Sciences 8 (05) (2015) 97.

[84] L.F. Polania, R.E. Carrillo, M. Blanco-Velasco, K.E. Barner, Compressed sensing based method for ECG compression, in: 2011 IEEE International Conference on Acoustics, Speech and Signal Processing (ICASSP), IEEE, 2011, pp. 761–764.

[85] A. Mishra, F.N. Thakkar, C. Modi, R. Kher, Selecting the most favorable wavelet for compressing ECG signals using compressive sensing approach, in: 2012 International Conference on Communication Systems and Network Technologies (CSNT), IEEE, 2012, pp. 128–132.

[86] A. Mishra, F. Thakkar, C. Modi, R. Kher, Comparative analysis of wavelet basis functions for ECG signal compression through compressive sensing, International Journal of Computer Science and Telecommunications 3 (5) (2012) 23–31.

[87] A. Singh, L. Sharma, S. Dandapat, Multi-channel ECG data compression using compressed sensing in eigenspace, Computers in Biology and Medicine 73 (2016) 24–37.

[88] V. Cambareri, M. Mangia, F. Pareschi, R. Rovatti, G. Setti, A case study in low-complexity ECG signal encoding: how compressing is compressed sensing?, IEEE Signal Processing Letters 22 (10) (2015) 1743–1747.

[89] G. Da Poian, C.J. Rozell, R. Bernardini, R. Rinaldo, G.D. Clifford, Matched filtering for heart rate estimation on compressive sensing ECG measurements, IEEE Transactions on Biomedical Engineering 65 (6) (2018) 1349–1358.

[90] M. Elgendi, A. Mohamed, R. Ward, Efficient ECG compression and QRS detection for e-health applications, Scientific Reports 7 (1) (2017) 459.

[91] A.M. Dixon, E.G. Allstot, A.Y. Chen, D. Gangopadhyay, D.J. Allstot, Compressed sensing reconstruction: comparative study with applications to ECG bio-signals, in: 2011 IEEE International Symposium on Circuits and Systems (ISCAS), IEEE, 2011, pp. 805–808.

[92] A.M. Dixon, E.G. Allstot, D. Gangopadhyay, D.J. Allstot, Compressed sensing system considerations for ECG and EMG wireless biosensors, IEEE Transactions on Biomedical Circuits and Systems 6 (2) (2012) 156–166.

[93] R.E. Carrillo, L.F. Polania, K.E. Barner, Iterative algorithms for compressed sensing with partially known support, in: 2010 IEEE International Conference on Acoustics Speech and Signal Processing (ICASSP), IEEE, 2010, pp. 3654–3657.

[94] H. Mamaghanian, N. Khaled, D. Atienza, P. Vandergheynst, Compressed sensing for real-time energy-efficient ECG compression on wireless body sensor nodes, IEEE Transactions on Biomedical Engineering 58 (9) (2011) 2456–2466.

[95] D.H. Chae, Y.F. Alem, S. Durrani, R.A. Kennedy, et al., Performance study of compressive sampling for ECG signal compression in noisy and varying sparsity acquisition, in: ICASSP, 2013, pp. 1306–1309.

[96] L.F. Polania, R.E. Carrillo, M. Blanco-Velasco, K.E. Barner, Compressive sensing for ECG signals in the presence of electromyographic noise, in: 2012 38th Annual Northeast Bioengineering Conference (NEBEC), IEEE, 2012, pp. 295–296.

[97] F. Ren, R. Dorrace, W. Xu, D. Marković, A single-precision compressive sensing signal reconstruction engine on FPGAs, in: 2013 23rd International Conference on Field Programmable Logic and Applications (FPL), IEEE, 2013, pp. 1–4.

[98] D. Gangopadhyay, E.G. Allstot, A.M. Dixon, K. Natarajan, S. Gupta, D.J. Allstot, Compressed sensing analog front-end for bio-sensor applications, IEEE Journal of Solid-State Circuits 49 (2) (2014) 426–438.

CHAPTER

10 Multichannel ECG reconstruction based on joint compressed sensing for healthcare applications

Sushant Kumar, Bhabesh Deka, Sumit Datta

Department of Electronics and Communication Engineering, Tezpur University, Tezpur, Assam, India

10.1 Introduction

As our society is moving towards current demographic and lifestyle trends, we are constantly facing healthcare delivery crisis. The increased frequency of cardiac disorders due to world's population has increased rapidly and cardiovascular diseases (CVD) is a major cause of death worldwide.[1] Moreover, we need an efficient medical management system which can supervise the healthcare within sustainable costs and better infrastructure. An effective method for diagnosis of CVD is the continuous monitoring of the electrocardiogram (ECG) signals.

A new technology is developed called the wireless body area network (WBAN), a subset of the wireless body sensor network (WBSN), which is a self-organized network at human body scale consisting of multiple sensor nodes [1]. The major stages of WBAN-based e-healthcare system are sensing, signal processing and wireless communication. The sensing nodes are conventionally integrated into the WBSN to continuously acquire and process data from the body, and transmit them through the WBSN gateway for further investigation and follow ups from the e-health service providers. This technology promises to offer large-scale, cost-effective, continuous monitoring of physiological signals such as blood pressure, electroencephalography (EEG), glucose level, and electrocardiogram (ECG) and reporting the same wirelessly. However, these systems suffer in terms of limited power capacity, low level signal processing capability, and low storage capacity among others, as they are battery powered. Of these, energy efficiency remains one of the core issues that needs careful attention from scientists and design engineers for the success of e-healthcare.

1 https://www.who.int/news-room/fact-sheets/detail/cardiovascular-diseases-(cvds).

Compressive Sensing in Healthcare. https://doi.org/10.1016/B978-0-12-821247-9.00015-9

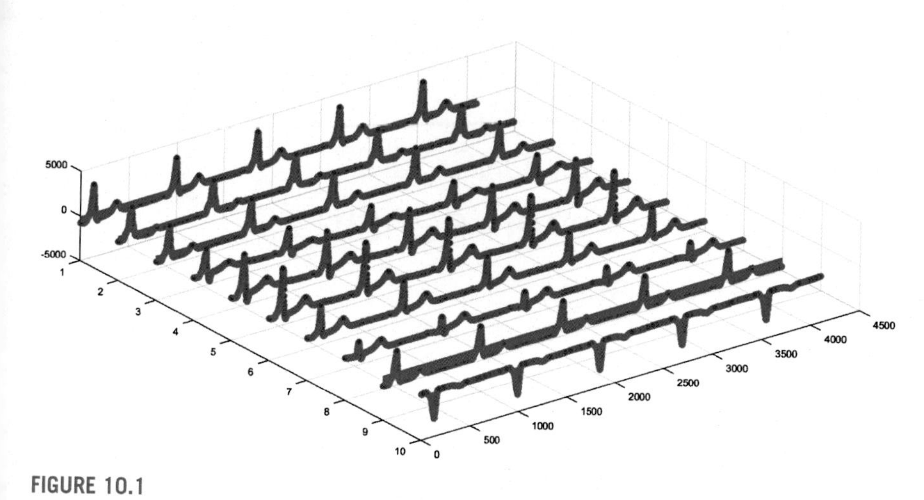

FIGURE 10.1

Ten-channel MECG signals showing spatio-temporal correlation structure in time domain.

The compressed sensing (CS) technique can be very useful for simultaneous sensing and reconstruction of sparse or compressible signal from a very few selected linear measurements by using a nonlinear optimization technique [2]. Recently, research in healthcare applications shows that CS-based data compression is a promising technique for low-complexity encoding, which can also be energy efficient for the WBAN-based e-healthcare monitoring system [3–6].

MECGs are the most favorable signals over single channel ECG due to their detailed clinical/pathological information in multiple leads for diagnosis and treatment of cardiac diseases [7]. Fig. 10.1 demonstrates inter-channel or spatial correlation among different channels, and intra-channel or temporal correlation within each channel of the MECG in the time domain. A similar correlation structure of MECG signals also exists in the wavelet domain. In fact joint amplitudes of wavelet coefficients follow this structure in a more prominent way [8, Fig. 2]. Therefore, in the CS-based recovery model of MECG, spatio-temporal correlations in the wavelet domain should be considered as a joint *a priori* information instead of either spatial or temporal correlation alone for better performance. Existing work on CS-based reconstruction of MECG described in the literature has exploited either temporal correlation [3–6,9,10] or spatial correlation [11–13], but a very few works have been reported that exploit spatial and temporal correlations simultaneously [8,14]. In this chapter, we propose a joint CS-based model that exploits spatio-temporal correlations of MECG, while modeling a block-sparsity structure of MECG in the wavelet domain. Furthermore, multi-scale information of wavelet coefficients is considered as an additional feature to weight detailed wavelet coefficients at different scales for the efficient recovery of high quality MECG from a relatively few compressed measurements.

Main contributions

The main contributions of this chapter are as follows.

- We propose a novel weighted block-sparsity-based joint compressed sensing model for MECG reconstruction. Weights to different sparse blocks or groups are determined according to their scales in the wavelet domain.
- Experiments are carried out using Physikalish-Technische Bundesanstalt (PTB) diagnostic MECG database. The performance of the proposed method is compared with existing CS-based MECG reconstruction techniques.
- Simulations are carried out to demonstrate suitability for e-healthcare applications. Successfully demonstrated benefits of the proposed method due to its reduced computational cost at varying number of CS measurements.

The rest of the chapter is organized as follows: Sect. 10.2 briefly discusses about the background and related works. Section 10.3 gives details of the proposed methodology. Simulation results are then detailed in Sect. 10.4. Finally, conclusions of the chapter are drawn in Sect. 10.5.

10.2 Background and related work

10.2.1 Compressed sensing-based MECG compression

We assume that compressed MECG measurements $\mathbf{Y}$ corresponding to different channels are being obtained using the model:

$$\mathbf{Y} = \mathbf{\Phi X} = \mathbf{\Theta} \boldsymbol{\alpha}, \tag{10.1}$$

where $\mathbf{X} \in \mathbb{R}^{N \times L}$ is the original MECG bundle, L represents the number of channels and N is the length of data in each channel. The matrix $\mathbf{\Theta} = \mathbf{\Phi \Psi}$ is called the measurement matrix formed by the selected orthonormal basis set, like wavelets, $\mathbf{\Psi} \in \mathbb{R}^{N \times N}$ and the sensing matrix $\mathbf{\Phi} \in \mathbb{R}^{M \times N}$ such that $M \ll N$. Solution of Eq. (10.1), i.e. $\boldsymbol{\alpha} \in \mathbb{R}^{N \times L}$ is the unknown sparse coefficient vector. Given $\mathbf{Y}$ and $\mathbf{\Theta}$, finding $\boldsymbol{\alpha}$ using the above equation is an ill-posed inverse problem, which can be solved efficiently using the CS theory.

10.2.1.1 Sensing matrix

In the following, we describe the implementation of a few sensing matrices:

Gaussian random matrix

The columns of a Gaussian random matrix (GRM) are independently sampled from the probability density function (pdf) of a normal distribution with mean zero and variance 1/M, which satisfies the restricted isometry property (RIP) [15]. Unfortunately, it is very difficult to generate random numbers in the hardware; even it is impossible to store all possible random numbers in the hardware platform due to the memory constraints. Therefore, in practice, it is preferable to use pseudorandom sensing instead of a full random matrix. However, it is also highly time consuming and not efficient for hardware implementation [3].

Sparse binary matrix

The entries of the sparse binary matrix (SBM) have all zeros except exactly d nonzero entries in each column, which satisfies a modified restricted isometric property RIP_p [3]. In [3,11], authors used the SBM for the CS-based ECG compression. It is an efficient sensing matrix to encode ECG signals and leads to fast computations and low memory requirements, which helps to minimize the energy cost of the encoder. The execution time of sparse binary matrices is less than the GRM [3].

10.2.2 Compressed sensing based MECG reconstruction

There are generally two types of CS models for ECG reconstruction, the single measurement vector (SMV) model for single channel ECG (SECG) signals and we also have the multiple measurement vector (MMV) model employed for MECG signals. The MMV model [16] is an extension of the conventional SMV model and deals with multiple signals sharing joint structures. MECG signals have joint sparsity structure on an orthonormal basis, $\boldsymbol{\Psi}$ as discussed above. We have received L different compressed measurements from different sensors nodes, respectively, following the model

$$\mathbf{Y} = \boldsymbol{\Theta}\boldsymbol{\alpha} + \mathbf{E} \tag{10.2}$$

where $\mathbf{Y} = [y_1, y_2, ..., y_L] \in \mathbb{R}^{M \times L}$ is the compressed data, $\boldsymbol{\alpha} = [\alpha_1, \alpha_2, ..., \alpha_L] \in \mathbb{R}^{N \times L}$ is the solution matrix, and $\mathbf{E} = [e_1, e_2, ...] \in \mathbb{R}^{M \times L}$ is the unknown noise or measurement error matrix. Now, $\mathbf{X}$ can be recovered/decoded from the compressed measurements $\mathbf{Y}$ by solving the standard basis pursuit denoising (BPDN) problem [17–19]:

$$\min_{\boldsymbol{\alpha}} \frac{1}{2} \|\mathbf{Y} - \boldsymbol{\Theta}\boldsymbol{\alpha}\|_2^2 + \lambda \|\boldsymbol{\alpha}\|_1, \tag{10.3}$$

where $\|\boldsymbol{\alpha}\|_1$ is the l_1-norm of $\boldsymbol{\alpha}$, λ is a regularization parameter to achieve a trade-off between sparsity and data consistency.

The MMV CS model would be more beneficial than the SMV CS model for CS-based MECG signal reconstruction in WBAN applications. The authors in [11] exploited the spatial correlation of MECG signals by using a mixed-norm minimization (MNM)-based joint CS (JCS) reconstruction technique. Their model is also compared to the SMV CS model for ECG signals in terms of power efficiency and reconstruction quality. In [12], the authors present a weighted mixed-norm minimization (WMNM)-based JCS reconstruction for MECG signals. The WMNM technique improves the reconstruction performance in terms of the percentage root-mean-square difference (PRD) and SNR to be compared to [11]. In [20], the authors present a prior weighted mixed-norm minimization (PWMNM) method for JCS reconstruction of MECG signals by exploiting spatial correlation as well as additional prior information. The PWMNM technique improves the reconstruction performance in terms of

PRD and SNR compared to [11,12]. The authors in [21] successfully demonstrated block-sparsity recovery of fetal ECG (FECG) signals using the ℓ_2/ℓ_{1-2} mixed-norm minimization. It is reported that the ℓ_2/ℓ_{1-2} norm is highly effective for sparse signal recovery. They consider temporal correlation only while inducing block sparsity in FECG. In [8], the authors present a block-sparsity-based JCS reconstruction for MECG signals by using spatio-temporal sparse Bayesian learning (STSBL) [22]. Their model performs better than the JCS techniques [11,12] in terms of reduced PRD levels at the same CR value, but PRD values are higher than the PWMNM. In [13], authors present a weighted $l_{1,2}$ minimization (WL12M) method for the JCS reconstruction of MECG by exploiting inter-channel correlation and multiscale prior knowledge in the wavelet domain. The WL12M technique improves reconstruction performance in terms of PRD compared to the weighted l_1 minimization (WLM) [5] and block-sparse Bayesian learning (BSBL). However, these methods ignore the temporal correlation structure of MECG that is present within each channel.

In this work, we have employed a JCS model which takes advantages of spatial as well as temporal correlations simultaneously while modeling a block-sparsity structure of MECG in the wavelet domain.

10.3 Proposed method

In a WBAN-based system for e-healthcare applications, we are supposed to continuously acquire MECG signals using multiple sensors nodes and transmit them in real-time to a remote cloud or server for further processing. We aim to efficiently compress MECG signals without loss of their clinical/pathological information. In this work, we propose a block-sparsity-based joint compressive sensing (JCS) reconstruction of MECG signals.

In order to induce block-sparsity property in α, we exploit spatial as well as temporal correlations of MECG in the wavelet domain. Multi-level *a priori* information of wavelet coefficients may be exploited to reinforce the importance of detailed wavelet coefficients according to their detail levels by introducing a new multi-level weighting approach, while obtaining the block-sparsity structure of MECG. The weights are determined according to the positions of coefficients in the wavelet decomposition for different blocks or groups; assigning a single weight for coefficients belonging to one level. Weights decrease as we go from finer scales to coarser scales. The importance of multi-level weighting is that it speeds up the convergence rate of the JCS reconstruction besides enhancing the reconstruction performance.

The proposed weighted block-sparsity based compressed sensing model of MECG reconstruction using l_2/l_{1-2} minimization is given by

$$\min_{\boldsymbol{\alpha}} \frac{1}{2}\|\boldsymbol{\Theta}\boldsymbol{\alpha} - \mathbf{Y}\|_2^2 + \lambda \left(\sum_{g_i=1}^{p} \omega_{g_i} \left\|\boldsymbol{\alpha}_{g_i}\right\|_2 - \|\boldsymbol{\alpha}\|_2 \right), \tag{10.4}$$

where the second term indicates the l_2/l_{1-2}-norm, λ is a positive regularization parameter to achieve a trade-off between sparsity and data consistency. The ith block index $g_i \in \{g_1, g_2, ..., g_p\}$ consists of indices of coefficients in each block of MECG in the wavelet domain. So, $\boldsymbol{\alpha}_{g_i}$ denotes the sub-vector of $\boldsymbol{\alpha}$ indexed by g_i. Let us assume that there are p blocks with block size $b = N/p$ in $\boldsymbol{\alpha}$. If $\boldsymbol{\alpha}$ has at most s nonzero blocks, we refer to it as the s block-sparse signal. An efficient block-sparse representation improves the signal reconstruction further at a given number of compressed measurements. The weights ω_{g_i} are calculated as follows [5]:

$$\omega_{g_i} = \begin{cases} 0, & \text{if } \alpha^k_{g_i} \in \alpha^k_{L_0}, \\ 2^{-(J-1-j)}, & \text{if } \alpha^k_{g_i} \in \alpha^k_{H_j}. \end{cases} \tag{10.5}$$

Now, reconstruction problem in Eq. (10.4) may be simplified after rearranging its terms as reported in [23]. We reproduce it here as follows:

$$\boldsymbol{\alpha}^{k+1} = \text{argmin} \frac{1}{2} \|\boldsymbol{\Theta\alpha} - \mathbf{Y}\|_2^2 + \lambda \sum_{g_i=1}^{p} \omega_{g_i} \left\|\boldsymbol{\alpha}_{g_i}\right\|_2 + \langle \mathrm{v}^k, \boldsymbol{\alpha} \rangle \tag{10.6}$$

where $v^k = -\lambda\boldsymbol{\alpha}^k/\|\boldsymbol{\alpha}^k\|_2$ when $\|\boldsymbol{\alpha}^k\|_2 \neq 0$ and $v^k = 0$ otherwise. In order to solve the above problem using the alternating direction method of multipliers (ADMM) algorithm [24], we rewrite the problem by introducing a new variable $\mathbf{b}$ as follows:

$$\min_{\alpha, b} \underbrace{\frac{1}{2} \|\boldsymbol{\Theta\alpha} - \mathbf{Y}\|_2^2 + \langle v^k, \boldsymbol{\alpha} \rangle}_{f(\boldsymbol{\alpha})} + \underbrace{\lambda \sum_{g_i=1}^{p} \omega_{g_i} \left\|b_{g_i}\right\|_2}_{g(\mathbf{b})}, \text{ subject to } \boldsymbol{\alpha} - \mathbf{b} = 0. \tag{10.7}$$

The augmented Lagrangian form of the above problem may be defined as

$$\mathcal{L}_\beta(\boldsymbol{\alpha}, \mathbf{b}, \mathbf{z}) = f(\boldsymbol{\alpha}) + g(\mathbf{b}) + \mathbf{z}^T(\boldsymbol{\alpha} - \mathbf{b}) + \frac{\beta}{2} \|\boldsymbol{\alpha} - \mathbf{b}\|_2^2, \tag{10.8}$$

where β is a penalty parameter and z denotes the Lagrangian multiplier. According to the ADMM, we need to solve the following subproblems iteratively:

$$\boldsymbol{\alpha}^{t+1} = \underset{\alpha}{\text{argmin}}\ f(\boldsymbol{\alpha}) + \frac{\beta}{2} \left\|\boldsymbol{\alpha}^t - (\mathbf{b} - \mathbf{z}^t/\beta)\right\|_2^2, \tag{10.9}$$

$$\mathbf{b}^{t+1} = \underset{b}{\text{argmin}}\ g(\mathbf{b}) + \frac{\beta}{2} \left\|\mathbf{b} - (\boldsymbol{\alpha}^{t+1} - \mathbf{z}^t/\beta)\right\|_2^2, \tag{10.10}$$

$$\mathbf{z}^{t+1} = \mathbf{z}^t + \beta(\boldsymbol{\alpha}^{t+1} - \mathbf{b}^{t+1}). \tag{10.11}$$

We called the above algorithm the weighted block-sparsity-based MECG (WB-MECG) reconstruction. Steps of the proposed algorithm are summarized in Algorithm 10.1. After reconstruction of the wavelet coefficients $\boldsymbol{\alpha}$, one can get back the MECG signals by applying the inverse wavelet transform.

Algorithm 10.1 WBMECG reconstruction.

Input: $\mathbf{Y}$, $\mathbf{\Phi}$, $\boldsymbol{\omega}$, p, λ, β
Ensure: $\boldsymbol{\alpha}^* \in \mathbb{R}^{N\times L}$
Initialize: $v^{(0)} = 0$ and $k = 0$
  while not converge **do**
    Initialize: $b^{(0)} = 0$, $z^{(0)} = 0$ and $t = 0$
    while not converge **do**
      $\boldsymbol{\alpha}^{t+1} = (\Phi^T\Phi + \beta I)^{-1}(\Phi^T\mathbf{Y} - v^{(k)} + \beta(b^{(t)} - z^{(t)}))$
      for $i = 1, 2, ..., p$ **do**

$$b_{g_i}^{(t+1)} = \begin{cases} (1 - \frac{\lambda\omega_{g_i}}{\beta\|\theta_{g_i}\|_2})\theta_{g_i}, & \|\theta_{g_i}\|_2 > \frac{\lambda}{\beta} \\ 0, & \text{otherwise} \end{cases}$$

        where $\theta = \boldsymbol{\alpha}^{(t+1)} + z^{(t)}$
      end for
      $z^{(t+1)} = z^{(t)} + \boldsymbol{\alpha}^{(t+1)} - b^{(t+1)}$
      $t = t + 1$
    end while
    $\boldsymbol{\alpha}^{(k+1)} = \boldsymbol{\alpha}^{(t)}$

$$v^{(k+1)} = \begin{cases} -\frac{\lambda\boldsymbol{\alpha}^{(k+1)}}{\|\boldsymbol{\alpha}^{(k+1)}\|_2}, & \boldsymbol{\alpha}^{(k+1)} \neq 0 \\ 0, & \text{otherwise} \end{cases}$$

    $k = k + 1$
  end while
Output: $\boldsymbol{\alpha}^* = \boldsymbol{\alpha}^k$

10.4 Results and discussion

10.4.1 Experimental setup

All experiments are performed in the MATLAB® environment on a PC with 3.40 GHz Intel-i7 core CPU, 10 GB of RAM. We use the PTB diagnostic MECG database from [25] to verify the performance of the proposed method. The database consists of 15-channel MECG from 290 patients and each signal is sampled at $f_s = 1$ kHz with 16-bit resolution. In particular, we select dataset "s0029lrem" of the PTB database for our simulations. The ECG matrix $\mathbf{X} \in \mathbb{R}^{512\times 8}$ is built from eight fundamental ECG channels of the above dataset. All other channels of the standard 12-lead ECG may be obtained from these fundamental channels. We choose two types of sensing matrices, i.e. Gaussian random and sparse binary sensing matrices [3] for computing $\mathbf{\Phi}$. A Daubechies-6 ("db6") wavelet [26] has been used as the sparsifying bases $\mathbf{\Psi}$ for the sparse representation of MECG signals. All results are averaged over 50 independent trials with different realizations of the sensing matrix in each trial. The dependent ECG channels may be obtained by using the following relations:

$$\begin{aligned} Lead\ III &= Lead\ II - Lead\ I, \\ Lead\ aVR &= (Lead\ I + Lead\ II)/2, \\ Lead\ aVL &= Lead\ I - Lead\ II/2, \\ Lead\ aVF &= Lead\ II - Lead\ I/2. \end{aligned}$$

We have also compared our results with some of the existing MECG reconstruction techniques. MATLAB codes of sparse Bayesian learning exploiting temporal correlation in multiple measurement vectors (tMSBL) [27] and the FOCal underdetermined system solver exploiting temporal correlation in multiple measurement vectors (tMFOCUSS) [28] algorithms are downloaded.[2] We use the YALL1 toolbox[3] to implement MBP.

10.4.2 Performance metrics

The performance of the proposed reconstruction algorithm is evaluated by different quality measures in terms of percentage root-mean-square difference (PRD) and signal-to-noise ratio (SNR) [11]. For the ECG reconstruction quality, in [29] the authors had classified different values of PRD and SNR based on the signal quality perceived by specialists. They are defined as follows:

$$PRD(\%) = \frac{\|x_{orig} - x_{rec}\|_2}{\|x_{orig}\|_2} \times 100 \tag{10.12}$$

$$SNR_{dB} = -20\log_{10}(0.01PRD), \tag{10.13}$$

where $\mathbf{x}_{orig}$ and $\mathbf{x}_{rec}$ represent the original and reconstructed signals, respectively. Benchmarks of the quality of CS-based ECG reconstruction, Table 10.1 shows the range of PRD and SNR values for different signal qualities [29].

Table 10.1 Range of PRD and SNR for different signal qualities [29].

Signal quality	PRD (%)	SNR (dB)
Very good	0–2	>34
Very good or Good	2–9	34–20.9
Not possible to determine	>9	<20.9

Table 10.2 Average PRD values at varying number of measurements in each channel for GRM.

M	Method	Channels							
		I	**II**	**V1**	**V2**	**V3**	**V4**	**V5**	**V6**
64	**Proposed**	**12.93**	5.959	4.001	**4.351**	10.39	**4.078**	3.528	**5.539**
	BSBL	15.14	3.573	3.624	4.843	4.666	4.609	3.395	8.251
	tMSBL	79.69	81.57	85.59	81.23	84.33	83.70	79.69	63.47
	MBP	88.40	93.57	95.77	91.76	95.28	95.57	91.10	76.20
	tMFOCUSS	77.12	80.79	84.18	80.51	83.51	83.42	79.02	64.08
96	**Proposed**	**9.653**	**2.670**	**1.582**	**2.306**	7.320	**2.268**	**1.587**	**3.140**
	BSBL	11.15	2.675	2.670	3.113	2.564	3.261	2.542	5.381
	tMSBL	10.25	3.543	3.267	3.798	3.385	3.819	3.160	4.253
	MBP	72.94	75.98	78.61	75.65	78.27	77.99	73.77	60.39
	tMFOCUSS	38.89	39.66	41.60	39.80	41.17	40.93	38.72	31.12
128	**Proposed**	**7.762**	**1.719**	**1.149**	**1.668**	4.398	**1.616**	**1.082**	**2.129**
	BSBL	8.444	1.782	1.674	2.064	1.526	2.143	1.492	2.748
	tMSBL	9.103	2.149	1.562	2.025	1.599	2.080	1.581	2.818
	MBP	16.49	13.30	13.45	13.75	13.86	14.12	12.58	11.46
	tMFOCUSS	18.71	17.49	18.38	17.63	18.18	18.09	17.07	13.83
160	**Proposed**	**6.524**	**1.381**	**0.882**	**1.193**	2.868	**1.278**	**0.912**	**1.783**
	BSBL	6.559	1.482	1.085	1.539	1.144	1.554	1.021	1.908
	tMSBL	7.991	1.774	1.082	1.374	1.035	1.429	1.172	2.203
	MBP	7.008	2.428	2.223	2.886	2.561	2.935	2.130	3.091
	tMFOCUSS	10.57	8.589	8.949	8.652	8.894	8.907	8.324	6.907
192	**Proposed**	5.466	**1.165**	**0.668**	**0.838**	1.647	**0.926**	**0.759**	1.502
	BSBL	5.451	1.245	0.777	1.110	0.803	1.100	0.760	1.478
	tMSBL	7.062	1.537	0.879	1.059	0.786	1.114	0.962	1.908
	MBP	6.016	1.596	1.292	1.666	1.349	1.651	1.245	2.164
	tMFOCUSS	7.626	5.486	5.624	5.433	5.548	5.565	5.241	4.513
224	**Proposed**	4.880	**0.999**	**0.573**	**0.689**	1.202	**0.772**	0.637	1.272
	BSBL	4.699	1.038	0.607	0.838	0.600	0.849	0.606	1.217
	tMSBL	6.212	1.364	0.755	0.870	0.642	0.939	0.838	1.717
	MBP	5.183	1.295	0.918	1.152	0.916	1.192	0.956	1.736
	tMFOCUSS	6.343	4.163	4.277	4.137	4.231	4.256	3.994	3.465

10.4.3 Performance evaluation

Averaged PRD values per channel using the proposed algorithm, the BSBL, the tMSBL, the tMFOCUSS, and the MBP at different number of measurements for eight channels of PTB MECG database are shown in Tables 10.2 and 10.3, respectively,

[2] http://dsp.ucsd.edu/~zhilin/.

[3] http://yall1.blogs.rice.edu/.

Table 10.3 Average PRD values at varying number of measurements in each channel for SBM.

M	Method	Channels							
		I	II	V1	V2	V3	V4	V5	V6
64	**Proposed**	**13.61**	3.830	3.743	5.552	9.639	**3.641**	**2.710**	**5.266**
	BSBL	15.76	3.612	3.548	4.899	4.570	4.653	3.374	8.277
	tMSBL	79.29	81.31	85.62	81.31	84.36	83.63	79.46	62.98
	MBP	77.01	79.61	83.27	81.26	83.43	83.23	77.46	63.00
	tMFOCUSS	71.90	74.01	77.92	74.75	77.25	76.95	72.42	57.95
96	**Proposed**	**9.570**	2.819	**1.833**	**2.617**	7.505	**2.254**	**1.516**	**2.997**
	BSBL	11.67	2.466	2.608	3.010	2.420	3.059	2.468	5.297
	tMSBL	10.47	3.760	3.592	4.170	3.724	4.094	3.278	4.222
	MBP	48.66	48.49	50.59	48.88	50.41	50.17	46.93	38.09
	tMFOCUSS	33.62	32.78	34.62	33.00	34.11	33.81	31.95	25.39
128	**Proposed**	**7.656**	1.851	**1.131**	**1.633**	4.894	**1.655**	**1.148**	**2.269**
	BSBL	8.562	1.845	1.637	1.994	1.505	2.128	1.539	2.714
	tMSBL	8.985	2.209	1.537	2.066	1.616	2.068	1.542	2.874
	MBP	9.353	3.789	4.059	5.010	4.536	4.859	3.461	4.260
	tMFOCUSS	14.28	11.98	12.75	12.29	12.57	12.51	11.74	9.607
160	**Proposed**	**6.405**	**1.425**	**0.890**	**1.207**	3.359	**1.246**	**0.878**	**1.738**
	BSBL	7.033	1.533	1.125	1.517	1.097	1.503	0.988	1.908
	tMSBL	8.010	1.848	1.102	1.417	1.066	1.453	1.182	2.302
	MBP	7.064	2.112	1.951	2.616	2.212	2.521	1.742	2.798
	tMFOCUSS	10.30	7.828	8.204	7.911	8.091	8.060	7.579	6.353
192	**Proposed**	**5.551**	**1.165**	**0.709**	**0.889**	2.185	**0.986**	0.762	1.531
	BSBL	5.577	1.266	0.789	1.136	0.845	1.146	0.762	1.515
	tMSBL	7.309	1.571	0.894	1.067	0.798	1.134	0.977	1.964
	MBP	6.062	1.557	1.283	1.699	1.385	1.681	1.238	2.145
	tMFOCUSS	7.848	5.323	5.487	5.274	5.406	5.417	5.118	4.406
224	**Proposed**	4.833	**1.001**	**0.576**	**0.713**	1.291	**0.773**	0.655	1.288
	BSBL	4.787	1.051	0.621	0.855	0.600	0.849	0.603	1.208
	tMSBL	6.441	1.373	0.746	0.892	0.647	0.944	0.816	1.660
	MBP	5.261	1.266	0.923	1.191	0.929	1.194	0.934	1.729
	tMFOCUSS	6.262	3.787	3.878	3.731	3.800	3.827	3.610	3.177

for GRM and SBM. The proposed algorithm performs superior over the tMSBL, the tMFOCUSS, and the MBP with the improvement of average PRD at almost all levels of measurements for different channels. This is because the proposed algorithm improves the reconstruction quality by exploiting spatio-temporal correlations along with multi-scale prior information simultaneously.

Averaged PRD and SNR over eight channels are also plotted in Figs. 10.2, 10.3, 10.4, and 10.5, respectively, for different numbers of CS measurements for the GRM and SBM sensing matrix.

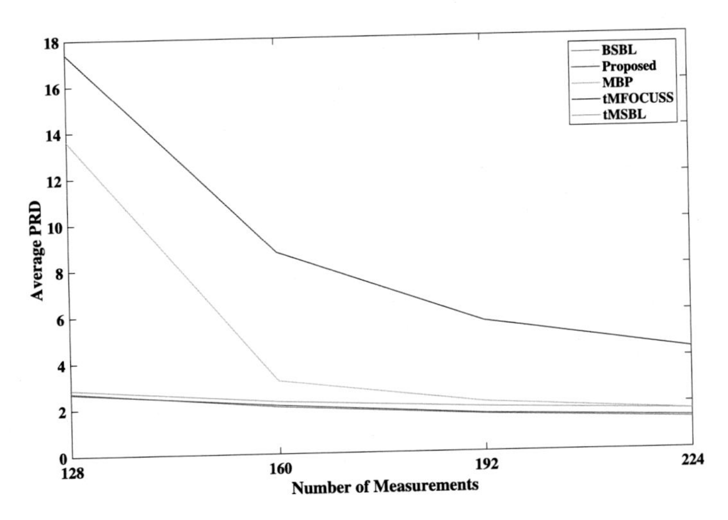

FIGURE 10.2

Average PRD at different numbers of measurements for different algorithms and GRM sensing matrices.

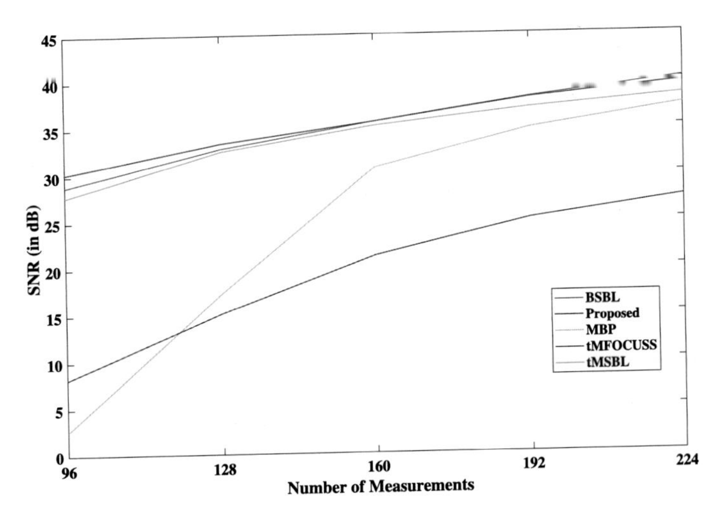

FIGURE 10.3

Average SNR at different numbers of measurements for different algorithms and GRM sensing matrices.

10.4.4 Healthcare application perspective

We also compare the signal quality of the proposed algorithm and the BSBL algorithm from an healthcare application viewpoint. In healthcare applications, the diagnosis/clinical information in terms of R-R interval, QRS interval, ST segment,

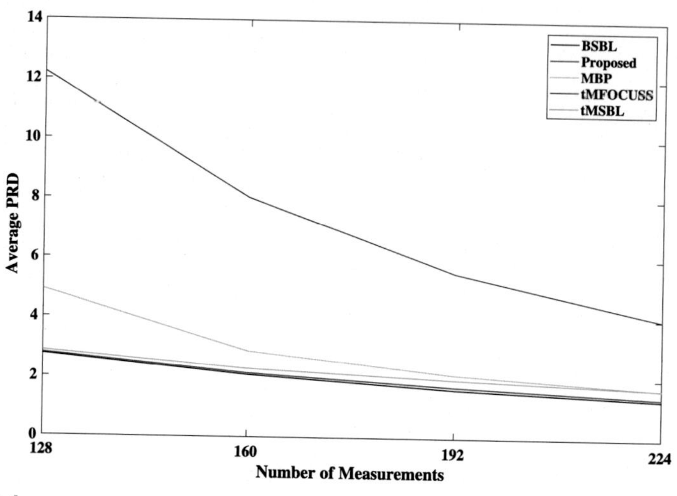

FIGURE 10.4

Average PRD at different numbers of measurements for different algorithms and SBM sensing matrices.

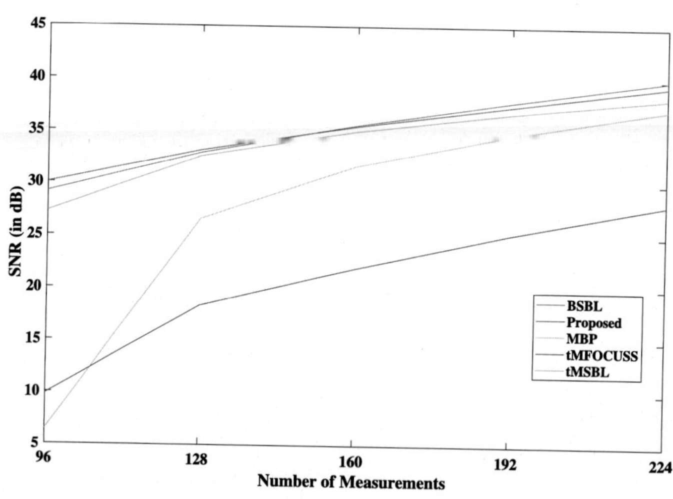

FIGURE 10.5

Average SNR at different numbers of measurements for different algorithms and SBM sensing matrices.

and R amplitude, are preserved in the recovered MECG signal. We use the SBM sensing matrix of 96×512 to compress the "s0029lrem" MECG dataset. We plot the 8-channel reconstruction of MECG signals by BSBL and the proposed method. Results are shown in Figs. 10.6, 10.7, respectively. Here, we observe that the recon-

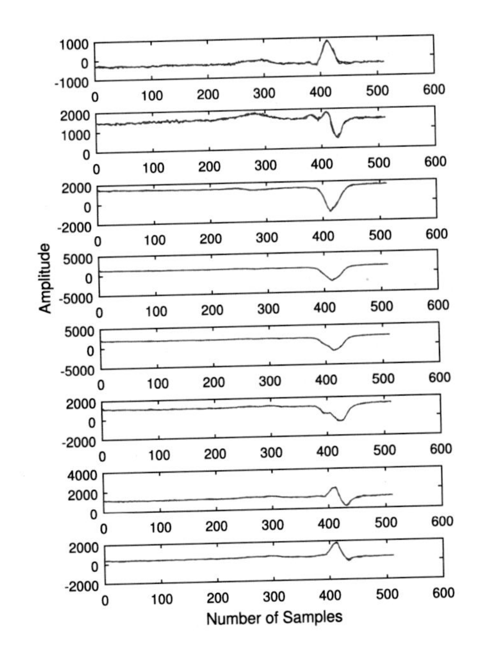

FIGURE 10.6

Recovered results of BSBL for SDM sensing matrix

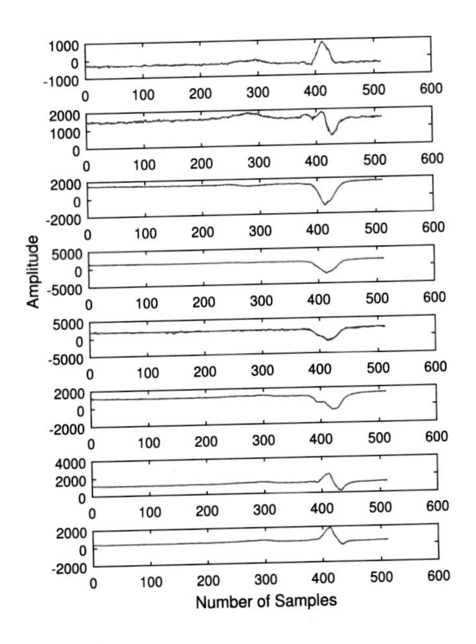

FIGURE 10.7

Recovered results of the proposed SBM sensing matrix.

structed signals using the proposed algorithm are better than the BSBL algorithm. The diagnostic features present in all segments are very well preserved in the case of the proposed method.

10.4.5 Power saving

From the point of view of energy saving in healthcare applications, power consumption in both compression and wireless transmission should be minimum. The proposed method has low complexity in sensing compared to the existing CS-based methods. We implement the SBM with each column having zero entries except four entries of 1s at random locations for our experiments. The other columns are also similar with four nonzero entries at random locations and independent of the rest. So, the proposed method helps to save more than 88.57% energy in the sensing process of the JMCS [11] which uses a number of d = 35 of 1s in each column.

10.5 Conclusion

In this chapter, we propose a block-sparsity-based weighted l_2/l_{1-2}-norm minimization for MECG reconstruction. It is observed that the SBM is more suitable for the WBAN-based real-time transmission of MECG signals than the GRM. It is an effective way to improve the power efficiency and the lifetime of the network, which meets the growing demand for more efficient e-healthcare monitoring system. Exploiting spatial-temporal correlations and the multi-level prior information simultaneously in the wavelet domain plays a significant role in the MECG reconstruction using a mixed norm compared to the spatial or temporal correlation of MECG signals alone.

References

[1] H. Cao, V. Leung, C. Chow, H. Chan, Enabling technologies for wireless body area networks: a survey and outlook, IEEE Communications Magazine 47 (12) (2009) 84–93.

[2] E.J. Candes, M.B. Wakin, An introduction to compressive sampling, IEEE Signal Processing Magazine 25 (2) (Mar. 2008) 21–30.

[3] H. Mamaghanian, N. Khaled, D. Atienza, P. Vandergheynst, Compressed sensing for real-time energy-efficient ECG compression on wireless body sensor nodes, IEEE Transactions on Biomedical Engineering 58 (9) (Sep. 2011) 2456–2466.

[4] L.F. Polania, R.E. Carrillo, M. Blanco-Velasco, K.E. Barner, Exploiting prior knowledge in compressed sensing wireless ECG systems, IEEE Journal of Biomedical and Health Informatics 19 (2) (Mar. 2015) 508–519.

[5] J. Zhang, Z. Gu, Z.L. Yu, Y. Li, Energy-efficient ECG compression on wireless biosensors via minimal coherence sensing and weighted l_1 minimization reconstruction, IEEE Journal of Biomedical and Health Informatics 19 (2) (Mar. 2015) 520–528.

[6] Z. Zhang, T. Jung, S. Makeig, B.D. Rao, Compressed sensing for energy-efficient wireless telemonitoring of noninvasive fetal ECG via block sparse Bayesian learning, IEEE Transactions on Biomedical Engineering 60 (2) (Feb. 2013) 300–309.

[7] B. Surawicz, T. Knilans, Chou's Electrocardiography in Clinical Practice, 6th edition, Elsevier, Canada, 2008.
[8] A. Singh, S. Dandapat, Block sparsity-based joint compressed sensing recovery of multi-channel ECG signals, Healthcare Technology Letters 4 (2) (2017) 50–56.
[9] L.F. Polania, R.E. Carrillo, M. Blanco-Velasco, K.E. Barner, Compressed sensing based method for ECG compression, in: IEEE International Conference on Acoustics, Speech and Signal Processing (ICASSP), May 2011, pp. 761–764.
[10] L. Polania, K. Barner, A weighted l_1 minimization algorithm for compressed sensing ECG, in: IEEE International Conference on Acoustics, Speech and Signal Processing (ICASSP), 2014, pp. 4413–4417.
[11] H. Mamaghanian, G. Ansaloni, D. Atienza, P. Vandergheynst, Power-efficient joint compressed sensing of multi-lead ECG signals, in: IEEE International Conference on Acoustics, Speech and Signal Processing (ICASSP), May 2014, pp. 4409–4412.
[12] A. Singh, S. Dandapat, Weighted mixed-norm minimization based joint compressed sensing recovery of multi-channel electrocardiogram signals, Computers & Electrical Engineering 53 (2016) 203–218.
[13] J. Zhang, Z.L. Yu, Z. Gu, Y. Li, Z. Lin, Multichannel electrocardiogram reconstruction in wireless body sensor networks through weighted $l1,2$ minimization, IEEE Transactions on Instrumentation and Measurement 67 (9) (2018) 2024–2034.
[14] S. Kumar, B. Deka, S. Datta, Block-sparsity based compressed sensing for multichannel ECG reconstruction, in: Pattern Recognition and Machine Intelligence, Springer International Publishing, 2019, pp. 210–217.
[15] E.J. Candes, T. Tao, Decoding by linear programming, IEEE Transactions on Information Theory 51 (12) (2005) 4203–4215.
[16] S.F. Cotter, B.D. Rao, K. Engan, K. Kreutz-Delgado, Sparse solutions to linear inverse problems with multiple measurement vectors, IEEE Transactions on Signal Processing 53 (7) (Jul. 2005) 2477–2488.
[17] S. Chen, D. Donoho, M. Saunders, Atomic decomposition by basis pursuit, SIAM Journal on Scientific Computing 20 (1) (1998) 33–61.
[18] B. Deka, S. Datta, Compressed Sensing Magnetic Resonance Image Reconstruction Algorithms: A Convex Optimization Approach, Springer Series on Bio- and Neurosystems, Springer, Singapore, 2019.
[19] T. Cabral, M. Khosravy, F. Dias, H. Monteiro, M. Lima, L. Silva, R. Naji, C. Duque, Compressive sensing in medical signal processing and imaging systems, in: Sensors for Health Monitoring, Academic Press, 2019, pp. 69–92.
[20] A. Singh, S. Dandapat, Exploiting multi-scale signal information in joint compressed sensing recovery of multi-channel ECG signals, Biomedical Signal Processing and Control 29 (2016) 53–66.
[21] W. Wang, J. Wang, Z. Zhang, Block-sparse signal recovery via l_2/l_{1-2} minimisation method, IET Signal Processing 12 (4) (2018) 422–430.
[22] Z. Zhang, T. Jung, S. Makeig, Z. Pi, B.D. Rao, Spatiotemporal sparse Bayesian learning with applications to compressed sensing of multichannel physiological signals, IEEE Transactions on Neural Systems and Rehabilitation Engineering 22 (6) (Nov. 2014) 1186–1197.
[23] P. Yin, Y. Lou, Q. He, J. Xin, Minimization of l_{1-2} for compressed sensing, SIAM Journal on Scientific Computing 37 (1) (2015) 536–563.
[24] S. Boyd, N. Parikh, E. Chu, B. Peleato, J. Eckstein, Distributed optimization and statistical learning via the alternating direction method of multipliers, Foundations and Trends in Machine Learning 3 (1) (2011) 1–122.
[25] A.L. Goldberger, L.A. Amaral, L. Glass, J.M. Hausdorff, P.C. Ivanov, R.G. Mark, J.E. Mietus, G.B. Moody, C.K. Peng, H.E. Stanley, PhysioBank, PhysioToolkit, and PhysioNet: components of a new research resource for complex physiologic signals, Circulation 101 (23) (2000) 215–220.
[26] I. Daubechies, Ten Lectures on Wavelets, Society for Industrial and Applied Mathematics, Philadelphia, PA, USA, 1992.
[27] Z. Zhang, B.D. Rao, Sparse signal recovery with temporally correlated source vectors using sparse Bayesian learning, IEEE Journal of Selected Topics in Signal Processing 5 (5) (Sep. 2011) 912–926.

[28] Z. Zhang, B.D. Rao, Iterative reweighted algorithms for sparse signal recovery with temporally correlated source vectors, in: IEEE International Conference on Acoustics, Speech and Signal Processing (ICASSP), 2011, pp. 3932–3935.
[29] Y. Zigel, A. Cohen, A. Katz, The weighted diagnostic distortion (WDD) measure for ECG signal compression, IEEE Transactions on Biomedical Engineering 47 (11) (Nov. 2000) 1422–1430.

CHAPTER

11 Neural signal compressive sensing

Denise Fonseca Resende[a], Mahdi Khosravy[b], Henrique L.M. Monteiro[a], Neeraj Gupta[c], Nilesh Patel[c], Carlos A. Duque[a]

[a]*Department of Electrical Engineering, Federal University of Juiz de Fora, Juiz de Fora, Brazil*
[b]*Media Integrated Communication Laboratory, Graduate School of Engineering, Osaka University, Suita, Osaka, Japan*
[c]*Computer Science and Engineering, Oakland University, Rochester, MI, United States*

11.1 Introduction

A promising method for acquiring and reconstructing a signal with potential benefits is compressive sensing (CS). This method takes advantage of the sparsity of a signal in a given domain to decrease the number of samples required to reconstruct the same signal. As an advantage, it generally requires a smaller number of samples than Nyquist sampling, which has been the fundamental principle of data acquisition in signal processing [1].

Considering the Nyquist theorem, the sampling rate must be at least greater than twice the maximum signal frequency, i.e. the Nyquist rate. Clearly, troubles occur when the signal has a very broad bandwidth, requiring a very high sampling rate and resulting in the large amounts of data measured. In contrast, CS can acquire the signal with less samples, representing a sparse signal, and uses mathematical techniques such as L1 optimization to reconstruct the original signal.

In recent years, interest in outpatient monitoring of bioelectric signals has increased. This monitoring has major advantages, such as increased patient mobility, constant patient observation and reduced health-care costs. An important clinical tool used for monitoring and enhancing brain knowledge is the electroencephalogram (EEG); this equipment is to record the electrical activity of the brain and is capable of diagnosing neurological diseases or disorders.

Compressive sensing has drawn much attention of researchers in the fields of biomedical signal processing and imaging systems [2]. Besides its widely increased applications, it has still a great potential for application. Among others, CS has a great potential to be applied to agriculture machinery [3,4], signal processing [5,6], blind signal processing [7–12], acoustic OFDM [13], electrocardiogram processing [14–20], power quality analysis [21–23], power line communications [24], public transportation system [25], power system planing [26,27], location-based services

Compressive Sensing in Healthcare. https://doi.org/10.1016/B978-0-12-821247-9.00016-0

[28], text data processing [29], smart environment [30], human motion analysis [31], texture analysis [32], image adaptation [33], image enhancement [34,35], medical image processing [36–38], human-robot interaction [39], telecommunications [40–45], sentiment mining [46], data mining [47,48], and software intensive systems [49].

Medical applications include monitoring or diagnosis of Alzheimer's, epilepsy, fall monitoring in the elderly and sleep disorders, for example. For the convenience of patients, the recent research focus is on developing mobile EEG technology, particularly data capture to monitor clinical events such as seizures. However, there is a need to store a large amount of EEG data and this entails the problem of mobile device batteries.

One of the key challenges is processing large amounts of EEG data. The CS technique can help in the data transmission efficiency by wireless methods when a portable embedded system is used if sampling compression is applied before wireless transmission.

11.2 Compressed sensing theory – a brief review

Compressive sensing (CS) or compressive sampling is an emerging technique for acquiring and reconstructing a digital signal with potential benefits in many applications. The CS method takes advantage of a sparse signal in a specific domain to significantly reduce the number of samples needed to reconstruct the signal [1].

There is a difference between the signal change rate and the signal information rate, this is the concept of compressive sensing [50,51].

For working with the digital domain signal in wireless transmission, for example, the first concept is taken into account according to traditional Nyquist sampling [52].

Nyquist sampling means that it is essential to acquire the signal at a rate at more than twice the frequency of the original signal. A usual compression algorithm would then be used to all such collected samples and eliminate any redundancy, providing a smaller number of bits to represent the signal.

On the other hand, compressive detection exploits the rate of information within a given signal. Signal redundancy is eliminated during the sampling process, resulting in a lower efficient sampling rate [53].

Then consider an n-length vector $\boldsymbol{x}$ being sampled as an m-length vector $\boldsymbol{y}$:

$$\boldsymbol{y} = \Phi \boldsymbol{x}, \tag{11.1}$$

where $\Phi \in \mathbb{R}^{m \times n}$ is the sensor assembly and m is defined as the sampling rate. The matrix Φ of random elements. Considering $m << n$, Eq. (11.1) is undetermined.

On the other hand, $\boldsymbol{x}$ can be considered by a set of sparse coefficients $\mu \in \mathbb{R}^n$, in determined conditions of dispersion-inducing bases $\Psi \in \mathbb{R}^{n \times n}$:

$$\boldsymbol{x} = \Psi \boldsymbol{u}. \tag{11.2}$$

Consequently, considering (11.1) and (11.2), the sparse vector, $\boldsymbol{u}$, can be defined by

$$\boldsymbol{y} = \Phi\Psi\boldsymbol{u} = \Theta_{m\times n}\boldsymbol{u}, \tag{11.3}$$

on what, $\Theta_{m\times n} = \Phi\Psi$ is a $m \times n$ matrix MxN called measurement matrix.

For a discrete signal, the compressed signal $\boldsymbol{y}$ can be defined by

$$\hat{\boldsymbol{y}} = Q_b(\boldsymbol{y}) \tag{11.4}$$

where $Q_b(.)$ is the quantization function considering b bits.

The vector u, defined in (11.2), is known as a sparse vector. It can be estimated by minimization of the l_0 norm. However, the l_0 norm is an NP-hard problem and requires non-linear metaheuristic optimization approaches [20,54–64]. A very common solution is replacement by the l_1 norm minimization. Then, the estimative can be considered by

$$\hat{\boldsymbol{u}} = \min \|\boldsymbol{u}\|_1 \quad \text{s.t.} \quad \|\hat{\boldsymbol{y}} - \Theta\boldsymbol{u}\| < \epsilon \tag{11.5}$$

where ϵ is the reconstruction margin of error. The minimization of l_1 approximates the optimal minimization l_0, besides being convex, and can be resolved within polynomial time. Therefore, the rebuilt signal, $\hat{\boldsymbol{x}}$, is recovered considering the following equation:

$$\hat{\boldsymbol{x}} = \Psi\hat{\boldsymbol{u}}. \tag{11.6}$$

11.3 Compressive sensing for energy-efficient neural recording

Implantable devices for recording, neural decoding, and peak detection belong to the most important dealings for investigating and monitoring neural movement within a designated brain area [53,65,66]. Tools such as multiple electrode arrays (MEAs) have been widely used by neuroscientists to monitor neural activities. However, the patient's quality of life must be taken into consideration, so wireless and implantable sensor networking technologies are extensively being explored. Systems such as this comprise an electroencephalography (EEG) signal sampled sensor and a remote processing center. But a problem with the total of data in the wireless transmission is the sensor design to reduce the energy consumed.

One solution to this problem would be compressive sensing (CS) because this technique can reduce the data transmitted. The complete system includes a low power front end (CS) and a remote server [67].

The front end is a compound with three main components, an arbitrary coding module, a quantization module, and a coarse-grained screening module. As shown in Fig. 11.1 the signal $\boldsymbol{x}$ from the gross dimension sensor n is compacted into m dimensions in the random coding module $\boldsymbol{y}$ [67].

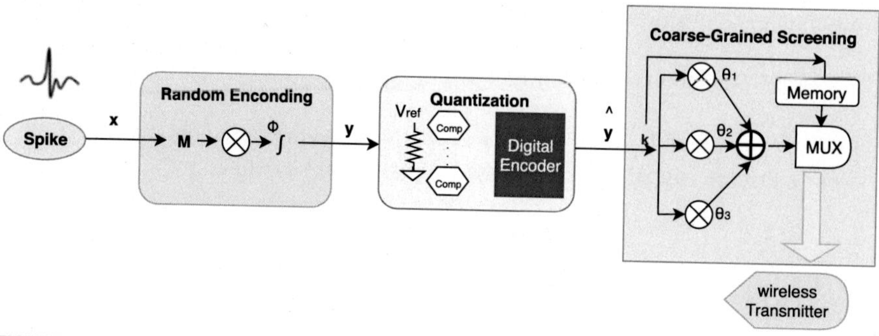

FIGURE 11.1

The block diagram of the selective CS structure for wireless implantable neural decoding.

The random coding module has m branches, each branch is a random combination for acquisition, where each branch completes this process.

In the quantization module, each comparator issues a binary decision to compare the signal with a voltage. Another element in this step is the digital encoder that arranges the ending and a $\hat{y}$ confidence score.

If the score is greater than the pre-quantization output, comparing these decisions, the coarse-grained screening module analyzes the set threshold. The wireless transmitter sends the final forecast to the distant server. If not, it transmits $\hat{y}$ compacted samples to the server in fine-grained form by a deep learning algorithm.

11.4 Compressive sensing in seizure detection

A very important part of the human body is the brain, which is responsible for controlling the central nervous system [68]. The physiological electroencephalogram (EEG) is a device capable of capturing signals as regards the state of the brain from which information is extracted. These signals are widely used for the analysis of different types of brain activities.

The third most common neurological disorder that affects more than 50 million people worldwide is epilepsy. It is a chronic neurological disorder characterized by repeated seizures. Fear of sudden and inexplicable death in epilepsy (SUDEP) is always a concern for epilepsy patients; it is believed that about a quarter of deaths in people with epilepsy are attributable to SUDEP [69]. Therefore, continually monitoring convulsive patients is not a trivial task, as such hospital monitoring is expensive and also frustrating for patients. As a promising market solution for seizure detection comes portable home monitoring [70,71].

Portable and wireless devices are each ever more utilized. The estimative is that one has 5 billion devices worldwide. However, a critical issue of using such devices is the energy expended by these devices [72].

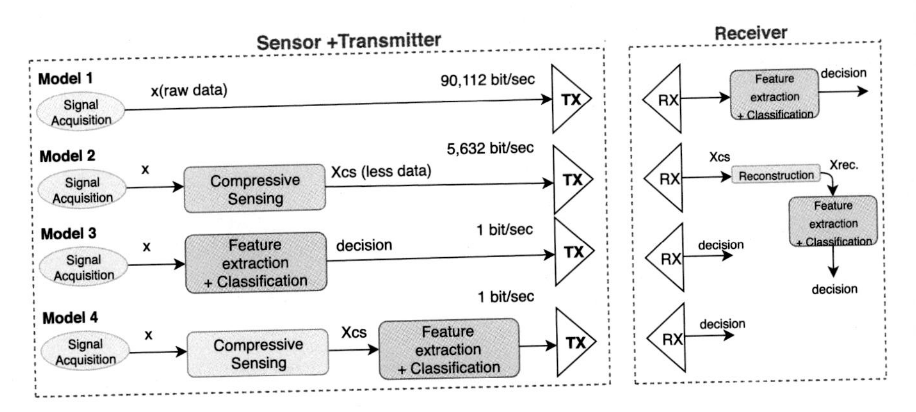

FIGURE 11.2

Four different system models that can be used in bio-sensor applications. We assume that the TX power scales with the data rate.

Power consumption in portable home monitoring systems for seizure detection is still a concern as models use wireless systems and use batteries as a power source. In view of this, the high energy cost of transmitting some data and the limited bandwidth of radios require an application of some data compression technique to reduce power consumption and data transfer rate [73].

Despite the great development of portable devices, power consumption is still a major concern. This is because of the wireless method to be utilized in data transmission, requiring the use of batteries. Thus, to reduce the consumption in data transmission, it is necessary to apply data compression techniques.

As different block diagrams show in Fig. 11.2, these blocks represent systems used for biomedical monitoring sensors. Assuming the TX power scales with data rate [69], analyzing the models, one can observe the advantages and disadvantages of each one. For example, model 1 shows a system that captures the AFE signal and sends the data through a transmitter. The disadvantage of this method is the power consumption due to the uncompressed transmission of data.

In model 2, the signal is compressed using the CS technique before it is transmitted. Despite reducing the transmission rate, this system is not yet suitable for wearable appliances, because it consumes a considerably high energy. Although the power consumption is still high, the disadvantage is the need of a reconstruction algorithm inserted into the receiving system.

Model 3 consumes less power than models 1 and 2 as it processes the entire signal at the sensor node and sends 1 bit/s, which concerns the final decision, through the transmitter and it may also be in sleep mode during processing.

Model 4 is proposed using the CS technique, which compresses the sensor data locally on it. This approach has some advantages to satisfy the reduction of system power consumption.

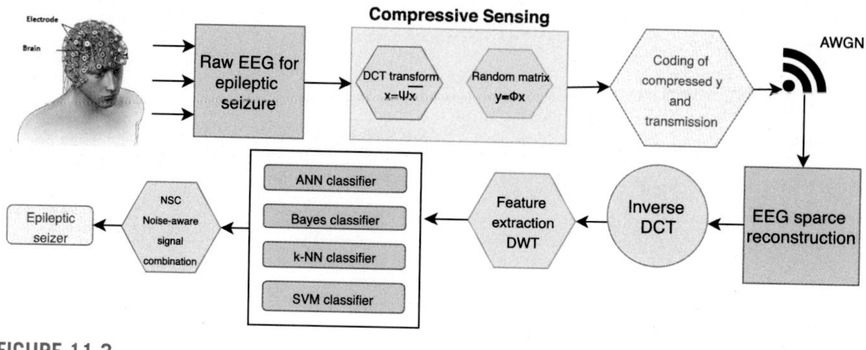

FIGURE 11.3

EEG-based epileptic seizure framework.

- The final decision is sent after the signal processing is done on the sensor.
- It is a more efficient system because the reconstruction algorithm is not utilized.
- Using CS, fewer input samples are used for resource extraction than the system shown in Model 3.

Fig. 11.3 shows another methodology, using a unified structure in which the EEG signal is compressed using a CS technique in the sensor sending through two types of channels [68]. In the first channel, it sends the signal through a silent channel, while in the second channel we have a Gaussian white noise additive wireless (AWGN) method.

The structured architecture in Fig. 11.3 shows that the system model is composed of two parts: transmitter and receiver. The transmitter has samples of raw EEG signals, represented by $\boldsymbol{x}$, ψ is the DCT basis for different amounts of m to get the $\hat{\boldsymbol{x}}$ compressed data that will be transmitted over silent channels and will be noisy. It will be considered that thermal noise using the AWGN model on the receiving side is the most used model to represent thermal noise.

From Fig. 11.3 it is possible to see that who receives a signal of size M can retrieve the data from the EEG using a DCT inverter (iDCT). A problem in the optimization algorithm with certain constraints is solved for CS.

11.5 Compressive sensing in Alzheimer's disease

Alzheimer's disease (AD) is considered the most common and fastest-growing neurological disease in the world [74]. One way to control and monitor patients, or to study the increased prevalence of AD is to use biomarkers for early diagnosis.

The annual conversion rate of normality to Alzheimer's dementia (AD) ranges from 0.2% to 4%, while that of mild cognitive impairment MCI to AD is between 6% and 25% [75–77]. It is an open question, with relevant clinical implications, whether or not ischemic cardiomyopathy (ICM) is a prodromal stage of AD [78].

A relatively inexpensive and effective way to detect early dementia and AD is the electroencephalogram (EEG), although it does not meet the specificity prescribed for clinical use. Portable wireless sensor-based EEG systems are used for monitoring, even with the disadvantage of battery life. Systems such as these require intensive recording of massive EEG data, increasing the need for efficient compression techniques; this compression may help to reduce data, but can also discriminate brain states to distinguish patients with different diseases, such as AD, individuals with mild cognitive impairment, cranial cruciate ligament (CCL) and healthy and normal elderly. An approach using wavelets and compressive sensing (CS) shows the relationships between the compression ratio and entropic measurements of EEG complexity, the reduction of which is considered the hallmark of AD.

11.6 Compressive electroencephalography (EEG) sensor design

For the use of EEG wireless sensors, efficient data representation and low power consumption should be obtained [79]. Several techniques are used for this type of application, such as principal component analysis (ICA) and neural networks. These techniques can improve the system reducing the noise and data volume.

One way to establish robust performance, considering the constraints of this application, is to adopt efficient sampling schemes and detection protocols, improving sensor and data efficiency [80]. This should be carefully considered when is considered wireless applications for storing a lot of data. The procedure used is based on compression sampling and multiplexing. Usually, the bases used are the Wavelet or Fourier series.

The EEG sensor consists of several pseudorandom sampling elements, providing signals with different spatial resolutions. To obtain the signals of each channel, a pseudorandom vector generator is used for control. The output signal is the result of summing the signals of each channel within a time window after each sample is encoded with a Bernoulli number. With this methodology you can reduce (1) the number of samples and the sampling rate, (2) throughput and power consumption, and (3) computational complexity of processing.

The proposed detection protocol in Ref. [80] is shown in Fig. 11.4. As stated earlier, the sensor sampling rates are controlled by a binary random vector generator. Thus, if the random vector generator output is ‘1’, the corresponding ADC channel is enabled and if the generator output is ‘0’, the corresponding ADC channel is disabled. The Bernoulli encoder is used for data acquisition and reconstruction, giving each sample a random appearance before the system provides the output. With compressed samples, data can be segmented into parts and used for further processing such as detection and classification and can be used for signal modeling and data compression. This is expected to be more efficient in acquiring EEG data and lower power consumption.

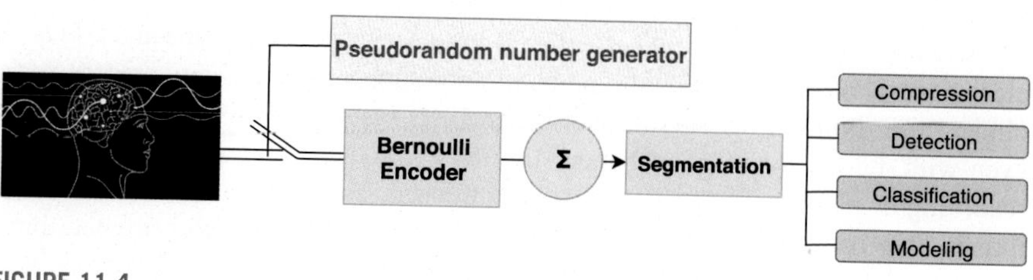

FIGURE 11.4

Illustration of the incoherent sensing protocol.

11.7 Compressive sensing in neural recording

One of the tools most used by neuroscientists to understand and study brain activity is neural recording microsystems consisting of one or more electrodes implanted in the cortex capable of collecting spikes generated by individual neurons, allowing to analyze their function and connectivity concerning brain circuits and its role in cognition [81,82].

However, one of the main challenges is the ability to transmit low power signal data. Spikes have a bandwidth of up to 10 kHz and amplitude ranging from 50 μV to 500 μV, using, for example, a Multi-Electrode Array (MEA), which contains hundreds of electrodes and is used to capture the brain's neuronal activities. To transmit this chipless data, which is of the order of megabytes per second, requires a power consumption of mV [83,84]. Using CS data is compressed before it is transmitted, reducing transmission power consumption.

Since the peak of each neuron has a unique shape, a dictionary used to increase the compression ratio (CR) can be built, capable of maintaining high-quality recovery in the structure of the CS [85]. The peak of each neuron has a characteristic shape and amplitude, and peak trains recorded on a single electrode contain peaks of several nearby neurons. Thus, peak waveforms can learn a signal-dependent dictionary and represent sparse peaks similar to those recorded on the same electrode. This technique allows CS to achieve comparable CR quality and recovery to the method based on extremely efficient wavelet transformation.

However, some changes may occur in the burning process. Thus, the dictionary used must be adaptable to accommodate the changes caused by the signal. Without this adaptation, the signal recovery quality would decrease over time as the dictionary would lose its ability to represent sparse peaks. One solution to this problem would be a closed-loop CS neural recording system [84], where the system includes application-specific integrated circuits (ASICs) containing four recording electrodes and compression circuits, a recovery algorithm off-chip real-time signal retrieval and, most importantly, a QE retrieval method that provides ASIC adaptive closed-loop feedback for better work between CR and retrieval quality [84], as illustrated by Fig. 11.5.

The system is used as a learned dictionary for CS neural recording. This system has a high compression ratio when applied to raw neural signals (CR > 10.6) and

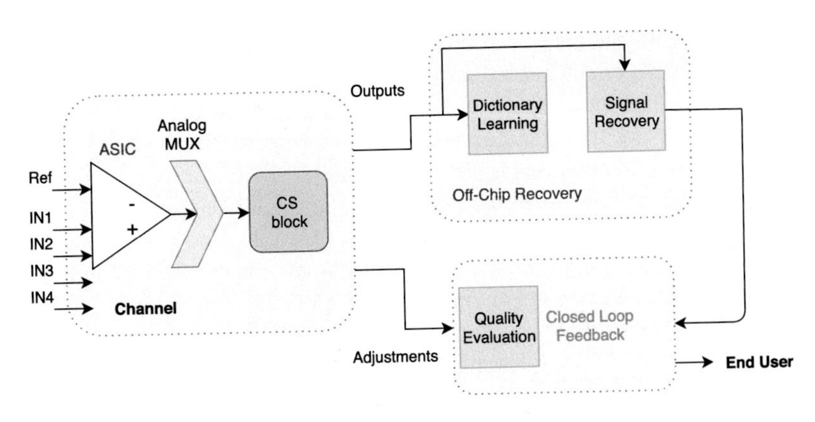

FIGURE 11.5

Summary of the neural recording ASIC, the off-chip recovery system and the closed-loop feedback linking two sub-systems.

detected peaks (CR > 16). This system has extremely low power (< 0.83 μW per electrode) and has a very small area (< 0.11 mm^2). In addition, it can be sized and integrated into large recording arrays using thousands of electrodes on the compaction sensor.

11.8 Compressive sensing for fall prevention

Due to its fragility, many elderly people are at risk of falling and consequently suffering from multiple pathologies. As scary as it sounds, one-third of people over 65 report falls, and this has become one of the leading causes of death either directly from falling or in its aftermath in older people. Health costs related to medical treatments are a major expense and need to be mitigated with low-cost solutions using preventive measures.

Biomedical technologies that can add to fall prevention and that do not disturb the individual's daily life can use sensors in plastic, paper or even fabric, so it will be possible to combine wireless communication and computing technology. Wearable care devices can be worn by individuals in a comfortable, non-invasive manner.

New wearable care systems, considering the risks associated with the fall, are under development. These devices can detect the risk of a fall by monitoring EEG signals and giving a warning before the actual fall happens [86].

In addition, the underfoot pressure distribution, cadence and stride length are associated with the risk of falling. To monitor the pressure under the feet and submit the fall risk assessment the Smart-Insole system can be used. In addition to providing fall risk assessment, this system can predict the fall potential by analyzing the inappropriate use of canes by the elderly and disabled.

For an efficient system, two procedures are required: one is the wireless data transmission between the client and the server, and the other is server-side data mining. It is important that the time of data transmission is fast, or the system response, because time is critical for the prevention of falls, if the information to send that will serve to predict the fall of the individual is slow. CS theory is an innovative method for high-efficiency data processing ideal for portable systems and for the study in question, fall prevention for the elderly [87].

Using the compression sensor, you can sub-sample the actual data with $m << n$ measurements and estimate $\boldsymbol{x}$. For example, consider a discrete signal $\boldsymbol{x}$, represented by a $n \times 1$ vector. In terms of a vector base, this signal can be represented by an $n \times 1$ basis $\Psi_i (i = 1, ..., n)$. Thus, to express the $\boldsymbol{x}$ as a linear combination of base vectors with coefficients α_i, the $n \times n$ basis matrix $\Psi = [\Psi_1, \Psi_2, ..., \Psi_N]$ is used:

$$\boldsymbol{x} = \sum_{i=1}^{n} \alpha_i \psi_i \tag{11.7}$$

where α_i, represented in orthogonal space or domain Ψ, is the weighting coefficient vector, which represents the bio-signal $\boldsymbol{x}$, represented in the temporal domain. Generally, the biological signal is represented in the k-sparse domain ($k << n$). Thus, the α_i vector has only large coefficients of k, making the $\boldsymbol{x}$ signal compressible.

A representation of the compressed signal can be acquired directly by the compression sensor, representing all useful information without having to use all n samples. Considering a linear measurement process, the calculation of $m < n$ internal product between $\boldsymbol{x}$ and the measurement vector can be defined as

$$\boldsymbol{y}_j = \langle \boldsymbol{x}, \boldsymbol{\phi}_j \rangle \quad \text{or} \quad \boldsymbol{y} = \Phi \boldsymbol{x} = \Phi \psi \boldsymbol{\alpha} = \Theta \boldsymbol{\alpha} \quad (\Theta \in \mathbb{R}^{m \times n}, \boldsymbol{y} \in \mathbb{R}^m). \tag{11.8}$$

Therefore, the bio-signal x can be represented by a set of compressed measures y_j, where the original sign x can be exclusively reconstructed by measures y_j. In addition, the basis for fall prevention analysis can be provided, enabling detection through only small-scale y_j measurements. As shown in Fig. 11.6, Ref. [86] presents a compression sensor-based biological signal processing system.

The original biological signals from the input go through the projection and measurement stages; at these stages the original signal transforms to the Ψ domain and is subsequently measured by the ϕ matrix. From the gross bio-signal $\boldsymbol{x}$ a small-scale representation of the $\boldsymbol{y}$ data is extracted, but this is equivalent. Consequently, the $\boldsymbol{y}$

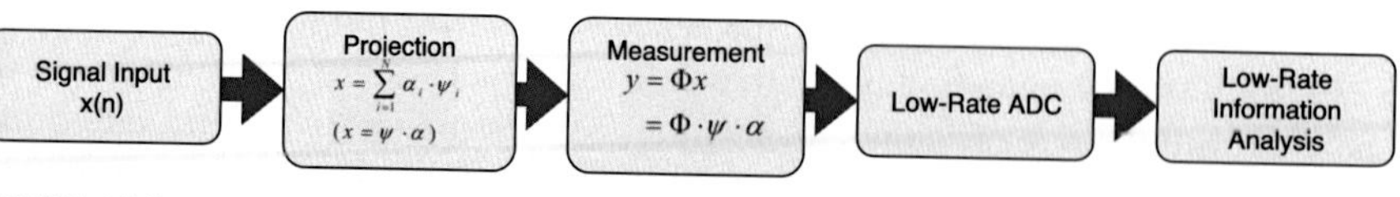

FIGURE 11.6

Overview of bio-signal acquisition system based on compressive sensing.

sign is sparse and has only m nonzero elements, if $k \geq m \log n/m$, one samples randomly from $\boldsymbol{y}$ providing enough information to reconstruct the signal with a minimal error [88].

By applying valid pattern recognition to low rate measurements rather than to the original sample signal, it is possible to detect the signature of an imminent fall through the bio-signal data. Based on the pattern recognition results a potential fall risk prediction can be provided and preventive measures can be taken to help the elderly recover their physical balance.

11.9 EEG compressive sensing for person identification

Identification systems of more than one method are used to ensure that only the legitimate user can access certain services. Several systems require reliable personal recognition schemes to confirm or determine the identity of a person requesting their services [89].

Biometric recognition is performed through its physiological and/or behavioral characteristics. Thus, biometric identification can also be done by EEG signals. This is a relatively new process that can distinguish people by comparing each individual's EEG signals [90].

Identifying people using sparse techniques, wearable devices that are of high quality and reliable for the consumer is due in the future. In addition to these characteristics, EEG identification must also be adjusted to the ubiquitous environment. The points to be followed by identification by EEG signals are: (i) reliable processing: shall include accuracy, security and robustness; (ii) real-time processing: response time must be fast; (iii) being wearable: the identification process should reduce unnecessary channels, avoiding additional energy and manufacturing costs [91].

Reference [91] gives a schematic of the four main elements in the measurement system configuration as shown in Fig. 11.7. All four steps are integrated into the complete interacting, sparse EEG detection system. This makes the application suitable for web-enabled applications.

The electrodes are sparsely distributed at selected positions according to the specific function of the cerebral cortex and the applications. An example of this type of application is the acquisition of a visual signal from the EEG. The electrodes in the visual cortex are distributed or activated, while the electrodes in the other cortex are deactivated or not used. For the reduction of information in data transmission and analysis, improving the dynamic performance of the EEG sensor, three considerations can be adopted:

1. Reduce the number of electrodes sparsely as defined above.
2. Add a data reduction block to reduce the sampling rate or perform discontinuous EEG recordings; see Fig. 11.7.
3. Block for data compression (CS) in order to minimize streaming data on the web; see Fig. 11.7.

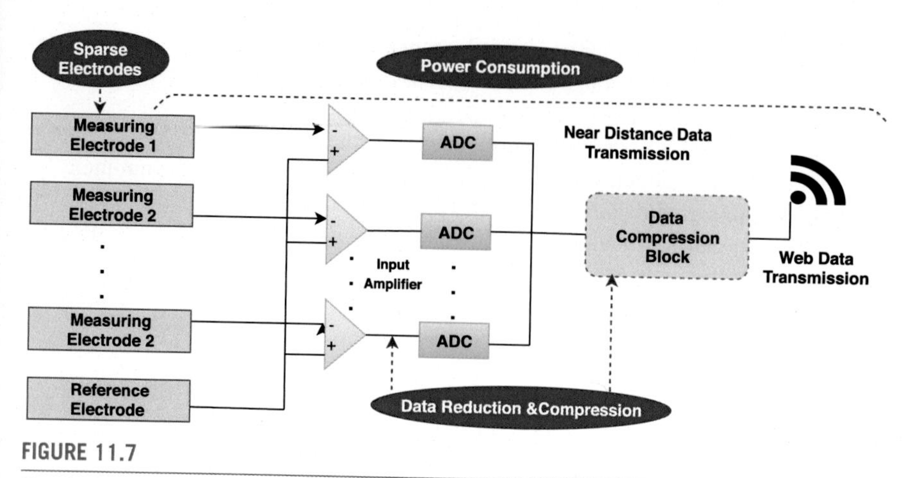

FIGURE 11.7

Fundamental elements of web-enabled sparse EEG sensing.

Data compression can be done in the data reduction block or in the data compression block, both located separately in the web-enabled front-end measurement system [92].

Power consumption of the running basic system is provided by the input amplifier, ADC, remote data transmission (per channel) and web data transmission, respectively. This power consumption is reduced by using the sparse structure with data reduction and compression blocks. In transmission, the front-end detection system must have an interconnected hardware and software structure to perform detection and decision. This final step of the project consists of three layers: detection layer, transmission layer, and decision layer. The decision and the corresponding command are passed from the decision layer to the detection layer, completing the feedback and control process. Using sparse EEG compression detection, it is possible to demonstrate the feasibility of identifying people.

11.10 Compressive sensing wireless neural recording

Wireless neural recording systems capable of capturing brain-implanted or scalp-related spikes, although subject to energy constraints, can increase understanding of brain function and increase user viability. Therefore, it is important to establish a way to reduce energy in the applications of this system. Thus, it should be noted that the power consumption is directly related to the radio transmitter, and it is necessary to reduce the data rate to have a lower power consumption [93].

To establish this data reduction we use the CS technique, which corresponds to the signal width and not to its bandwidth [94]. The application of CS in this system occurs in order to accentuate its main components, quantifying the information content by

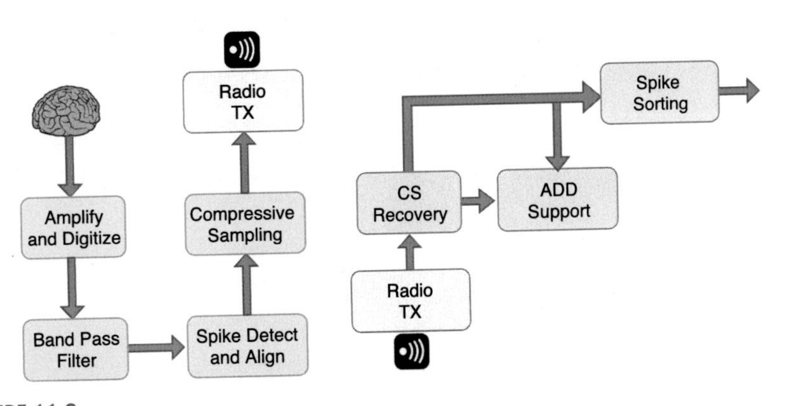

FIGURE 11.8

Schematic representation of a compressive wireless neural recording system, with half *in vivo* and half *ex vivo*.

estimating the number of significant coefficients. This reduces communication and bandwidth costs.

Reference [93] gives a schematic diagram of the compressive neural system as shown in Fig. 11.8. The implanted device has band filtering, peak detection and chip alignment steps to extract action potential waveforms. Analyzing the system, it is apparent that the waveform peak windows are sequentially encoded through a compression sensor block and then transmitted by a low-frequency radio.

To recover the spikes a "learned union of supports" is used in order to increase the sparsity in the reconstruction. The basis for peak classification is the small difference between the peaks, which depends on the source neurons. In the signal recovery process using a compression sensor, the signal must be on a weighted basis, where the weight indices are a union of the recovered peak support assemblies. The compressed sensor can provide high-quality reconstructions of the peak morphology, but power consumption is higher when compared to sending only peak citation features [95].

11.11 Compressive sensing in portable EEG systems

The need for human-computer interactions intrinsically requires some form of human physiological monitoring. A widely used tool for this is the electroencephalogram (EEG), capable of recording the signals on the micro-volt scale. This technology, however, for the user to have mobility and comfort, must be discreet, comfortable and durable. The main problem of the wireless EEG method, which theoretically concerns a more discreet and portable device, is the size and life of the battery, where the main consumption occurs in the wireless transmitter. Therefore, it is desirable to compress the raw EEG data in real time to reduce the amount of data to be transmitted and thus to increase battery operating life.

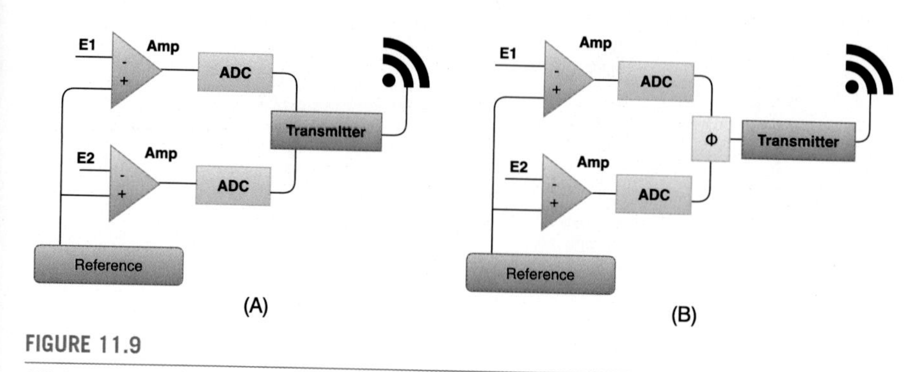

FIGURE 11.9

(A) A standard EEG unit. (B) An EEG unit incorporating compressive sensing.

A modern development in compression theory is compressive sensing (CS), capable of performing online data compression. This technique, as more channels are used, and as many systems generally use 128 or more channels, improves the compression detection scheme and can lead to a significant reduction in overall power consumption [52]. Fig. 11.9 shows a simplified power model of an EEG device with and without CS.

A simple model of an EEG system is shown in Fig. 11.9. In this system, there is an input instrumentation amplifier in order to amplify the small captured EEG signals. After the amplifier, an analog to digital converter (ADC) converts the signal to transmit it. Thus the C power consumption for channel (P_{sys}) is found:

$$P_{sys} = C(P_{Amp} + P_{ADC} + J f_s R) \tag{11.9}$$

where P_x is the power consumption of transmitter block from Fig. 11.9A, f_s is the sampling frequency, R is the number of bits per sample, J is the bit transmission network and $J F_s R$ provides the transmitter power consumption. It is assumed that the bandwidth limitation of the EEG signal is inserted into the instrumentation amplifier.

Modifications to incorporate detection using CS are illustrated in Fig. 11.9B. The detection process is implemented in the discrete domain and only one block is required to generate the ϕ measurements. This matrix is used to select a random group and form $\mathbf{y}$. The elements in ϕ represent a pseudorandom sequence after a specific probability distribution.

With these considerations, you can find the power consumption on each channel ($P_{sys_{cs}}$):

$$P_{sys_{cs}} = C(P_{Amp} + P_{ADC}) + P_{RNG} + P_{DSP} + P_{Sync} + J(\frac{m}{n} C f_s R + S). \tag{11.10}$$

In this configuration, the amplifier and power consumption remain unchanged. However, extra hardware, containing three components, is required: a random number generator (P_{RNG}), used to generate an array ϕ; a (P_{DSP}) microcontroller, used to perform matrix multiplications, and a (P_{Sync}) synchronization unit, used to prevent transmission of ϕ arrays. Concerning the last term, a matrix can be reconstructed at the receiver based on the pseudorandom sequence. Regardless of the number of system channels, only one of these blocks is required.

The feasibility of compression sensor-based systems in low power portable EEG equipment is evaluated by comparing (11.9) and (11.10). This analysis quantifies the energetic feasibility of using the sensor to reduce data online.

11.12 Compressed and distributed sensing of neuronal activity

A key neural communication mechanism is 'spikes' or peak trains, which are used to relay, process and store information in the central nervous system. Understanding spike information is a fundamental goal of neuroscience [65].

Peak trains for motor systems are said to carry important information about the execution of movements; therefore, they are indispensable in the development of neuroprotetics and in brain–machine interface (BMI) technology to help people suffering from serious diseases [96,97].

Peak trains of motor cortical neurons are recorded over a very short range, usually between 100–200 ms, and cortically controlled IMC systems are based essentially on instantaneous 'spike' decoding.

Fig. 11.10 shows instant decoding, which is the process step where amplification and filtering can be observed after binary peak detection and classification is performed [65]. After this process, a Gaussian function is applied to filter the peaks, producing a smooth estimate of the instantaneous firing rate. All of these steps must

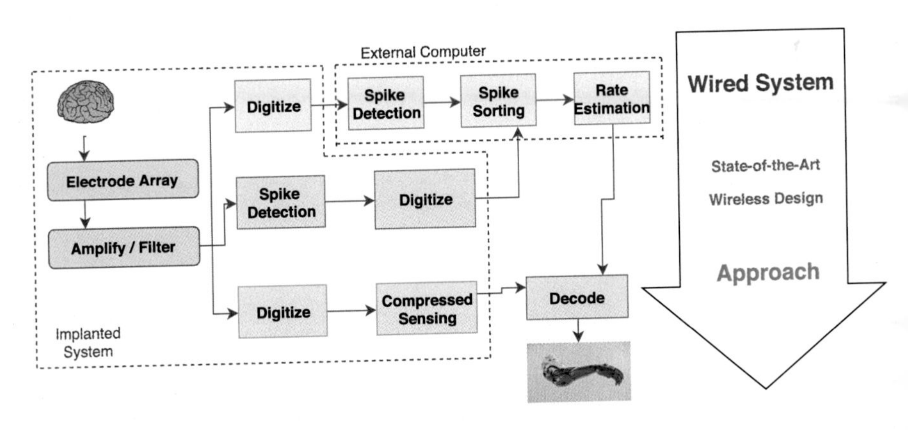

FIGURE 11.10

Schematic diagram of a typical data flow in a neuro-motor prosthetic application. Ensemble neural recordings are first amplified and filtered prior to telemetry transmission to the outside world.

be performed within the movement preparation period to allow the subject to experience natural motor behavior.

The next step in the process requires that the peak be classified and analyzed in two modes: one of training and one of runtime. Based on certain characteristics peaks are detected and aligned in training mode. During runtime, to know which neuronal class it belongs to the resources of an observed peak are compared to the stored resources. This identification/classification process must have a wired brain connection to allow neural data to be streamed to a computer.

Fig. 11.10 shows the sparse representation diagram for better unit separation when compared with classical PCA techniques. In this case, it is not necessary to completely reconstruct the time domain peak waveforms. This is because performance can be achieved using a peak rating. Thus, there is a significant reduction in the computational and communication costs of implantable neural prosthetic systems. Under these conditions, performance and potential use in clinical applications can be significantly improved.

11.13 Summary

This chapter reviews compressive sensing in neuro-signal acquisition. Because of the high efficiency of compressive sensing in data sampling with a much smaller need for sensory devices, much less memory storage, higher data transmission rate, many times less power consumption, etc., it has been used in different fields of healthcare engineering. This chapter reviews the compressive sensing application aspects in neuro-signal sampling concerning the designs for energy efficiency, seizure detection, Alzheimer disease analysis, electroencephalography, neural recording, fall prevention, EEG-based person identification, wireless neural recording, and portable EEG systems.

References

[1] D. Craven, B. McGinley, L. Kilmartin, M. Glavin, E. Jones, Compressed sensing for bioelectric signals: a review, IEEE Journal of Biomedical and Health Informatics 19 (2) (2014) 529–540.

[2] T.W. Cabral, M. Khosravy, F.M. Dias, H.L.M. Monteiro, M.A.A. Lima, L.R.M. Silva, R. Naji, C.A. Duque, Compressive sensing in medical signal processing and imaging systems, in: Sensors for Health Monitoring, Elsevier, 2019, pp. 69–92.

[3] S. Gupta, M. Khosravy, N. Gupta, H. Darbari, N. Patel, Hydraulic system onboard monitoring and fault diagnostic in agricultural machine, Brazilian Archives of Biology and Technology 62 (2019).

[4] S. Gupta, M. Khosravy, N. Gupta, H. Darbari, In-field failure assessment of tractor hydraulic system operation via pseudospectrum of acoustic measurements, Turkish Journal of Electrical Engineering & Computer Sciences 27 (4) (2019) 2718–2729.

[5] M.H. Sedaaghi, R. Daj, M. Khosravi, Mediated morphological filters, in: 2001 International Conference on Image Processing, 2001, Proceedings, vol. 3, IEEE, 2001, pp. 692–695.

[6] M. Khosravy, N. Gupta, N. Marina, I.K. Sethi, M.R. Asharif, Morphological filters: an inspiration from natural geometrical erosion and dilation, in: Nature-Inspired Computing and Optimization, Springer, Cham, 2017, pp. 349–379.

[7] M. Khosravy, M.R. Asharif, K. Yamashita, A PDF-matched short-term linear predictability approach to blind source separation, International Journal of Innovative Computing, Information & Control (IJICIC) 5 (11) (2009) 3677–3690.

[8] M. Khosravy, M.R. Asharif, K. Yamashita, A theoretical discussion on the foundation of Stone's blind source separation, Signal, Image and Video Processing 5 (3) (2011) 379–388.

[9] M. Khosravy, M.R. Asharif, K. Yamashita, A probabilistic short-length linear predictability approach to blind source separation, in: 23rd International Technical Conference on Circuits/Systems, Computers and Communications (ITC-CSCC 2008), Yamaguchi, Japan, ITC-CSCC, 2008, pp. 381–384.

[10] M. Khosravy, M.R. Alsharif, K. Yamashita, A PDF-matched modification to Stone's measure of predictability for blind source separation, in: International Symposium on Neural Networks, Springer, Berlin, Heidelberg, 2009, pp. 219–222.

[11] M. Khosravy, Blind source separation and its application to speech, image and MIMO-OFDM communication systems, PhD thesis, University of the Ryukyus, Japan, 2010.

[12] M. Khosravy, M. Gupta, M. Marina, M.R. Asharif, F. Asharif, I. Sethi, Blind components processing a novel approach to array signal processing: a research orientation, in: 2015 International Conference on Intelligent Informatics and Biomedical Sciences, ICIIBMS, 2015, pp. 20–26.

[13] M. Khosravy, N. Punkoska, F. Asharif, M.R. Asharif, Acoustic OFDM data embedding by reversible Walsh–Hadamard transform, AIP Conference Proceedings 1618 (2014) 720–723.

[14] M.H. Sedaaghi, M. Khosravi, Morphological ECG signal preprocessing with more efficient baseline drift removal, in: 7th IASTED International Conference, ASC, 2003, pp. 205–209.

[15] N. Dey, S. Samanta, X.-S. Yang, A. Das, S.S. Chaudhuri, Optimisation of scaling factors in electrocardiogram signal watermarking using cuckoo search, International Journal of Bio-Inspired Computation 5 (5) (2013) 315–326.

[16] M. Khosravy, M.R. Asharif, M.H. Sedaaghi, Morphological adult and fetal ECG preprocessing: employing mediated morphology, IEICE Technical Report, IEICE 107 (2008) 363–369.

[17] N. Dey, A.S. Ashour, F. Shi, S.J. Fong, R.S. Sherratt, Developing residential wireless sensor networks for ECG healthcare monitoring, IEEE Transactions on Consumer Electronics 63 (4) (2017) 442–449.

[18] M. Khosravi, M.H. Sedaaghi, Impulsive noise suppression of electrocardiogram signals with mediated morphological filters, in: 11th Iranian Conference on Biomedical Engineering, ICBME, 2004, pp. 207–212.

[19] N. Dey, S. Mukhopadhyay, A. Das, S.S. Chaudhuri, Analysis of P-QRS-T components modified by blind watermarking technique within the electrocardiogram signal for authentication in wireless telecardiology using DWT, International Journal of Image, Graphics and Signal Processing 4 (7) (2012) 33.

[20] M. Khosravy, N. Gupta, N. Patel, T. Senjyu, C.A. Duque, Particle swarm optimization of morphological filters for electrocardiogram baseline drift estimation, in: Applied Nature-Inspired Computing: Algorithms and Case Studies, Springer, 2020, pp. 1–21.

[21] E. Santos, M. Khosravy, M.A. Lima, A.S. Cerqueira, C.A. Duque, A. Yona, High accuracy power quality evaluation under a colored noisy condition by filter bank ESPRIT, Electronics 8 (11) (2019) 1259.

[22] E. Santos, M. Khosravy, M.A. Lima, A.S. Cerqueira, C.A. Duque, ESPRIT associated with filter bank for power-line harmonics, sub-harmonics and inter-harmonics parameters estimation, International Journal of Electrical Power & Energy Systems 118 (105) (2020) 731.

[23] K. Melo, M. Khosravy, C. Duque, N. Dey, Chirp code deterministic compressive sensing: analysis on power signal, in: 4th International Conference on Information Technology and Intelligent Transportation Systems, IOS Press, 2020, pp. 125–134.

[24] A.A. Picorone, T.R. de Oliveira, R. Sampaio-Neto, M. Khosravy, M.V. Ribeiro, Channel characterization of low voltage electric power distribution networks for PLC applications based on measurement campaign, International Journal of Electrical Power & Energy Systems 116 (105) (2020) 554.

[25] M. Foth, R. Schroeter, J. Ti, Opportunities of public transport experience enhancements with mobile services and urban screens, International Journal of Ambient Computing and Intelligence (IJACI) 5 (1) (2013) 1–18.

[26] N. Gupta, M. Khosravy, N. Patel, T. Senjyu, A bi-level evolutionary optimization for coordinated transmission expansion planning, IEEE Access 6 (2018) 48455–48477.
[27] N. Gupta, M. Khosravy, K. Saurav, I.K. Sethi, N. Marina, Value assessment method for expansion planning of generators and transmission networks: a non-iterative approach, Electrical Engineering 100 (3) (2018) 1405–1420.
[28] M. Yamin, A.A.A. Sen, Improving privacy and security of user data in location based services, International Journal of Ambient Computing and Intelligence (IJACI) 9 (1) (2018) 19–42.
[29] C.E. Gutierrez, P.M.R. Alsharif, M. Khosravy, P.K. Yamashita, P.H. Miyagi, R. Villa, Main large data set features detection by a linear predictor model, AIP Conference Proceedings 1618 (2014) 733–737.
[30] C. Castelfranchi, G. Pezzulo, L. Tummolini, Behavioral implicit communication (BIC): communicating with smart environments, International Journal of Ambient Computing and Intelligence (IJACI) 2 (1) (2010) 1–12.
[31] G.V. Kale, V.H. Patil, A study of vision based human motion recognition and analysis, International Journal of Ambient Computing and Intelligence (IJACI) 7 (2) (2016) 75–92.
[32] S. Hemalatha, S.M. Anouncia, Unsupervised segmentation of remote sensing images using FD based texture analysis model and isodata, International Journal of Ambient Computing and Intelligence (IJACI) 8 (3) (2017) 58–75.
[33] M. Khosravy, N. Gupta, N. Marina, I. Sethi, M. Asharifa, Perceptual adaptation of image based on Chevreul–Mach bands visual phenomenon, IEEE Signal Processing Letters 24 (5) (2017) 594–598.
[34] A.S. Ashour, S. Samanta, N. Dey, N. Kausar, W.B. Abdessalemkaraa, A.E. Hassanien, Computed tomography image enhancement using cuckoo search: a log transform based approach, Journal of Signal and Information Processing 6 (03) (2015) 244.
[35] M. Khosravy, N. Gupta, N. Marina, I. Sethi, M. Asharif, Brain action inspired morphological image enhancement, in: Nature-Inspired Computing and Optimization, Springer, Cham, 2017, pp. 381–407.
[36] M. Khosravy, M.R. Asharif, M.H. Sedaaghi, Medical image noise suppression: using mediated morphology, IEICE, Technical Report 107 (461) (2008) 265–270.
[37] N. Dey, A.S. Ashour, A.S. Ashour, A. Singh, Digital analysis of microscopic images in medicine, Journal of Advanced Microscopy Research 10 (1) (2015) 1–13.
[38] M. Khosravy, M.R. Asharif, M.H. Sedaaghi, Medical image noise suppression using mediated morphology, in: IEICE Tech. Rep., IEICE, 2008, pp. 265–270.
[39] B. Alenljung, J. Lindblom, R. Andreasson, T. Ziemke, User experience in social human–robot interaction, in: Rapid Automation: Concepts, Methodologies, Tools, and Applications, IGI Global, 2019, pp. 1468–1490.
[40] M. Khosravy, M.R. Alsharif, M. Khosravi, K. Yamashita, An optimum pre-filter for ICA based multi-input multi-output OFDM system, in: 2010 2nd International Conference on Education Technology and Computer, vol. 5, IEEE, 2010, pp. V5–129.
[41] F. Asharif, S. Tamaki, M.R. Alsharif, H. Ryu, Performance improvement of constant modulus algorithm blind equalizer for 16 QAM modulation, International Journal on Innovative Computing, Information and Control 7 (4) (2013) 1377–1384.
[42] M. Khosravy, M.R. Alsharif, K. Yamashita, An efficient ICA based approach to multiuser detection in MIMO OFDM systems, in: Multi-Carrier Systems & Solutions 2009, Springer, 2009, pp. 47–56.
[43] M. Khosravy, M.R. Alsharif, B. Guo, H. Lin, K. Yamashita, A robust and precise solution to permutation indeterminacy and complex scaling ambiguity in BSS-based blind MIMO-OFDM receiver, in: International Conference on Independent Component Analysis and Signal Separation, Springer, 2009, pp. 670–677.
[44] M. Khosravy, A blind ICA based receiver with efficient multiuser detection for multi-input multi-output OFDM systems, in: The 8th International Conference on Applications and Principles of Information Science (APIS), Okinawa, Japan, 2009, 2009, pp. 311–314.
[45] M. Khosravy, S. Kakazu, M.R. Alsharif, K. Yamashita, Multiuser data separation for short message service using ICA, SIP, IEICE Technical Report 109 (435) (2010) 113–117.
[46] M. Baumgarten, M.D. Mulvenna, N. Rooney, J. Reid, Keyword-based sentiment mining using Twitter, International Journal of Ambient Computing and Intelligence (IJACI) 5 (2) (2013) 56–69.

[47] C.E. Gutierrez, M.R. Alsharif, K. Yamashita, M. Khosravy, A tweets mining approach to detection of critical events characteristics using random forest, International Journal of Next-Generation Computing 5 (2) (2014) 167–176.
[48] N. Kausar, S. Palaniappan, B.B. Samir, A. Abdullah, N. Dey, Systematic analysis of applied data mining based optimization algorithms in clinical attribute extraction and classification for diagnosis of cardiac patients, in: Applications of Intelligent Optimization in Biology and Medicine, Springer, 2016, pp. 217–231.
[49] P. Sosnin, Precedent-oriented approach to conceptually experimental activity in designing the software intensive systems, International Journal of Ambient Computing and Intelligence (IJACI) 7 (1) (2016) 69–93.
[50] M. Lustig, D. Donoho, J. Santos, J. Pauly, Compressed sensing MRI, IEEE Signal Processing (2008).
[51] E.J. Candès, M.B. Wakin, An introduction to compressive sampling [a sensing/sampling paradigm that goes against the common knowledge in data acquisition], IEEE Signal Processing Magazine 25 (2) (2008) 21–30.
[52] A.M. Abdulghani, A.J. Casson, E. Rodriguez-Villegas, Quantifying the feasibility of compressive sensing in portable electroencephalography systems, in: International Conference on Foundations of Augmented Cognition, Springer, 2009, pp. 319–328.
[53] C. Song, A. Wang, F. Lin, J. Xiao, X. Yao, W. Xu, Selective CS: an energy-efficient sensing architecture for wireless implantable neural decoding, IEEE Journal on Emerging and Selected Topics in Circuits and Systems 8 (2) (2018) 201–210.
[54] M. Khosravy, N. Gupta, N. Patel, T. Senjyu, Frontier Applications of Nature Inspired Computation, Springer, 2020.
[55] N. Gupta, M. Khosravy, O.P. Mahela, N. Patel, Plant biology-inspired genetic algorithm: superior efficiency to firefly optimizer, in: Applications of Firefly Algorithm and Its Variants, Springer, 2020, pp. 193–219.
[56] C. Moraes, E. De Oliveira, M. Khosravy, L. Oliveira, L. Honório, M. Pinto, A hybrid bat-inspired algorithm for power transmission expansion planning on a practical Brazilian network, in: Applied Nature-Inspired Computing: Algorithms and Case Studies, Springer, 2020, pp. 71–95.
[57] G. Singh, N. Gupta, M. Khosravy, New crossover operators for real coded genetic algorithm (RCGA), in: 2015 International Conference on Intelligent Informatics and Biomedical Sciences (ICIIBMS), IEEE, 2015, pp. 135–140.
[58] N. Gupta, N. Patel, B.N. Tiwari, M. Khosravy, Genetic algorithm based on enhanced selection and log-scaled mutation technique, in: Proceedings of the Future Technologies Conference, Springer, 2018, pp. 730–748.
[59] N. Gupta, M. Khosravy, N. Patel, I. Sethi, Evolutionary optimization based on biological evolution in plants, Procedia Computer Science 126 (2018) 146–155.
[60] J. Kaliannan, A. Baskaran, N. Dey, A.S. Ashour, M. Khosravy, R. Kumar, ACO based control strategy in interconnected thermal power system for regulation of frequency with HAE and UPFC unit, in: International Conference on Data Science and Application (ICDSA-2019), in: LNNS, Springer, 2019.
[61] N. Gupta, M. Khosravy, N. Patel, O. Mahela, G. Varshney, Plants genetics inspired evolutionary optimization: a descriptive tutorial, in: Frontier Applications of Nature Inspired Computation, Springer, 2020, pp. 53–77.
[62] M. Khosravy, N. Gupta, N. Patel, O. Mahela, G. Varshney, Tracing the points in search space in plants biology genetics algorithm optimization, in: Frontier Applications of Nature Inspired Computation, Springer, 2020, pp. 180–195.
[63] N. Gupta, M. Khosravy, N. Patel, S. Gupta, G. Varshney, Evolutionary artificial neural networks: comparative study on state of the art optimizers, in: Frontier Applications of Nature Inspired Computation, Springer, 2020, pp. 302–318.
[64] N. Gupta, M. Khosravy, N. Patel, S. Gupta, G. Varshney, Artificial neural network trained by plant genetics-inspired optimizer, in: Frontier Applications of Nature Inspired Computation, Springer, 2020, pp. 266–280.
[65] M. Aghagolzadeh, K. Oweiss, Compressed and distributed sensing of neuronal activity for real time spike train decoding, IEEE Transactions on Neural Systems and Rehabilitation Engineering 17 (2) (2009) 116–127.

[66] M.S. Lewicki, A review of methods for spike sorting: the detection and classification of neural action potentials, Network: Computation in Neural Systems 9 (4) (1998) R53–R78.
[67] A. Wang, C. Song, X. Xu, F. Lin, Z. Jin, W. Xu, Selective and compressive sensing for energy-efficient implantable neural decoding, in: 2015 IEEE Biomedical Circuits and Systems Conference (BioCAS), IEEE, 2015, pp. 1–4.
[68] K. Abualsaud, M. Mahmuddin, M. Saleh, A. Mohamed, Ensemble classifier for epileptic seizure detection for imperfect EEG data, The Scientific World Journal 2015 (2015).
[69] A. Jafari, A. Page, C. Sagedy, E. Smith, T. Mohsenin, A low power seizure detection processor based on direct use of compressively-sensed data and employing a deterministic random matrix, in: 2015 IEEE Biomedical Circuits and Systems Conference (BioCAS), IEEE, 2015, pp. 1–4.
[70] G. Ullah, S.J. Schiff, Models of epilepsy, Scholarpedia 4 (7) (2009) 1409.
[71] A. Page, S.P.T. Oates, T. Mohsenin, An ultra low power feature extraction and classification system for wearable seizure detection, in: 2015 37th Annual International Conference of the IEEE Engineering in Medicine and Biology Society (EMBC), IEEE, 2015, pp. 7111–7114.
[72] K. Abualsaud, M. Mahmuddin, R. Hussein, A. Mohamed, Performance evaluation for compression-accuracy trade-off using compressive sensing for EEG-based epileptic seizure detection in wireless tele-monitoring, in: 2013 9th International Wireless Communications and Mobile Computing Conference (IWCMC), IEEE, 2013, pp. 231–236.
[73] F. Chen, A.P. Chandrakasan, V.M. Stojanovic, Design and analysis of a hardware-efficient compressed sensing architecture for data compression in wireless sensors, IEEE Journal of Solid-State Circuits 47 (3) (2012) 744–756.
[74] F.C. Morabito, D. Labate, A. Bramanti, F. La Foresta, G. Morabito, I. Palamara, H.H. Szu, Enhanced compressibility of EEG signal in Alzheimer's disease patients, IEEE Sensors Journal 13 (9) (2013) 3255–3262.
[75] A. Abbott, Cognition: the brain's decline: treating cognitive problems common in elderly people requires a deeper understanding of how a healthy brain ages, Nature 492 (7427) (2012).
[76] M.P. Mattson, Pathways towards and away from Alzheimer's disease, Nature 430 (7000) (2004) 631.
[77] C. Babiloni, R. Ferri, G. Binetti, F. Vecchio, G.B. Frisoni, B. Lanuzza, C. Miniussi, F. Nobili, G. Rodriguez, F. Rundo, et al., Directionality of EEG synchronization in Alzheimer's disease subjects, Neurobiology of Aging 30 (1) (2009) 93–102.
[78] R.C. Petersen, G.E. Smith, S.C. Waring, R.J. Ivnik, E.G. Tangalos, E. Kokmen, Mild cognitive impairment: clinical characterization and outcome, Archives of Neurology 56 (3) (1999) 303–308.
[79] J.P. Carmo, N.S. Dias, H.R. Silva, P.M. Mendes, C. Couto, J.H. Correia, A 2.4-GHz low-power/low-voltage wireless plug-and-play module for EEG applications, IEEE Sensors Journal 7 (11) (2007) 1524–1531.
[80] Q. Hao, F. Hu, A compressive electroencephalography (EEG) sensor design, in: SENSORS, 2010 IEEE, IEEE, 2010, pp. 318–322.
[81] S. Mitra, J. Putzeys, C.M. Lopez, C.A. Pennartz, R.F. Yazicioglu, 24 channel dual-band wireless neural recorder with activity-dependent power consumption, Analog Integrated Circuits and Signal Processing 83 (3) (2015) 317–329.
[82] C.M. Lopez, A. Andrei, S. Mitra, M. Welkenhuysen, W. Eberle, C. Bartic, R. Puers, R.F. Yazicioglu, G.G. Gielen, An implantable 455-active-electrode 52-channel CMOS neural probe, IEEE Journal of Solid-State Circuits 49 (1) (2013) 248–261.
[83] J. Zhang, Y. Suo, S. Mitra, S.P. Chin, S. Hsiao, R.F. Yazicioglu, T.D. Tran, R. Etienne-Cummings, An efficient and compact compressed sensing microsystem for implantable neural recordings, IEEE Transactions on Biomedical Circuits and Systems 8 (4) (2014) 485–496.
[84] J. Zhang, S. Mitra, Y. Suo, A. Cheng, T. Xiong, F. Michon, M. Welkenhuysen, F. Kloosterman, P.S. Chin, S. Hsiao, et al., A closed-loop compressive-sensing-based neural recording system, Journal of Neural Engineering 12 (3) (2015) 036005.
[85] Y. Suo, J. Zhang, R. Etienne-Cummings, T.D. Tran, S. Chin, Energy-efficient two-stage compressed sensing method for implantable neural recordings, in: 2013 IEEE Biomedical Circuits and Systems Conference (BioCAS), IEEE, 2013, pp. 150–153.

[86] W. Xu, F. Gong, L. He, M. Sarrafzadeh, Wearable assistive system design for fall prevention, in: Joint Workshop on High Confidence Medical Devices, Software, Systems & Medical Device Plug-and-Play Interoperability (HCMDSS/MDPnP'11), Chicago, USA, 2011, pp. 1–8.
[87] D.L. Donoho, et al., Compressed sensing, IEEE Transactions on Information Theory 52 (4) (2006) 1289–1306.
[88] E.J. Candes, J.K. Romberg, T. Tao, Stable signal recovery from incomplete and inaccurate measurements, Communications on Pure and Applied Mathematics 59 (8) (2006) 1207–1223.
[89] A.K. Jain, A. Ross, S. Prabhakar, et al., An introduction to biometric recognition, IEEE Transactions on Circuits and Systems for Video Technology 14 (1) (2004).
[90] M. Poulos, M. Rangoussi, V. Chrissikopoulos, A. Evangelou, Person identification based on parametric processing of the EEG, in: ICECS'99, Proceedings of ICECS'99, 6th IEEE International Conference on Electronics, Circuits and Systems (Cat. No. 99EX357), vol. 1, IEEE, 1999, pp. 283–286.
[91] Y. Dai, X. Wang, X. Li, Y. Tan, Sparse EEG compressive sensing for web-enabled person identification, Measurement 74 (2015) 11–20.
[92] A.J. Casson, D.C. Yates, S.J. Smith, J.S. Duncan, E. Rodriguez-Villegas, Wearable electroencephalography, IEEE Engineering in Medicine and Biology Magazine 29 (3) (2010) 44–56.
[93] Z. Charbiwala, V. Karkare, S. Gibson, D. Markovic, M.B. Srivastava, Compressive sensing of neural action potentials using a learned union of supports, in: 2011 International Conference on Body Sensor Networks, IEEE, 2011, pp. 53–58.
[94] E. Candes, T. Tao, Decoding by linear programming, IEEE Transactions on Information Theory 51 (12) (2005) 4203–4215.
[95] V. Karkare, S. Gibson, D. Markovic, A 130-μW, 64-channel neural spike-sorting DSP chip, IEEE Journal of Solid-State Circuits 46 (5) (2011) 1214–1222.
[96] L.R. Hochberg, M.D. Serruya, G.M. Friehs, J.A. Mukand, M. Saleh, A.H. Caplan, A. Branner, D. Chen, R.D. Penn, J.P. Donoghue, Neuronal ensemble control of prosthetic devices by a human with tetraplegia, Nature 442 (7099) (2006) 164.
[97] D.M. Taylor, S.I.H. Tillery, A.B. Schwartz, Direct cortical control of 3d neuroprosthetic devices, Science 296 (5574) (2002) 1829–1832.

CHAPTER

12 Level-crossing sampling: principles, circuits, and processing for healthcare applications

Nassim Ravanshad, Hamidreza Rezaee-Dehsorkh

Faculty of Electrical and Biomedical Engineering, Sadjad University of Technology, Mashhad, Iran

12.1 Introduction

Restriction of the power consumption is the main issue in design of most of the ambulatory data-acquisition systems, which are faced in several applications such as portable biopotential signal acquisition. The whole process or some parts of the conditioning (amplification and filtering), transmission and processing must be powered with a battery with a limited size, wireless power transmission or even self-powered techniques, none of which can provide a large amount of power in long time intervals. In most of these systems a huge amount of data from multiple sources must be recorded and processes simultaneously. Most of the power is consumed in transmitting and processing of these data and the power consumption of these blocks are directly proportional to the data volume which is provided by the analog-to-digital converter (ADC).

The analog-to-digital converter is one of the main blocks in most of the conventional data-acquisition systems which converts the amplified and filtered input signal from the analog domain to the digital domain in order to be transmitted to and/or processed by digital processors as shown in Fig. 12.1. Analog to digital conversion is conventionally composed of three main steps which are sampling, quantization and coding. In the first step, some samples are taken from the input signal, which makes a discrete-time version of the continuous-time input signal. A conventional uniform (synchronous) sampling scheme is shown in Fig. 12.2. Each sample is defined by a sample amplitude and a sampling instant. The second step is to round off the sample amplitude and instant to some specified quantization levels and so demonstrating the samples with a limited accuracy. This makes the storing of the digital data possible. An example of the amplitude quantization is shown in Fig. 12.2. In the last step, binary representation of the instant and/or amplitude of the samples is provided to be analyzed by the digital processors.

The volume of the data which is provided by the ADC is related to the type of the sampling used in the analog-to-digital conversion procedure. Synchronous sam-

Compressive Sensing in Healthcare. https://doi.org/10.1016/B978-0-12-821247-9.00017-2

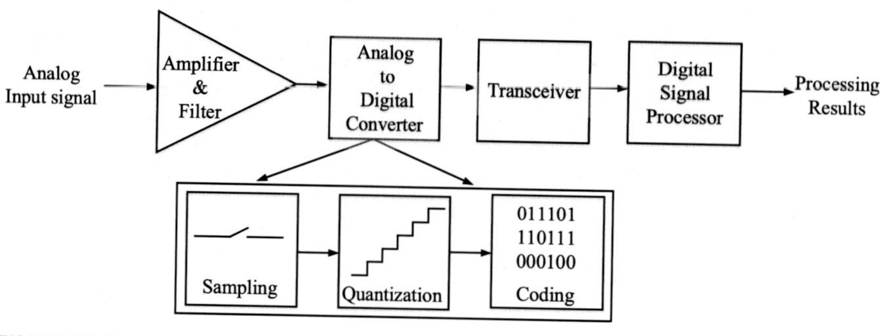

FIGURE 12.1

A conventional data-acquisition system.

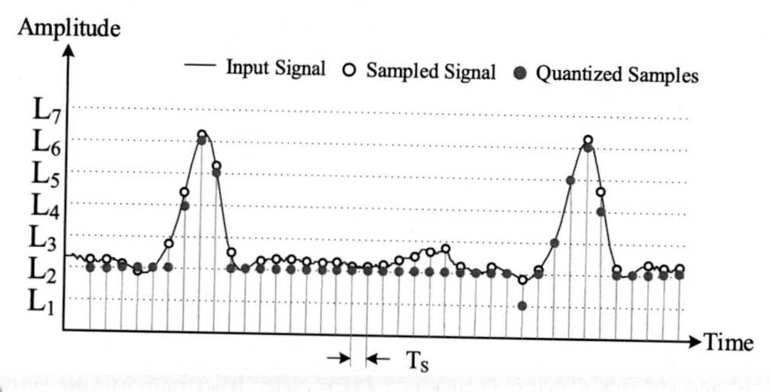

FIGURE 12.2

A conventional uniform (synchronous) sampling scheme.

pling is used as the most common sampling scheme in which the samples are taken uniformly in time with a constant rate. This rate is specified based on the Nyquist theorem to at least twice the maximum important frequency content of the input signal. Although this type of sampling has some advantages such as simplicity, it does not seem to provide an efficient volume of data in the case of the sparse and burst-like signals.

Sparse signals contain small durations of activity and are inactive in the rest of the times. These signals can be faced in many applications such as temperature, speech, accelerometer, and several biopotential signals such as heart and neural signals. Fig. 12.3A shows examples of these signals. Since the constant sampling rate is specified based on the active parts of the input signal in synchronous sampling, too many useless samples are taken from the inactive parts of the signal which usually does not contain any meaningful information. Transferring and processing of this huge amount of useless data means a complete waste of the power and memory.

Amplitude quantization is performed in conventional synchronous ADCs. As can be observed in Fig. 12.2, the accuracy of the samples is limited to the number of the quantization levels. Maximum of 2^M quantization levels can be considered for

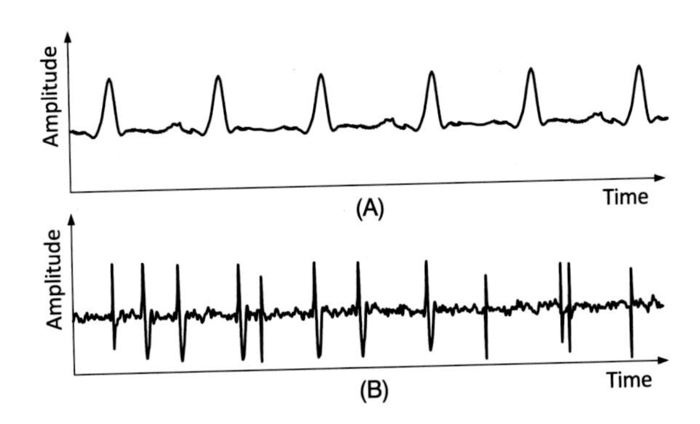

FIGURE 12.3

(A) Sample ECG signal. (B) Sample neural signal.

an M-bit representation of the sample amplitude, and so the accuracy is limited to the value of M. This means larger data volume (bits/s) is provided if more accurate samples are required, which is a serious disadvantage of synchronous sampling.

Level-crossing sampling can be considered as an alternative sampling scheme which results in a more efficient number of samples in the case of sparse signals. That is because of the fact that in this non-uniform sampling scheme, no sample is taken from the inactive parts of the signal and the sampling rate is directly proportional to the activity of the input signal. Also, based on the nature of the scheme, the input signal can be reconstructed from the two single-bit output signals of the level-crossing analog-to-digital converter (LCADC), which means a considerable decrease in the ADC output data volume and so a significant decrease in the transmitter power consumption.

One of the other advantages of the level-crossing analog-to-digital conversion is that no amplitude quantization is performed and so the accuracy of the sample's amplitude is not restricted by the ADC resolution. This means that the amplitude code length can be decreased especially in the case of the burst-like signals. A neural-spike signal, as an example of burst-like signals, is shown in Fig. 12.3B. Acquisition of data from these signals does not require a wide dynamic range, but the samples are needed to be accurate. The dynamic range of an ADC is defined as the ratio between the maximum and the minimum amplitude that can be detected by the converter.

Sparse and burst-like signals can be faced in many applications such as pressure, temperature and accelerometer sensors, speech, heart and neural signals. Therefore, applications such as implantable biosensors are perfect candidates for utilizing level-crossing sampling because of the nature of their signals and the importance of decreasing their power consumption.

Level-crossing sampling is first proposed in [1] as a data-compression technique. The first LCADC is proposed in [2] which is followed by a design procedure and circuit implementation in [3]. Research on the topic is continued in two fields. The first is proposing appropriate structures of LCADCs for different signal specification.

Several structures have been proposed for LCADCs in which the common tradeoff between the speed and power consumption can be observed [4–17]. The second field is working on processing the non-uniform samples comes from the LCADCs. It is shown that conventional signal processing techniques may need to be changed and modified to be efficient in this case [18–26].

This chapter starts by describing the basics of the level-crossing sampling scheme and the data-compression capability of this sampling type is shown in the case of some biopotential signals in the following. The two different types of the level-crossing sampling are described and several structures which are proposed for implementing LCADCs are presented and compared. Also, some requirements that should be considered for perfect recording of the level-crossing sampled data are described and processing of level-crossing sampled data is studied in this chapter.

12.2 Basics of level-crossing sampling

A real ECG signal is sampled with both the conventional synchronous and the level-crossing schemes in Fig. 12.4A and B, respectively. As can be observed in this figure, for an M-bit resolution, 2^M comparison levels are selected uniformly in the converter input amplitude range ($2 \cdot A_{FS}$) and a sample is taken when the input signal crosses one of these levels. Therefore, no sample is taken from the inactive parts of the signal and when the amplitude of the signal variations is less than an LSB, in which LSB is defined as follows:

$$LSB = \frac{2 \cdot A_{FS}}{2^M}. \tag{12.1}$$

Also, the activity of the input signal defines the sampling rate, which is an effective way for decreasing the number of useless samples and so the wasted power consumption in the case of the sparse signals. As can be observed in Fig. 12.4B, the amplitude of each sample is 1 LSB apart from the previous sample. Therefore, in level-crossing sampling scheme the amplitude of the output samples can be reconstructed a single-bit signal (UD), which is shown in Fig. 12.4C. This signal demonstrates the direction of the variations of the input signal; '0' for decreasing and '1' for increasing. Also, since the samples are not uniformly spaced in time, the sampling instants should be also recorded. This is done by using a second single-bit output signal (Token). As can be observed in Fig. 12.4C, this signal is a pulse which is activated in each sampling instant, remains active for a small time interval and deactivated before the next sample arises. This is the other advantage of the level-crossing sampling in which each sample can be demonstrated with two bits, which is much less than the corresponding value in synchronous ADCs (M).

The other fact that can be observed from Fig. 12.4A and B is that, although the value of M specifies the accuracy of the samples in synchronous sampling and so it must be selected larger than the required resolution in synchronous ADCs, it specifies the number of samples in the active parts of the signal in level-crossing sampling

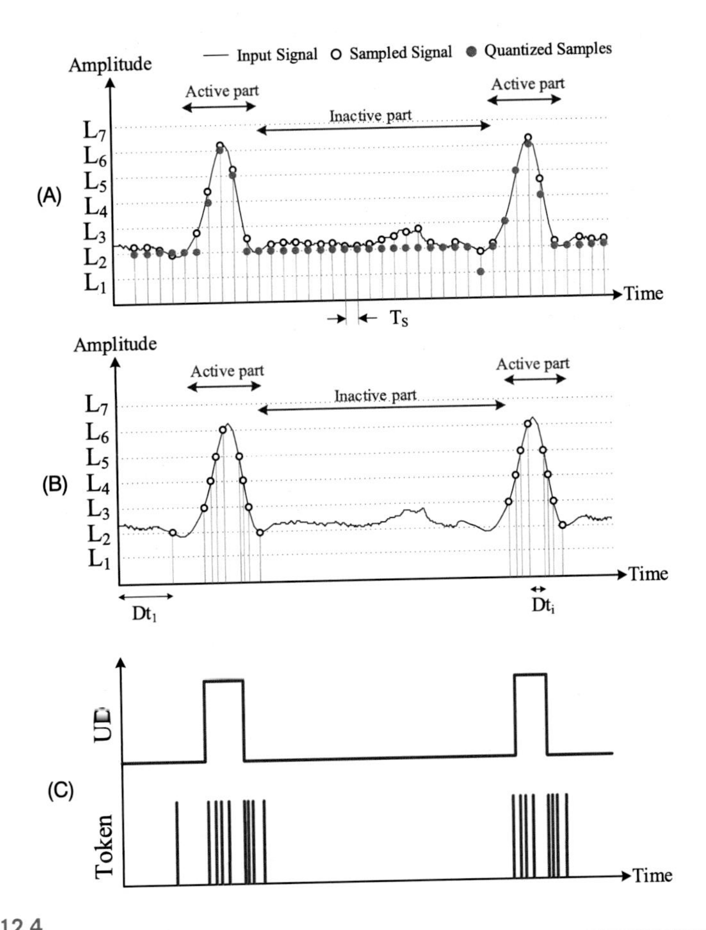

FIGURE 12.4

(A) Conventional synchronous sampling, (B) level-crossing sampling, (C) LCADC output signals.

and can be decreased considerably especially in the case of the burst-like signals. This may reduce the processing burden of the processor and so the processing power consumption.

Level crossings can be identified by comparing the input signal with the comparison levels. In order not to compare the input signals with all the comparison levels simultaneously and so saving the power consumption in the LCADC, two methods are mainly used for implementing the level-crossing sampling.

In the first method, which is called the floating-window scheme, the input signal is only compared with two specific comparison levels which surround the input signal at each moment of the time. These levels are called the upper and the lower reference levels (V_H and V_L). As can be observed in Fig. 12.5A, a sample is taken from the input signal when the ascending input signal passes V_H or the descending input signal

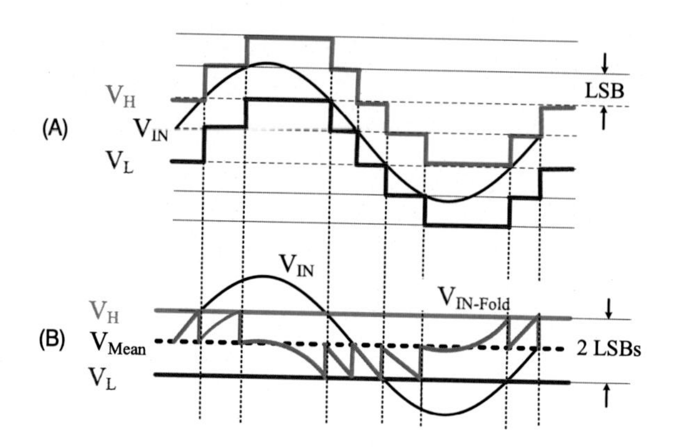

FIGURE 12.5

Two main methods for implementing level-crossing sampling: (A) floating window, (B) fixed window.

passes V_L. The gap between V_H and V_L is called the comparison window. In order to keep the input signal inside the comparison window, in the floating-window scheme, the values of V_H and V_L are increased (decreased) by 1 LSB at each sampling instant if the input signal crosses V_H (V_L).

Making the moving values of V_H and V_L may add some sort of complexity to the LCADC structures which results in higher power consumption and occupied silicon area compared to the conventional synchronous ADCs as discussed in the following But as the amplitude of the input signal has the specific 1 LSB change compared to the previous sample, it seems simpler to move the input signal to the comparison window instead of moving the comparison window based on the variation of the input signal. This is the idea behind the second method which is called the fixed-window scheme.

In the fixed-window scheme, the value of V_H and V_L are fixed and set around the input-signal DC value (V_{Mean}) as shown in Fig. 12.5B. A revised version of the input signal ($V_{IN\text{-}Fold}$) is made in this scheme which follows the input signal variations but it is folded by ± 1 LSB in each sampling instant such that it remains inside the comparison window. Folding the input signal can be implemented simpler than moving the reference levels and so the fixed-window LCADCs usually consume less power and occupy less area compared to the floating-window LCADCs. But the folding mechanism may result in some error which is accumulated during the ADC operation and so the fixed-window LCADCs are usually less accurate than their floating-window counterparts.

The other point that should be considered is the value of the comparison window. The comparison window is usually fixed at a value of more than 1 LSB in order to prevent toggling around a specific level as shown in Fig. 12.6A. This value also determines the minimum input signal amplitude which can be sensed by the LCADC. Although this value is conventionally set to 2 LSBs, it can be increased in some applications in order to omit some parts of the background noise as shown in Fig. 12.6B.

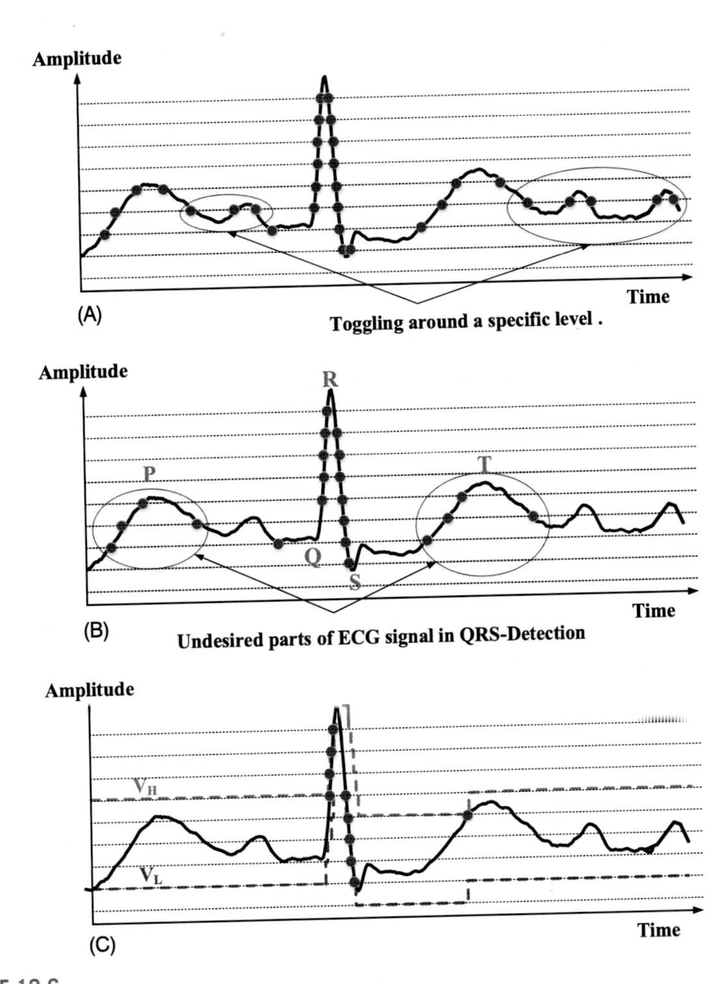

FIGURE 12.6

Level-crossing sampled ECG signal for comparison window of (A) 1 LSB, (B) 2 LSBs, and (C) 4 LSBs.

Also, by increasing the comparison window, some unwanted parts of the signal can be ignored as shown in Fig. 12.6C, in which major parts of the P-waves and the T-waves are omitted in an ECG signal. P-waves and the T-waves have destructive effects in detecting the actual position of the QRS complexes.

12.3 Design considerations in level-crossing sampling

In this section, several considerations are studied that should be made in the design of a proper level-crossing ADC (LCADC) for a specified application. It is shown that a suitable resolution must be selected for a signal-reconstruction capability and in order to achieve a specific dynamic range. Also, the accuracy in recording the sampling instants should be specified based of the required accuracy of the analog-to-digital conversion. Besides, the minimum conversion speed can be specified based on the maximum variation speed of the input signal.

Some input-signal specifications must be known for applying these considerations in a specific application, which are the maximum amplitude ($A_{in,max}$), the minimum substantial amplitude ($A_{in,min}$), the largest frequency component ($f_{in,max}$), power spectral density (PSD), and probability density P(x). Based on these data, several LCADC specifications can be determined such as: LCADC resolution (M), sample-time accuracy (T_{timer}), sample-amplitude accuracy (q), and the maximum tolerable conversion delay (d_{max}).

12.3.1 Resolution

Based on the Shannon–Nyquist theorem, for proper reconstruction of the input signal in the synchronous sampling, the sampling frequency must be selected more than the Nyquist frequency ($f_{Nyqu} = 2f_{in,max}$). For non-uniform sampling scheme, it is proved that the average sampling frequency ($f_{S,avg}$) should be more than f_{Nyqu} for the proper reconstruction [27]. In level-crossing sampling, the average sampling frequency can be theoretically determined based on the resolution (M), the probability density (P(x)) and the maximum slope of the input signal (S_{max}). Therefore, as $f_{S,avg}$ is a direct function of M, the minimum value of M can be determined based on the Beutler criterion ($f_{S,avg} > 2f_{in,max}$) [1]

In order to show this relation, in Fig. 12.7, the average sampling frequency is plotted versus the LCADC resolution for four sample real biopotential signals of ECG, EEG, EMG and single-unit neural spikes. ECG is the signal originated from the electrical activity of the heart, EEG comes from the electrical activity of the brain and EMG produced by the electrical activity of the muscles. For drawing Fig. 12.7, these signals are applied to a behaviorally simulated LCADC. In this figure, for ECG, EMG and neural signals, which contain inactive parts, $f_{S,avg}$ is calculated in two cases of the whole signal and only the active parts of the signal. This is due to the fact that the information is only in the active parts of the signal and so the Beutler criterion must be applied to these sections. Applying the Beutler condition to the whole signal results in an inapplicable value for M and destroy the whole benefit of using level-crossing sampling.

Based on the graphs drawn for the active parts of the signal and the bandwidths of the signals, the minimum acceptable value of M is reported in Table 12.1. The obtained resolution is less than the conventional values used in corresponding synchronous ADCs (8 bits for ECG and 10 bits for Neural). Also, as can be observed in Fig. 12.7, difference between the $f_{S,avg}$ in the active parts and the whole signal is significant in the cases of EMG and Neural signal, which means the data-compression capability of the level-crossing sampling.

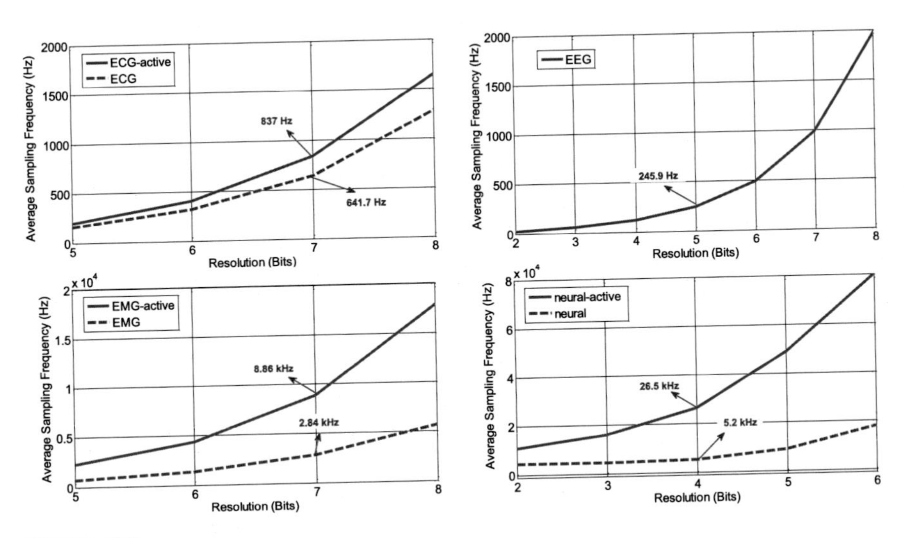

FIGURE 12.7

Average sampling frequency vs. LCADC resolution for ECG, EEG, EMG and single-unit recording neural signals in two cases of the whole signal and only the active parts of the signal.

Table 12.1 The minimum acceptable value of M for biopotential signals to satisfy the Beutler criterion.

Signal type	Δf (Hz)	Hardware number of bits
ECG	0.1–300	7
EEG	0.1–100	5
EMG	60–3k	7
Neural	5–10k	4

12.3.2 Dynamic range

Dynamic range (DR) is defined as the ratio of the maximum amplitude that can be applied to the converter input (A_{FS}) to the minimum amplitude that can be detected by the converter (A_P), as follows:

$$DR = \frac{A_{FS}}{A_P}. \tag{12.2}$$

In the level-crossing analog-to-digital converter, A_P is equal to the half of the comparison window. Comparison window, which is described in the previous section, is usually selected equal to couple of LSBs (G LSB). Therefore, using Eq. (12.1) for the LSB value, A_P can be obtained as:

$$A_P = \frac{1}{2} G \cdot \left(\frac{2 A_{FS}}{2^M} \right), \tag{12.3}$$

and so, DR is determined as:

$$DR = \frac{2^M}{G}. \tag{12.4}$$

Using this equation, the maximum size of the comparison window (G) can be obtained based on the minimum required dynamic range and the resolution obtained from the reconstruction criterion. By the way, the value of G cannot be less 1 LSB and it is better not to be less than 2 LSB as described in the previous section. If the value of G obtained by Eq. (12.4) is less than 1 LSB, then the value of M should be increased. This results in taking extra samples from the input signal which is in contrast to the aim of the level-crossing sampling.

Several solutions are proposed for solving this problem in the applications required large dynamic range. One of these methods is using adaptive resolution. In this method the comparison window is adjusted based of the input signal behavior such that the parts of the signal which have small amplitudes but contain valuable information, can be detected. In [10], this is done by estimating the slope of the signal. The comparison window is increased by increasing the slope of the signal. Using this method, no extra sample is taken from the large-amplitude high-speed parts of the signal. It may also help to soften the bandwidth limitation which is described in the following.

Adding dither to the input signal before applying it to the converter is the other method which can be used for increasing the dynamic range of the LCADC without increasing its resolution [8]. The operation of this method can be explained using a simple example. Suppose that we want to design a converter which can detect an input signal with an amplitude in the range of A_{min} to A_{max}. Based on Eq. (12.4), the minimum required value of M is equal to $\log_2(G \cdot A_{max}/A_{min})$. If we add a signal with an amplitude of d_{max} to the input, the required M is changed to $\log_2(G \cdot (A_{max} + d_{max})/(A_{min} + d_{max}))$. As a numeric example, suppose that the values of A_{max}, A_{min}, d_{max} and G are equal to $2^{10}\alpha$, α, $2^5\alpha$ and one respectively. Using these values, the value of M is equal to 10 bits and 5 bits before and after applying dither respectively. Although adding dither is effective in decreasing the value of M, this method suffers from several disadvantages. These include the complexity added by producing, adding and subtracting the dither, increasing the power of the error signal, and decreasing the maximum signal amplitude that can be applied to the converter input. The latter burdens a higher degree of noise immunity to the previous blocks.

As mentioned before and shown in Fig. 12.6A, the value of G cannot be selected equal to one in order to prevent toggling around a specific level at the presence of the noisy signals. This limits the minimum value of A_P to 1 LSB. To overcome this limitation, in [24] an LCADC is proposed in which the comparison window is increased by a small fraction of LSB (k'). In this floating-window structure, for an increasing signal, the upper reference level (V_H) for the current comparison window does not match the lower reference level (V_L) of the next comparison window and there is a gap of k' LSB between them. Therefore, even by choosing $G = 1$, noises with the amplitude less than $k'/2$ around a certain level do not detected by the converter. This method has two main disadvantages. First, some extra circuitry must be added to the converter structure which means additional power consumption and occupied silicon area. Second, the signal is distorted especially at the direction-changing points, which reduces the linearity of the converter. Also, similar solution can be done by adding hysteresis to the comparators.

12.3.3 Conversion delay

Conversion delay (d_{conv}), is the other characteristic, the maximum value of which, is limited to the input-signal specifications. d_{conv} is the delay of the converter in updating the comparison window by making the required new reference levels (V_H and V_L) at each sampling point in floating-window LCADCs. In fixed-window LCADCs, it is the time takes for the converter to fold the input signal. For proper operation of the LCADC, this time should be such short that no new level-crossing happens in this duration. Otherwise, as shown in Fig. 12.8, the reconstructed output signal cannot properly follow the input-signal variations and the signal is distorted. This means that the maximum variation of the input signal must be less than 1 LSB in a d_{conv} duration. In the other words:

$$S_{in,max} \leq \frac{LSB}{d_{conv}}, \quad (12.5)$$

in which $S_{in,max}$ is the maximum slope of the input signal. Based on the Bernstein theorem [28], with the maximum frequency component of $f_{in,max}$ and the maximum amplitude of $A_{in,max}$, the slope of the signal would be less than or equal to $2\pi f_{in,max} A_{in,max}$. Applying this value to Eq. (12.5) and replacing the value of LSB

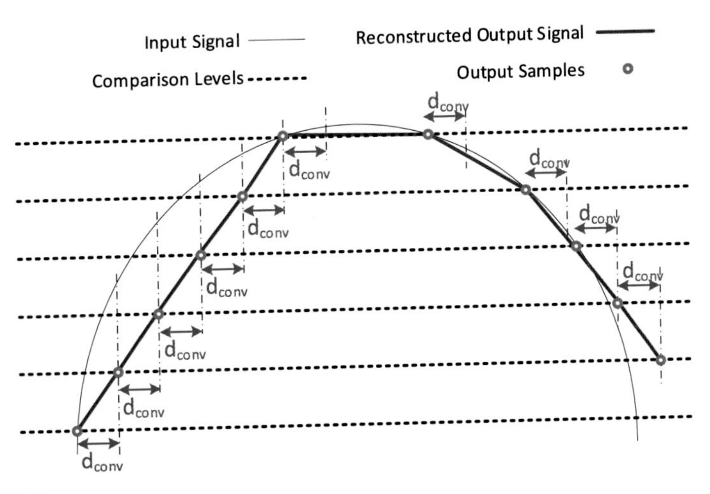

FIGURE 12.8

Distorted reconstructed signal due to a large conversion delay.

with $(2A_{FS}/2^M)$, the maximum tolerable value of d_{conv} can be obtained as follows:

$$d_{conv} \leq \frac{1}{\pi f_{in,max} 2^M} \left(\frac{A_{FS}}{A_{in,max}} \right). \tag{12.6}$$

This condition would not be a challenge in most of the biopotential signal acquisition applications. In most of these applications, $f_{in,max}$ and M would be less than or equal to 10 kHz and 10 bits, respectively, and so, for the worst case of $A_{FS} = A_{in,max}$, d_{conv} should be less than 31 ns. This value can be obtained simply in most of the LCADC structures in modern manufacturing technologies. By the way, LCADCs is going to be used in high-speed applications, this condition may limit the maximum acceptable value of M. If this value is lower than the minimum value of M obtained from the reconstruction or dynamic-range conditions, the problem can be fixed by using one of the methods described in the previous sub-section for decreasing the minimum value of M, such as an adaptive comparison window method or adding dither at the cost of more complexity.

12.3.4 Timer resolution and number of bits

One of the main differences between the level-crossing sampling and the conventional synchronous sampling is that, since the samples are not uniformly spaced in time, the sampling instants should also be recorded. As mentioned in the previous section, this is usually done by providing a token signal. However, sometimes in analyzing the data provided by an LCADC, the time interval between the consecutive samples are measured by a timer and provided as a digital code [20]. The resolution of the timer and the length of the digital codes provided by the timer are two other specifications that should be determined based on the input signal characteristics.

The limited resolution of the timer, acts as a variable delay in reporting the sampling instants and so makes some error. However, it can be shown that if the resolution of the timer be larger than the time interval between the consecutive samples, the reconstructed output signal cannot follow the variations of the input signal properly. This situation is shown in Fig. 12.9.

A simple counter is used as the timer in this figure which is triggered by the rising edge of a clock signal. The resolution of the timer is equal to the clock period in this situation. Different cases of latching and resetting of the timer is studied in [29]. The most reliable method is the synchronous operation, in which, if a level-crossing happens, the timer is synchronously reset in the next rising edge of the clock signal. It is shown that, for preventing the signal from compression in time, the counter should be synchronously reset to one instead of zero [29]. Both conditions of reset to one and reset to zero are shown in Fig. 12.9. It can be seen that, although the reconstructed output signal is much more accurate when the counter is reset to one, it cannot follow the input signal variations properly because the time interval between the consecutive samples is less than the clock period.

Based on the above description, the maximum acceptable resolution of the timer (T_{Timer}) is less than the minimum time interval between the consecutive samples and

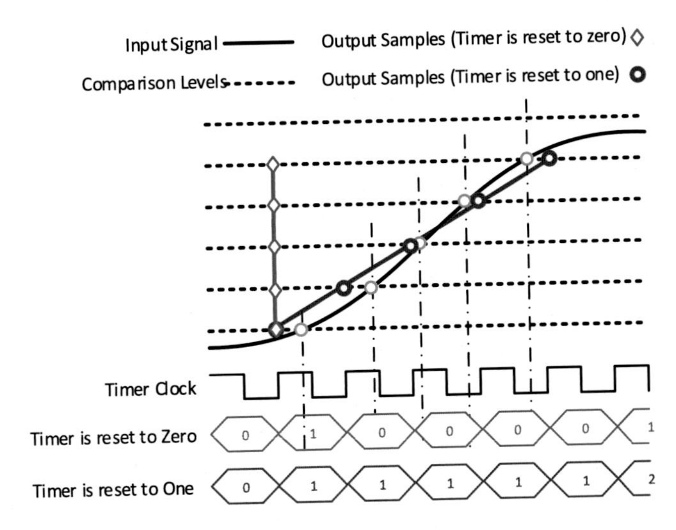

FIGURE 12.9

Effect of a large timer resolution on the LCADC operation and reconstructed signal.

so, it can be obtained based on the maximum slope of the input signal ($S_{in,max}$) as follows:

$$T_{Timer} \leq \frac{LSB}{S_{in,max}}. \tag{12.7}$$

Replacing the value of $S_{in,max}$ based on the descriptions presented in the previous sub-section, T_{Timer} can be obtained as follows:

$$T_{Timer} \leq \frac{1}{\pi f_{in,max} 2^M} \left(\frac{A_{FS}}{A_{in,max}} \right), \tag{12.8}$$

in which $f_{in,max}$, M, A_{FS} and $A_{in,max}$ represent the maximum frequency component of the input signal, LCADC resolution, LCADC full-scale input amplitude and the maximum amplitude of the input signal, respectively. In the worst case, A_{FS} is equal to $A_{in,max}$, and so the counter clock frequency ($f_{clk-Count}$) can be obtained:

$$f_{clk-Count} \geq \pi f_{in,max} 2^M. \tag{12.9}$$

The other specification of the timer, which is the length of the digital code provided by the timer (N_{Timer}), can be determined based on the maximum time interval between the consecutive samples (ΔT_{max}) and T_{Timer} as follows:

$$\Delta T_{max} \leq (2^{N_{Timer}} - 1) \cdot T_{Timer} \tag{12.10}$$

and so

$$N_{Timer} \geq \log_2 \left(\frac{\Delta T_{max}}{T_{Timer}} + 1 \right). \tag{12.11}$$

ΔT_{max} can be obtained by behavioral simulation on sample signals of a specific application. However, a larger ΔT_{max} is possible in sampling of other signals in similar applications. Therefore, an overflow mechanism must be considered in which an extra sample should be reported at the timer overflow condition. This may cause an error because no actual level-crossing happened at this point. By the way, as these extra samples are usually in the inactive parts of the signal, it may not cause an error in the following processing.

12.3.5 Accuracy of the converter

Even if all of the previously described conditions are considered, the accuracy of the converter may be limited by other factors. The accuracy of the amplitudes of the samples may be limited to the accuracy of making the reference levels (V_H and V_L) in the floating-window LCADCs. Switched-capacitor structures are usually used for making the required reference levels. The accuracy of these structures is usually limited by the matching on the capacitors, the non-ideal operation of the switches such as non-zero resistance at the on state, leakage at the off state, clock feed through and charge injection, thermal and flicker noise, and difference between the offset of the comparators. In the fixed-window LCADCs, the accuracy of the sample amplitude may be limited by a non-ideal folding mechanism, in which a single switch is simply connected the folded signal to the proper voltage. Therefore similar issues may corrupt this mechanism.

Accuracy of the sampling instants may also be limited due to the variation of the delay in recognizing the crossing point. This variation may result from several factors. One of them is the variation of the delay of the static comparators used for comparing the input or folded signal with the reference levels. Delay of a static comparator is a function of the slope of the input signal which changes from sample to sample. Dynamic comparators may also be used in order to reduce the power consumption in low-speed applications [14], [20]. In these structures, comparing is done with respect to the rising or falling edge of a clock signal. Since the crossing point can be in any point of the clock duration time cycle, there is a random delay in recognizing the sampling instants. This delay changes in the range of $[0 : T_{clk-Comp}]$, in which $T_{clk-Comp}$ is the period of the comparator trigger signal.

The other factor is the limited resolution of the timer which may be used for estimating the time interval between the consecutive samples or even the clock jitter. Based on the mechanism described in the previous sub-section, this delay may randomly change in the range of $[-T_{Timer}/2 : +T_{Timer}/2]$ in which T_{Timer} is the resolution of the timer.

Therefore, the signal-to-noise ratio can be defined for the level-crossing analog-to-digital converters as follows:

$$SNR_{dB} = 10 \cdot \log\left(\frac{P_{in}}{P_{Err-timer} + P_{Err-comp} + P_{Err-amp}}\right), \tag{12.12}$$

in which $P_{Err-timer}$, $P_{Err-comp}$, and $P_{Err-amp}$ means the power of the error due to the timer limited resolution, variation of the comparator delay and the limited accuracy of the comparison levels, respectively. Assigning an appropriate portion of the error power to each factor result in the efficient value of the timer resolution, the maximum acceptable delay for the comparators and the maximum error in making the reference levels.

Error in the time-domain can be moved to the amplitude domain using the following equation:

$$q_t = \frac{dV_{in}}{dt} \cdot t_{Err}, \tag{12.13}$$

in which t_{Err} and q_t represent the time error and its equivalent in the amplitude domain, respectively. Based on this equation, SNR can be rewritten as

$$SNR = 10\log\left(\frac{P\left(V_{in}\right)}{P\left(\frac{dV_{in}}{dt}\right)\cdot\left(E\left[d_{Timer}^2\right]+E\left[d_{comp}^2\right]\right)+E\left[q_{amp}^2\right]}\right), \tag{12.14}$$

in which $E[x]$ is expected value of random variable x, d_{Timer} and d_{comp} represent the variable delay due to the timer limited resolution and the comparator delay variation, respectively, and q_{amp} represents the random error in the amplitude of the samples. Suppose that q_{amp}, d_{Timer} and d_{comp} are random variables with uniform distribution in the range of $[-q/2 : q/2]$, $[-T_{Timer}/2 : T_{Timer}/2]$ and $[-\delta T_C/2 : \delta T_C/2]$, respectively, $E\left[q_{amp}^2\right]$, $E\left[d_{Timer}^2\right]$ and $E\left[d_{comp}^2\right]$ are equal to $q^2/12$, $T_{Timer}^2/12$ and $\delta T^2/12$, respectively. Using these values, SNR can be calculated as follows:

$$SNR = 10\log\left(\frac{P\left(V_{in}\right)}{P\left(\frac{dV_{in}}{dt}\right)\cdot\left(\frac{T_{Timer}^2}{12}+\frac{(\delta T_C)^2}{12}\right)+\frac{q^2}{12}}\right). \tag{12.15}$$

Based on the Bernstein theorem, $P(\frac{dV_{in}}{dt})$ would be less than $4\pi^2 f_{in,max}^2 A_{in,max}^2$. For a sine-wave input signal, this value is equal to $2\pi^2 f_{in,max}^2 A_{in,max}^2$. An interesting point that should be observed in Eq. (12.15) is that, in contrast to the synchronous ADCs in which the value of the $q = A_{FS}/2^M$ and so the SNR value is a function of the resolution of the ADC (M), the resolution of an LCADC does not have any direct effect of the accuracy of the converter.

One of the important notes that should be considered in calculating the SNR value in the case of a level-crossing sampled data is that the samples are not uniformly spaced in time. Therefore, in contrast to the synchronous sampling scheme, the power of the input error-free samples and the error signal cannot be determined without considering the time interval between the samples. The following equation must be used for calculating the energy (E_{Signal}) of a non-uniformly sampled signal:

$$E_{Signal} = \sum_{i=0}^{N-1} y_i^2 \Delta t_i. \tag{12.16}$$

In this equation y_i is the amplitude of the ith sample and Δt_i is the time interval between the current sample and the previous one. Most of the biopotential signal acquisition applications requires an accuracy between 7 to 10 effective number-of-bits (ENOB), in which ENOB is defined as follows:

$$ENOB = \frac{SNR_{dB} - 1.76}{6.02}. \tag{12.17}$$

In these accuracies, providing the required accuracy of the reference levels and the maximum required delay for the comparators would not be a challenge in the modern technologies. Therefore, if a timer used for estimating the sampling instants, the accuracy is mostly limited to timer resolution. In this case, using Eq. (12.15) the SNR and the ENOB values can be obtained for a sine-wave signal as follows:

$$SNR_{dB} = 20 \log \left(\frac{f_{Timer}}{f_{in}} \right) - 5.17, \tag{12.18}$$

$$ENOB = 3.32 \log \left(\frac{f_{Timer}}{f_{in}} \right) - 1.15, \tag{12.19}$$

in which $f_{Timer} = 1/T_{Timer}$.

The interesting point in Eq. (12.19) is that doubling the timer frequency results in a one bit increment in the ENOB of the LCADC. However, in oversampling synchronous ADCs, 0.5 bit increment of the ENOB can be obtained by doubling the sampling frequency.

12.4 Level-crossing analog-to-digital converters

As mentioned in Sect. 12.2, there are two main methods for implementing the level-crossing analog-to-digital converter (LCADC), which are named the fixed-window structures and the floating-window structures. As described in Sect. 12.2 and shown in Fig. 12.5, the value of the reference levels and so the comparison window are moved according to the variations of the input signal in floating-window structures such that the input signal remains inside the comparison window all the times in a normal operation. In the fixed-window scheme, the locations of the reference levels are fixed, but the input signal is folded such that it remains inside the comparison window.

Several structures have been proposed for implementing each of these methods. It can be seen that although the fixed-window structures are simpler and can be implemented with less power consumption, floating-window structures are usually more accurate than the fixed-window structures. This is due to the fact that the errors are integrated in fixed-window structures which distort the signal.

12.4.1 Floating-window LCADC

Two main structures, which are conventionally used for implementing the floating-window LCADCs, are shown in Fig. 12.10 and Fig. 12.11. The structure shown in Fig. 12.10A is one of the first implementations of the LCADC which is similar to a conventional synchronous Flash ADC [17]. For an M-bit resolution, the large number of $2^M - 1$ static comparators are used in this structure, using which the input signal is continuously compared with the comparison levels. Resistive ladder is usually used for making the comparison levels in this structure. A control logic is used for making

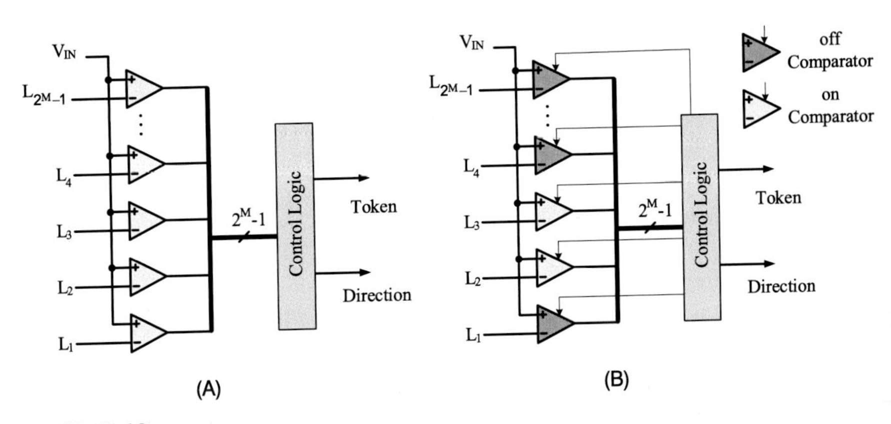

FIGURE 12.10

Flash-like variable-window structures for implementing level-crossing ADC.

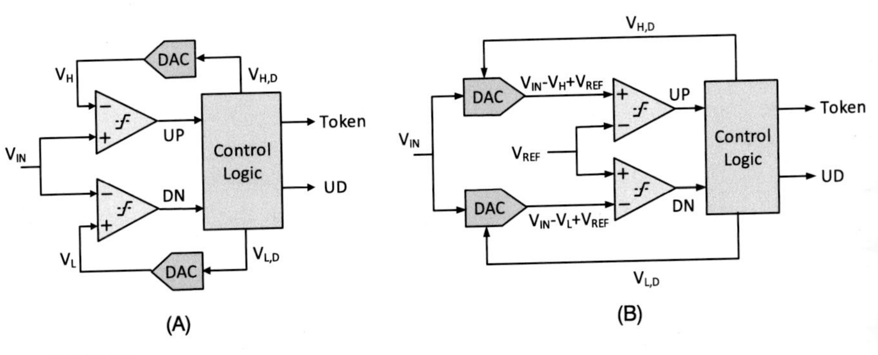

FIGURE 12.11

Two conventional structures for implementing floating-window LCADCs.

the required output signals from the comparators' outputs. This structure can be used in high-frequency applications. However, using a large number of static comparators and a resistive ladder, this structure consumes a high power and occupies a large area. Also, the mismatch between the resistors in the resistive ladder and the mismatch between the offsets of the comparators limits the accuracy of this converter.

As described in Sect. 12.2, the input signal is needed to be compared with only two comparison levels at any given point in time. Therefore, just two of the comparators which surround the input signal are needed to work and the others can be turned off. This structure is shown in Fig. 12.10B. In this structure, the control logic block decides which comparators should be "on" at any given point in time based on the input signal amplitude. In addition to the more complexity in the control logic, the other drawback of this structure is the limitation of the input signal frequency. This results from the time takes for deciding and turning on the suitable comparator after each level-crossing. This limitation can be softened by keeping more comparators around the input signal "on" at any given point in time at the expense of consuming more power.

The other floating-window structure is shown in Fig. 12.11A. In this structure, the reference levels are made based on the amplitude of the last sample using two digital-to-analog converters (DACs). The input signal is compared with these levels using two static comparators. The two resulted signals (UP and DN) are applied to a control logic circuit which provides the suitable LCADC output signals and the required digital representation of the reference levels ($V_{H,D}$ and $V_{L,D}$).

Digital-to-analog converters can be made with an ultra-low power consumption and a high accuracy using switched-capacitor structures. Also, this structure does not require any reference level and uses only two comparators. Therefore this structure consumes a low power and occupies a small area. The other advantage of this structure is that the comparison window can be set or even changed and tuned by changing the initial values of the reference voltages which are set by the control logic.

However, this structure has some disadvantages as follows. The first drawback is the closed-loop structure and the relatively large time takes for the DACs to update the value of the reference levels. This limits the variation speed and so the maximum frequency component of the input signal. Therefore, this structure can hardly be used in high-frequency applications.

The other disadvantage, which practically makes this structure non-applicable, is that the comparators must have a large input range because the reference levels are changed in the wide input amplitude range. Such comparators may consume a large power. This problem can be fixed by modifying the structure to the one shown in Fig. 12.11B. In this structure, the difference between the input signal and the reference levels are provided by a simple modification of the DAC structure. This difference is compared with a constant reference voltage (V_{REF}) in comparators which makes the design of the comparators much simpler. A sample schematic of the DAC is shown in Fig. 12.12, which contains an array of binary-weighted capacitors. The digital representation of the reference level is applied to L[0]-L[M-1] and loaded to the DAC by closing S_1 and S_2 switches in a small time interval after each level-

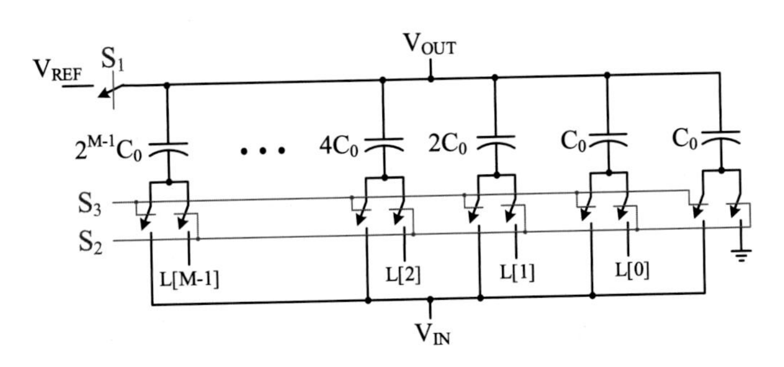

FIGURE 12.12

A binary-weighted switched-capacitor DAC used for implementing floating-window LCADCs.

crossing. By opening the S_1 and S_2 switches and closing the S_3 switches, the analog representation of the reference level is subtracted from the input signal (V_{IN}) and the result appears in the output (V_{OUT}).

Although the structure shown in Fig. 12.11B consumes a very small power compared to the one shown in Fig. 12.10A and can provide a larger accuracy compared to the one in Fig. 12.10B, it still consumes a large power compared to the conventional synchronous counterparts (such as the SAR ADC) because of using two continuous-time comparators which consume static power. A synchronous implementation of the structure shown in Fig. 12.11B is proposed in [29] which uses dynamic comparators which are triggered with an external clock signal. The frequency of the clock signal should be selected based on the restriction that should be considered for the minimum frequency of the timer in Eq. (12.9) which is described in the previous section. This LCADC is designed for the ECG-monitoring application and consumes an ultra-low power.

One of the issues that should be considered in design of most of the LCADC structures is the mismatch between the offsets of the comparators which may reduce the accuracy of the converter [10]. Complicated calibration techniques may be required for overcoming the problem [6,19]. A single-path structure is proposed in [15] which uses a single comparator for consecutively comparing the input signal with the two reference voltages and so the converter accuracy is not sensitive to the comparator offset. It has another benefit of using a single DAC and a single comparator which reduce the occupied area considerably. However, the delay of the system in comparing the input signal with each reference voltage is increased which limits the input signal variation speed. Therefore, this structure is suitable for low-frequency applications such as ECG-signal acquisition.

12.4.2 Fixed-window LCADCs

Fixed-window structures can be implemented simpler than the floating-window structures because the values of the reference levels are fixed in these LCADCs. Fig. 12.13 shows the schematic of a fixed-window LCADC. The required folding mechanism,

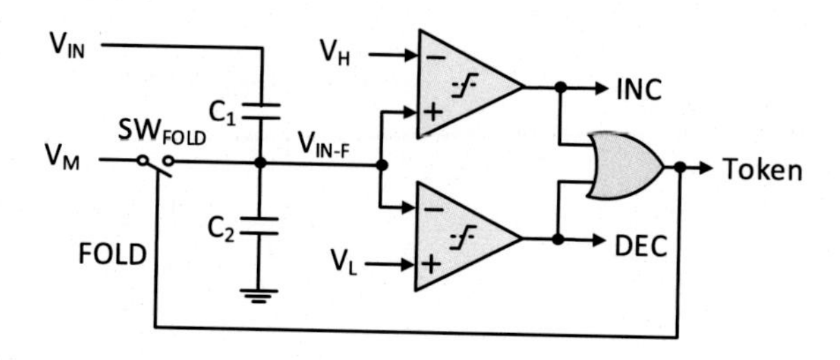

FIGURE 12.13

A conventional implementation of the fixed-window LCADC.

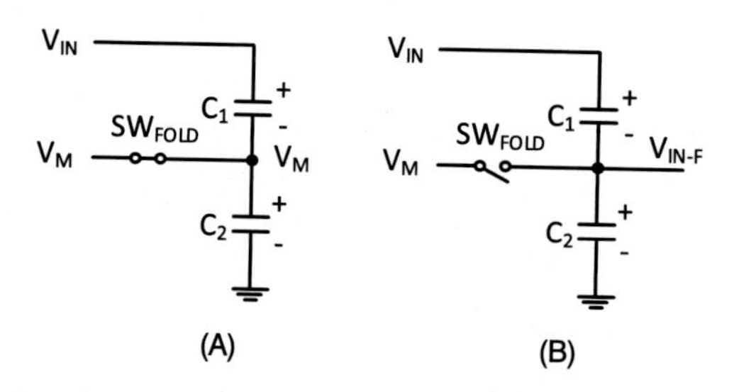

FIGURE 12.14

Two phases of the fixed-window LCADC circuit operation.

which is described in Sect. 12.2, is implemented using a simple switched-capacitor architecture composed of a single switch (SW_{Fold}) and two capacitors (C_1 and C_2). C_1 and C_2, which are in series, make a scaled version of the input signal ($V_{IN\text{-}F}$) which follows the variation of the input signal (V_{IN}). $V_{IN\text{-}F}$ can be folded to V_M by closing the SW_{Fold} after each level-crossing. The two phases of the circuit operation are shown in Fig. 12.14. Suppose that t_0 is the last instant that SW_{Fold} is closed, the relation between $V_{IN\text{-}F}$ and V_{IN} can be obtained by using charge-redistribution in which the charge in the center node is calculated in both phases (Q_{phase1} and Q_{phase2}) as follows:

$$\begin{aligned} Q_{phase1} &= C_2 V_M - C_1 \left(V_{IN}(t_0) - V_M \right), \\ Q_{phase2} &= C_2 V_{IN-F} - C_1 \left(V_{IN}(t) - V_{IN-F} \right). \end{aligned} \tag{12.20}$$

By equating Q_{phase1} and Q_{phase2}, $V_{IN\text{-}F}$ can be obtained as follows:

$$V_{IN-F} = \frac{C_1}{C_1 + C_2} \left(V_{IN}(t) - V_{IN}(t_0) \right) + V_M. \tag{12.21}$$

As can be observed from this equation, $V_{IN\text{-}F}$ is a scaled version of V_{IN} with a scale factor that can be adjusted with the values of C_1 and C_2. Two comparators are used for continuously comparing $V_{IN\text{-}F}$ with the fixed reference voltages of V_H and V_L. The comparison results are applied to an OR logic gate, the output of which is a token signal (FOLD) the activation of which denotes a level-crossing. The Fold signal is applied to the SW_{Fold} switch to trigger the folding mechanism. A synchronous implementation of this structure is also presented in [14] that uses dynamic comparators for decreasing the power consumption. This structure has the ability of tuning the size of the comparison window as well.

The structure of Fig. 12.13 has several benefits which are the pretty simple folding and controlling circuits and using only two comparators. These benefits make this structure very interesting for the low-power and low-area applications such as wearable and implantable biopotential acquisition systems. By the way, it has some important drawbacks which must be concerned. One of the main drawbacks of this structure is that the whole conversion process is done using $V_{IN\text{-}F}$ instead of the actual V_{IN} and so any error in making $V_{IN\text{-}F}$ will be accumulated while the conversion proceeds. Errors in making $V_{IN\text{-}F}$ may result from the delay of the comparator and the logic to make the FOLD signal, incomplete settling in the small folding duration, clock feedthrough and the charge injection of the SW_{Fold}, mismatch between the capacitors, error in making the values of V_H and V_L, and mismatch between the offset of the comparators. The resultant error is added to each sample. As the amplitude of each sample is reconstructed based on the amplitude of the previous samples, these errors will be accumulated and usually result in a drift in the output signal amplitude. Accurate results can be obtained by omitting this drift, using local amplitude data instead of the global amplitude data, or processing the slope of the signal instead of the amplitude of the signal which is common in the processing the biopotential signals. The other drawback is that two accurate reference voltages are required in this structure which may consume a large power or may occupy a large area to be implemented.

The other disadvantage of this structure is the limitation of the maximum frequency of the input signal which is resulted from the comparators delay in recognizing a level-crossing and the delay in completing the folding mechanism, as described and formulated in Sect. 12.3. This problem can be moderated by using an Operational Transconductance Amplifier (OTA), which facilitate the folding mechanism at the expense of higher power consumption as shown in Fig. 12.15 [7].

In this structure, a capacitive-coupled buffer is used as the first stage which makes the $V_{IN\text{-}F}$ in the output. The output of this buffer can be reset to V_M by closing the

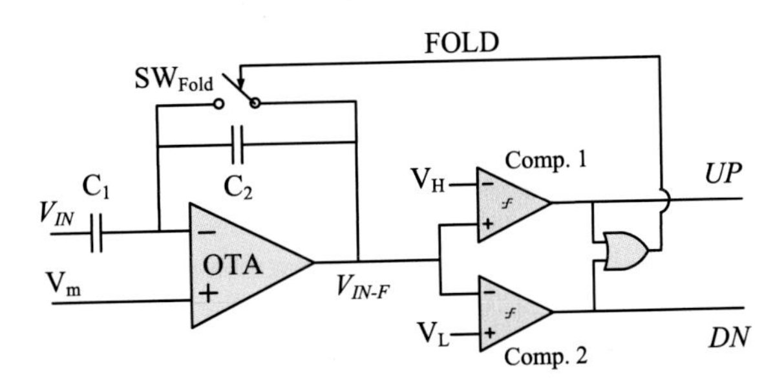

FIGURE 12.15

Implementation of the folding mechanism using a buffer in fixed-window LCADC.

SW_{Fold} switch for realizing the folding mechanism. In addition to faster operation, this structure has two other advantages. Despite the previous structure, the signal is not attenuated passing through the buffer. It can even be amplified if required. Also, the capacitors can be selected smaller than the ones required for the previous structure. In the structure shown in Fig. 12.13, the value of C_1 must be selected much larger than C_2 in order not to attenuate the input signal as can be observed from Eq. (12.21). This restriction is not needed to be applied in the structure shown in Fig. 12.15, since the buffer gain is equal to C_1/C_2 in this structure.

12.5 Processing of the level-crossing sampled data

It is shown in the previous sections that compared to conventional compressive-sensing techniques [30], level-crossing sampling is merged with the analog-to-digital conversion process and can be implemented simply without consuming a huge amount of power. It is shown in the following that by utilizing the special features of level-crossing sampling, compared to conventional processing techniques [31–33], level-crossing-sampled data can be processed using simple techniques. Several methods have been reported for processing the non-uniform level-crossing sampled data. If the processor is not restricted from the complexity and power consumption, uniform samples (in time) can be interpolated from the non-uniform samples [2]. Although the conventional synchronous processors can be used for processing these data, the accuracy of the samples may be reduced by the limited accuracy of the interpolation technique. A large number of samples should be taken from the signal in order to increase the interpolation accuracy which should be done by increasing the resolution of the LCADC. This may be corrupt the main advantage of using level-crossing sampling which is data compression and so this technique is not suggested for processing of the level-crossing sampled data.

The other technique which can be used for conventional synchronous transmitting, storing and processing of the non-uniform level-crossing sampled data is quantizing the time interval between the samples to a limited number of bits using a timer [20,21]. The processor can be synchronized to the timer clock and so it can utilize the simplicity and reliability of the synchronous digital system design.

This method has two main drawbacks. In addition to increasing the data volume by increasing the time-data code length, the accuracy of the samples may be decreased because of the limited resolution of the timer as described in Sect. 12.3. However, this is one of the simplest methods that can be used for processing of the level-crossing sampled data and its effectiveness is shown in several applications. For example, in [20] a QRS-detection system is proposed based on this technique. Utilizing the data compression capability of the level-crossing sampling in the case of the sparse burst-like ECG signal, this QRS detector consumes an ultra-low power and occupies a small area.

The other method is designing a tailor-made continuous-time digital signal processor which have the ability to process the non-uniform sampled data [19]. This method benefits from the complete data-compression capability of level-crossing sampling by utilizing the two single-bit LCADC output data of Token and UD (described in Sect. 12.2 and shown in Fig. 12.4). However, making special asynchronous processors may be a challenge.

References

[1] J. Mark, T. Todd, A nonuniform sampling approach to data compression, IEEE Transactions on Communications 29 (1) (Jan. 1981) 24–32.

[2] N. Sayiner, H. Sorensen, T. Viswanathan, A level-crossing sampling scheme for A/D conversion, IEEE Transactions on Circuits and Systems II: Analog and Digital Signal Processing 43 (4) (Apr. 1996) 335–339.

[3] E. Allier, G. Sicard, L. Fesquet, M. Renaudin, A new class of asynchronous A/D converters based on time quantization, in: Ninth International Symposium on Asynchronous Circuits and Systems, 2003, Proceedings, May 2003, pp. 196–205.

[4] M. Kurchuk, Y. Tsividis, Signal-dependent variable-resolution clockless A/D conversion with application to continuous-time digital signal processing, IEEE Transactions on Circuits and Systems I: Regular Papers 57 (5) (May 2010) 982–991.

[5] K. Kozmin, J. Johansson, J. Delsing, Level-crossing ADC performance evaluation toward ultrasound application, IEEE Transactions on Circuits and Systems I: Regular Papers 56 (8) (Aug. 2009) 1708–1719.

[6] Y. Li, D. Zhao, W.A. Serdijn, A sub-microwatt asynchronous level-crossing ADC for biomedical applications, IEEE Transactions on Biomedical Circuits and Systems 7 (2) (Apr. 2013) 149–157.

[7] W. Tang, A. Osman, D. Kim, B. Goldstein, C. Huang, B. Martini, V.A. Pieribone, E. Culurciello, Continuous time level crossing sampling ADC for bio-potential recording systems, IEEE Transactions on Circuits and Systems I: Regular Papers 60 (6) (Jun. 2013) 1407–1418.

[8] T. Wang, D. Wang, P.J. Hurst, B.C. Levy, S.H. Lewis, A level-crossing analog-to-digital converter with triangular dither, IEEE Transactions on Circuits and Systems I: Regular Papers 56 (9) (Sep. 2009) 2089–2099.

[9] C. Weltin-Wu, Y. Tsividis, An event-driven, alias-free ADC with signal-dependent resolution, in: 2012 Symposium on VLSI Circuits (VLSIC), Honolulu, HI, USA, IEEE, Jun. 2012, pp. 28–29.

[10] M. Trakimas, S.R. Sonkusale, An adaptive resolution asynchronous ADC architecture for data compression in energy constrained sensing applications, IEEE Transactions on Circuits and Systems I: Regular Papers 58 (5) (May 2011) 921–934.

[11] Y. Hou, K. Yousef, M. Atef, G. Wang, Y. Lian, A 1-to-1-kHz, 4.2-to-544-nW, multi-level comparator based level-crossing ADC for IoT applications, IEEE Transactions on Circuits and Systems II: Express Briefs 65 (10) (Oct. 2018) 1390–1394.

[12] J. Jimenez, S. Dai, J.K. Rosenstein, A microwatt front end and asynchronous ADC for sparse biopotential acquisition, in: 2017 IEEE 60th International Midwest Symposium on Circuits and Systems (MWSCAS), Boston, MA, USA, IEEE, Aug. 2017, pp. 503–506.

[13] T. Marisa, T. Niederhauser, A. Haeberlin, R.A. Wildhaber, R. Vogel, J. Goette, M. Jacomet, Pseudo asynchronous level crossing ADC for ECG signal acquisition, IEEE Transactions on Biomedical Circuits and Systems 11 (2) (Apr. 2017) 267–278.

[14] H. Teimoori, N. Ravanshad, H. Rezaee-Dehsorkh, Ultra-low-power fully-synchronous level-crossing analog-to-digital converter for biomedical signal acquisition, in: 2017 29th International Conference on Microelectronics (ICM), Beirut, IEEE, Dec. 2017, pp. 1–4.

[15] N. Ravanshad, H. Rezaee-Dehsorkh, R. Lotfi, A fully-synchronous offset-insensitive level-crossing analog-to-digital converter, in: 2016 IEEE 59th International Midwest Symposium on Circuits and Systems (MWSCAS), Abu Dhabi, United Arab Emirates, IEEE, Oct. 2016, pp. 1–4.

[16] Y. Hou, J. Qu, Z. Tian, M. Atef, K. Yousef, Y. Lian, G. Wang, A 61-nW level-crossing ADC with adaptive sampling for biomedical applications, IEEE Transactions on Circuits and Systems II: Express Briefs 66 (1) (Jan. 2019) 56–60.
[17] F. Akopyan, R. Manohar, A.B. Apsel, A level-crossing flash asynchronous analog-to-digital converter, in: 12th IEEE International Symposium on Asynchronous Circuits and Systems (ASYNC'06), Mar. 2006, 11 pp.–22.
[18] R. Agarwal, S. Sonkusale, Input-feature correlated asynchronous analog to information converter for ECG monitoring, IEEE Transactions on Biomedical Circuits and Systems 5 (5) (Oct. 2011) 459–467.
[19] B. Schell, Y. Tsividis, A continuous-time ADC/DSP/DAC system with no clock and with activity-dependent power dissipation, IEEE Journal of Solid-State Circuits 43 (11) (Nov. 2008) 2472–2481.
[20] N. Ravanshad, H. Rezaee-Dehsorkh, R. Lotfi, Yong Lian, A level-crossing based QRS-detection algorithm for wearable ECG sensors, IEEE Journal of Biomedical and Health Informatics 18 (1) (Jan. 2014) 183–192.
[21] F. Aeschlimann, E. Allier, L. Fresquet, M. Renaudin, Asynchronous FIR filters: towards a new digital processing chain, in: 10th International Symposium on Asynchronous Circuits and Systems, 2004, Proceedings, Crete, Greece, IEEE, 2004, pp. 198–206.
[22] Y. Li, K. Shepard, Y. Tsividis, A continuous-time programmable digital FIR filter, IEEE Journal of Solid-State Circuits 41 (11) (Nov. 2006) 2512–2520.
[23] C. Vezyrtzis, Y. Tsividis, Processing of signals using level-crossing sampling, in: 2009 IEEE International Symposium on Circuits and Systems, Taipei, Taiwan, IEEE, May 2009, pp. 2293–2296.
[24] Y. Hong, I. Rajendran, Y. Lian, A new ECG signal processing scheme for low-power wearable ECG devices, in: 2011 Asia Pacific Conference on Postgraduate Research in Microelectronics & Electronics, Macao, China, IEEE, Oct. 2011, pp. 74–77.
[25] N. Ravanshad, H. Rezaee-Dehsorkh, An event-based ECG-monitoring and QRS-detection system based on level-crossing sampling, in: 2017 Iranian Conference on Electrical Engineering (ICEE), Tehran, Iran, IEEE, May 2017, pp. 302–307.
[26] X. Zhang, Y. Lian, A 300-mV 220-nW event-driven ADC with real-time QRS detection for wearable ECG sensors, IEEE Transactions on Biomedical Circuits and Systems 8 (6) (Dec. 2014) 834–843.
[27] F.J. Beutler, Error-free recovery of signals from irregularly spaced samples, SIAM Review 8 (3) (Jul. 1966) 328–335.
[28] F. Marvasti, T. Lee, Analysis and recovery of sample-and-hold and linearly interpolated signals with irregular samples, IEEE Transactions on Signal Processing 40 (8) (Aug. 1992) 1884–1891.
[29] N. Ravanshad, H. Rezaee-Dehsorkh, R. Lotfi, Detailed study of the time estimation in level-crossing analog-to-digital converters, in: 2013 21st Iranian Conference on Electrical Engineering (ICEE), Mashhad, Iran, IEEE, May 2013, pp. 1–6.
[30] T.W. Cabral, M. Khosravy, F.M. Dias, H.L.M. Monteiro, M.A.A. Lima, L.R.M. Silva, R. Naji, C.A. Duque, Compressive sensing in medical signal processing and imaging systems, in: N. Dey, J. Chaki, R. Kumar (Eds.), Sensors for Health Monitoring, Academic Press, 2019, pp. 69–92.
[31] M.H. Sedaaghi, M. Khosravi, Morphological ECG signal preprocessing with more efficient baseline drift removal, presented at the 7th IASTED International Conference, 2003, pp. 205–209.
[32] M. Khosravi, M.H. Sedaaghi, Impulsive noise suppression of electrocardiogram signals with mediated morphological filters, presented at the 11th Iranian Conference on Biomedical Engineering, Tehran, Iran, Feb. 2004, pp. 207–212.
[33] M. Khosravy, M.R. Asharif, M.H. Sedaaghi, Morphological adult and fetal ECG preprocessing: employing mediated morphology, IEICE Technical Report 107 (461) (Jan. 18, 2008) 363–369.

CHAPTER

13 Compressive sensing of electroencephalogram: a review

Mateus M. de Oliveira[a], Mahdi Khosravy[b], Henrique L.M. Monteiro[a], Thales W. Cabral[a], Felipe M. Dias[a], Marcelo A.A. Lima[a], Leandro R. Manso Silva[a], Carlos A. Duque[a]

[a]*Department of Electrical Engineering, Federal University of Juiz de Fora, Juiz de Fora, Brazil*
[b]*Media Integrated Communication Laboratory, Graduate School of Engineering, Osaka University, Suita, Osaka, Japan*

13.1 Introduction

The human brain is made up of billions of neurons that command all parts of the human body and respond to stimuli from the external environment. In this context, every command or stimulus comes from information exchange between neurons through electrical signals. One of the possible approaches to capturing this electrical activity is to use an array of electrodes placed on the human scalp in an exam called an electroencephalogram (EEG) [1].

The EEG has tremendous diagnostic importance, and it is used to detect diseases such as Alzheimer's, epilepsy, and sleep problems. However, despite the great usefulness of this signal, it is affected by various types of noise, e.g., power line interference and muscular movement. Thus, several signal processing techniques must be applied so that the information contained in the EEG is preserved and can be used by health professionals [2].

Besides, a standard EEG signal measuring equipment has a large number of wires connecting the patient's scalp to a recording device. This type of equipment considerably limits patient movement and prevents long-term continuous monitoring of a patient. In order to overcome these issues, wearable EEG measuring equipment is a good alternative. However, there are several restrictions on the use of these sorts of equipment, and the high power consumption for wireless data transmission is the main one.

In this context, the theory of Compressive Sensing (CS) [3–5], proposed by Donoho, Candes, Romberg, and Tao in 2004, arose as a promising approach. The CS theory states that if a signal is sparse in any domain, it can be sampled at frequencies lower than the Nyquist–Shannon sampling frequency. Unlike other compression techniques, the CS approach can compress the signal during acquisition. Therefore,

Compressive Sensing in Healthcare. https://doi.org/10.1016/B978-0-12-821247-9.00018-4

it decreases the complexity of the measurement device, and it transfers all the demanding computational tasks to the device that receives the measurements, e.g., workstation. Therefore, CS allows for significant battery power saving [6]. Currently, CS has been deployed in a wide range of application from power quality analysis [7] to biomedical engineering [8]. Furthermore, there is a great potential of application of CS to agriculture equipment [9,10], signal processing [11,12], blind signal processing [13–18], acoustic OFDM [19], ECG processing [20–26], power quality analysis [7,27,28], power line communications [29], public transportation system [30], power system planning [31,32], location based services [33], text data processing [34], smart environment [35], human motion analysis [36], texture analysis [37], image adaptation [38], image enhancement [39,40], medical image processing [41–43], human-robot interaction [44], telecommunications [45–50], sentiment mining [51], data mining [52,53], software intensive systems [54], etc.

In this chapter, an overview of CS applications for EEG signals is presented. In Sect. 13.2, there is an overview of signal acquisition according to the Nyquist–Shannon paradigm. In Sect. 13.3, there is a brief introduction of EEG signals. In Sect. 13.4, a mathematical background of the CS theory is presented. In Sect. 13.5, there is a description of different applications of CS for EEG signals. Finally, the concluding remarks are conducted in Sect. 13.6.

13.2 Signal acquisition

Continuous-time signals have values for any given time [55]. In order to apply digital processing techniques to this signal, it is necessary to sample it first. The sampling process acquires the signal at equally spaced time steps. The distance between samples is defined as the sampling period (T_s), and its inverse is the sampling frequency (F_s). Therefore, high sampling frequencies require more computational complexity.

This section provides a brief overview of the traditional digital sampling approach. At first, a digital signal is introduced as the multiplication of a continuous-time signal by an impulse train. Then it is shown how to recover a signal after sampling. Subsequently, the Nyquist–Shannon sampling theorem is introduced, and it is shown what happens when the sampling frequency does not follow this theorem. Finally, a technique known as sample-hold is presented. This is the most used technique in digital sampling devices.

13.2.1 Pulse train sampling

In order to understand the concept of signal sampling, we consider a continuous-time signal $x(t)$ and a pulse train signal $\delta_T(t)$, defined in Eq. (13.1). The multiplication of these two signals results in the discrete-time form of the $x(t)$ signal. Fig. 13.1 shows

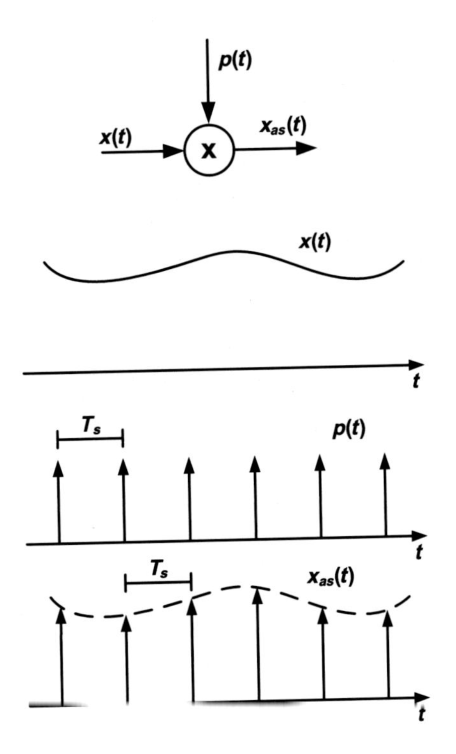

FIGURE 13.1

Representation of the signal sampling process, considering a function $x(t)$ multiplied by a pulse train function, resulting in the sampled signal $x_{as}(t)$.

this multiplication process and the obtained result. We have

$$\delta_T = \sum_{n=-\infty}^{\infty} \delta(t - nT). \tag{13.1}$$

The multiplication process can be mathematically described by Eq. (13.2):

$$x_{as}(t) = x(t) \sum_{n=-\infty}^{\infty} \delta(t - nT) = \sum_{n=-\infty}^{\infty} x(nT)\,\delta(t - nT). \tag{13.2}$$

The multiplication of two time domain signals results in a frequency domain convolution [55]. The convolution of the Fourier transform of the signal $x(t)$ and the Fourier transform of the signal $\delta_T(t)$ is shown in Eq. (13.3):

$$X_a(\omega) = \frac{X(\omega) * \Im\{\delta_T(t)\}}{2\pi}. \tag{13.3}$$

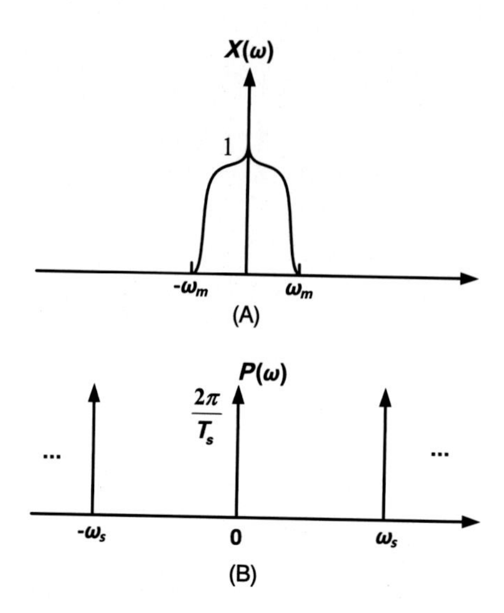

FIGURE 13.2

(A) $X(w)$ and (B) pulse train function.

The Fourier transform of the pulse train signal is a pulse train in the frequency domain where the space of each pulse is $\omega_s = \frac{2\pi}{T_s}$. This transformation is shown in Eq. (13.4):

$$\Im\{\delta_T(t)\} = \omega_s \sum_{k=-\infty}^{\infty} \delta(\omega - k\omega_s). \tag{13.4}$$

After substituting Eq. (13.4) into Eq. (13.3), we get

$$X_{as}(\omega) = \frac{1}{2\pi}\omega_s X(\omega) * \sum_{k=-\infty}^{\infty} \delta(\omega - k\omega_s). \tag{13.5}$$

Eq. (13.5) can be rewritten as

$$X_{as}(\omega) = \frac{1}{T}\sum_{k=-\infty}^{\infty} X(\omega - k\omega_s). \tag{13.6}$$

In Fig. 13.2A and 13.2B, the frequency domain representation of the signal $x(t)$ and the frequency domain pulse train are shown, respectively. Fig. 13.3 shows the convolution of these two signals.

13.2.2 Signal recovery

To reconstruct the original signal $x(t)$, the $X(\omega)$ spectrum must be recovered from $X_{as}(\omega)$. Therefore, it is necessary to guarantee that $\omega_m < \omega_s - \omega_m \Leftrightarrow \omega_s > 2\omega_m$

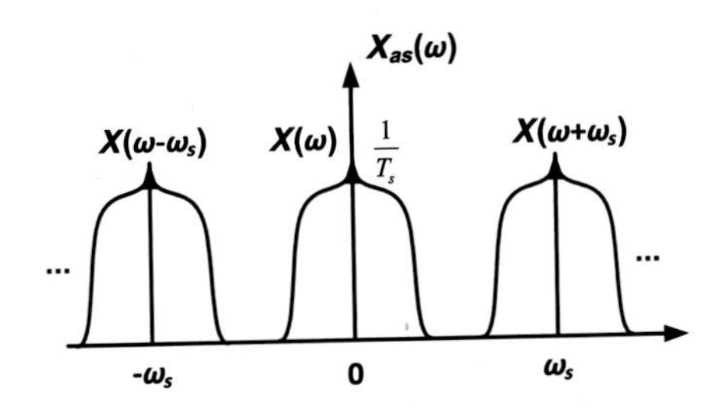

FIGURE 13.3

Sampled signal Fourier transform.

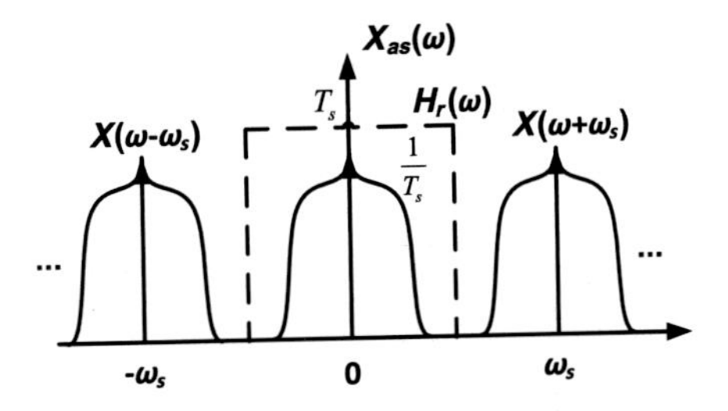

FIGURE 13.4

Spectrum of a sampled signal and response of a low-pass filter ideal for retrieving the original signal.

in order to avoid spectrum overlap. If there is no intersection, the central portion of $X_{as}(\omega)$ can be extracted using an ideal low-pass with a cutoff frequency filter between ω_m and $\omega_s - \omega_m$. The application of this filter is shown in Fig. 13.4.

13.2.3 Nyquist frequency

In order to guarantee that the spectrum of low-pass filter output is equal to $X(\omega)$, two conditions must be satisfied:

- $x(t)$ must be band limited, i.e., $-\omega_m < \omega < \omega_m$;
- $\omega_s > 2\omega_m \Leftrightarrow T < 2\pi/2\omega_m \Leftrightarrow T < 1/2f_m$.

The sampling frequency $\omega_{\text{nyquist-shannon}} = 2\omega_m$ is called the Nyquist–Shannon sampling frequency. The Nyquist–Shannon theorem states that if $\omega_s > \omega_{\text{nyquist-shannon}}$ the signal can be successfully reconstructed; otherwise, the reconstructed signal will lose information.

13.2.4 Aliasing

As stated in the previous section, if a signal is not sampled with a sampling frequency above the Nyquist frequency, overlap on the frequency spectrum will occur. Fig. 13.5

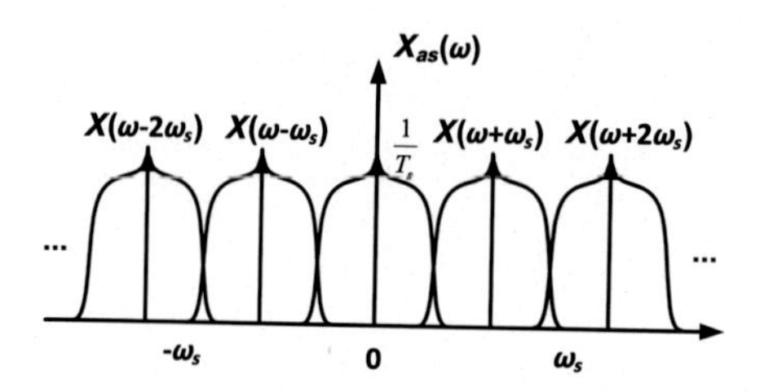

FIGURE 13.5

Spectrum of an aliasing sampled signal between shifted spectra.

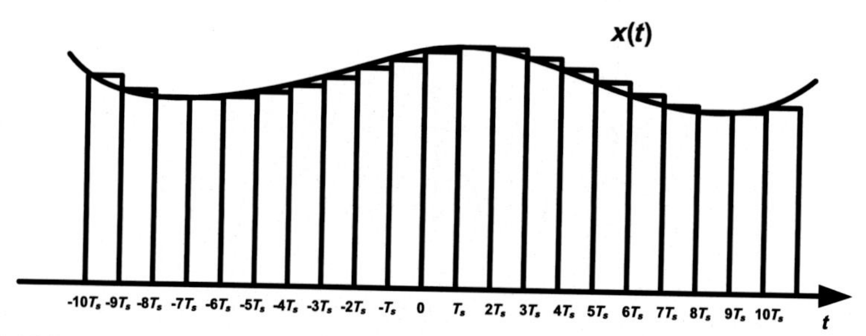

FIGURE 13.6

Continuous-time signal sampling using rectangular pulses.

shows this overlap. In such a case, it is not possible to reconstruct the signal $X(\omega)$ from the frequency spectrum $X_{as}(\omega)$. This effect is called aliasing.

13.2.5 Sample-hold

The pulse train signal assumes infinite value in an infinitesimal amount of time; therefore, it is impossible to reproduce this signal in the real world. In order to perform sampling, a common approach is to use rectangular pulses with a period of T_s. This sampling approach is called sample-hold. Fig. 13.6 shows the sampling process using the sample-hold approach.

The sample-hold approach can be described mathematically by Eq. (13.7), where the signal $p(t - nT_s)$ represents the periodic rectangular pulse.

$$x_{as}(t) = \sum_{n=-\infty}^{\infty} x(nT_s)\, p(t - nT_s). \tag{13.7}$$

The periodic rectangular pulse can be rewritten as shown in Eq. (13.8):

$$\sum_{n=-\infty}^{\infty} p(t - nT_s) = p(t) * \sum_{n=-\infty}^{\infty} \delta(t - nT_s). \tag{13.8}$$

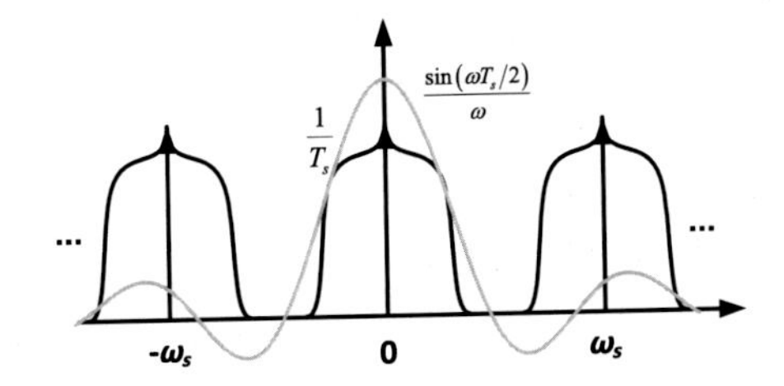

FIGURE 13.7

Spectrum obtained by the ideal sampling using sample-hold.

Substituting Eq. (13.8) into Eq. (13.7), we get

$$x_{as}(t) = p(t) * \sum_{n=-\infty}^{\infty} x(nT_s)\,\delta(t - nT_s). \tag{13.9}$$

The right term of Eq. (13.9) can be rewritten as shown in Eq. (13.10):

$$p(t) * \sum_{n=-\infty}^{\infty} x(nT_s)\,\delta(t - nT_s) = p(t) * \left(x(t) \sum_{n=-\infty}^{\infty} \delta(t - nT_s) \right). \tag{13.10}$$

Eq. (13.11) shows the Fourier transform of $x_{as}(t)$,

$$X_{as}(\omega) = \frac{\sin(\omega T_s/2)}{\omega} e^{-j\omega T_s/2} \cdot \frac{1}{T_s} \sum_{n=-\infty}^{\infty} X(\omega - k\omega_s). \tag{13.11}$$

Eq. (13.12) shows a simplification of Eq. (13.11):

$$X_{as}(\omega) = \frac{\sin(\omega T_s/2)}{\omega} e^{-j\omega T_s/2} X_a(\omega). \tag{13.12}$$

It is possible to see from Eq. (13.12) that if a signal is sampled using the sample-hold approach, its frequency spectrum will be different from sampling using the pulse train signal. However, this deformation can be corrected using a signal equalizer (Fig. 13.7). This equalizer frequency response is shown in Eq. (13.13):

$$H_{eq}(\omega) = \frac{\omega}{\sin(\omega T_s/2)}. \tag{13.13}$$

13.3 EEG signals

Electroencephalogram (EEG) is an examination that records the electrical activity of the brain using sensors placed on the scalp. This exam has been widely used as a clinical routine to diagnose physical or mental health problems such as epilepsy,

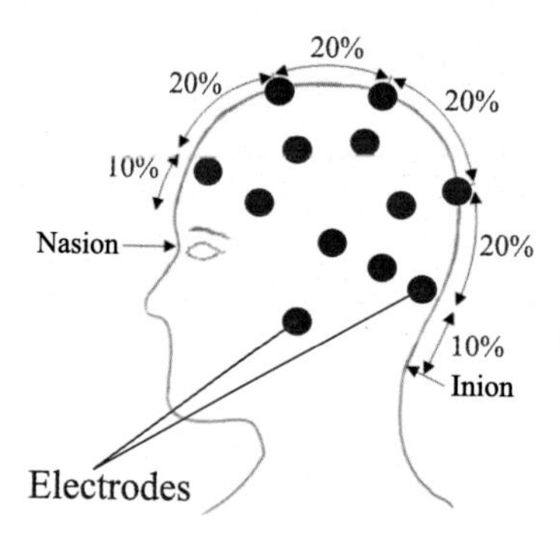

FIGURE 13.8

Electrode position according to the 10–20 system.

unconsciousness or dementia, and sleep disorders. Besides, EEG signals can also be used to build brain–computer interfaces (BCI) in order to assist persons with disabilities in the tasks of daily living [56–58].

Moreover, EEG signals are highly non-Gaussian, non-stationary, and nonlinear [1]. A typical EEG signal measured from the scalp will have an amplitude range from 10 μV to 100 μV and a frequency range from 1 Hz to 100 Hz. The brain activity of an abnormal person can be easily distinguished from a healthy person using signal processing methods.

There are two types of EEG recordings: (i) mono-polar (ii) bipolar. The mono-polar recording captures the voltage difference between an active electrode on the scalp and a reference electrode on the earlobe. Bipolar electrodes give the difference in voltage between two scalp electrodes. In general, the location of the electrodes for recording EEG signals are placed using the 10–20 system as recommended by the International Federation of Electroencephalographic Society for Electroencephalography and Clinical Neurophysiology (IFSECN) [59]. The 10–20 system is based on the relationship between the location of an electrode and the underlying area of the cerebral cortex. Fig. 13.8 shows electrode positions using the 10–20 system.

Five brain waves describe most of the useful information about the functional state of the human brain. These waves are: delta (0 to 4 Hz), theta (3.5 to 7.5 Hz), alpha (7.5 to 13 Hz), beta (13 to 26 Hz), and gamma (26 to 7 Hz). Delta waves have been found predominantly in infants up to 1-year-old and are related to the state of deep sleep. Theta waves were found in normal children and during sleepiness and sleep in adults. Alpha waves were found in relaxed adults. Beta waves are found predominantly in alert or anxious adults. Gamma waves are highly related to the brain's decision making mode [1,60].

One of the biggest challenges in collecting and analyzing EEG data is the amount of data that needs to be stored and processed. EEG activity collected from a single subject can correspond to hours of data. Efficient compression of EEG data can reduce the amount of data that needs to be stored and transmitted. Therefore, it is more

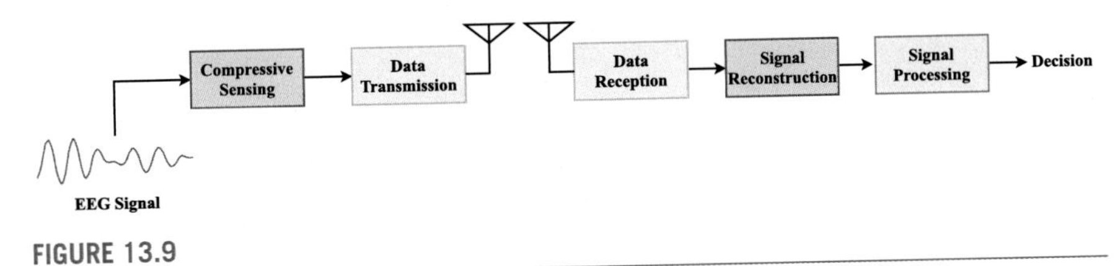

FIGURE 13.9

Diagram of a sparse compression and EEG signal processing system.

accessible to monitor EEG signals in real time and makes telehealthcare systems more feasible [56].

CS techniques have emerged as a potential feature for future continuous health monitoring devices because they can compress the signal during acquisition [56,61]. Therefore, CS is able to reduce the data to be transmitted and, consequently, decrease battery usage. A typical representation of an EEG monitoring system based on CS can be seen in Fig. 13.9.

13.4 Compressive sensing

CS technique was conceived by Donoho, Candès, Romberg and Tao [3–5]. It is a technique for compressing data in communication and information transmission processes.

Generally, coherent sampling capable of recovering the signal of interest utilizes the sampling frequency according to *Nyquist–Shannon* theorem. However, the CS technique allows a significant elimination in the number of samples used in the signal sampling process. The technique CS can be summarized in two steps: acquisition and reconstruction.

Initially, it is necessary to implement acquisition step which can be represented by Eq. (13.14), where **y** is the measured signal, **x** is the original signal, and $\mathbf{\Phi}$ is the measurement matrix. Eq. (13.14) assumes that the original signal is bigger than the measured signal, i.e., $M << N$.

$$\mathbf{y}_{M\times 1} = \mathbf{\Phi}_{M\times N}\mathbf{x}_{N\times 1}. \tag{13.14}$$

For the signal reconstruction, a sparse representation is applied according to (13.15),

$$\mathbf{x} = \sum_{i=1}^{N} s_i \Psi_i. \tag{13.15}$$

In such a case, **s** is a column vector $N \times 1$ and $\mathbf{\Psi}$ is the basis matrix $N \times N$. Furthermore, Eq. (13.15) is equivalent to $\mathbf{x} = \mathbf{\Psi}\mathbf{s}$. **x** and **s** are the same signal but with different representations. The first signal representation (**x**) is in the time domain, and

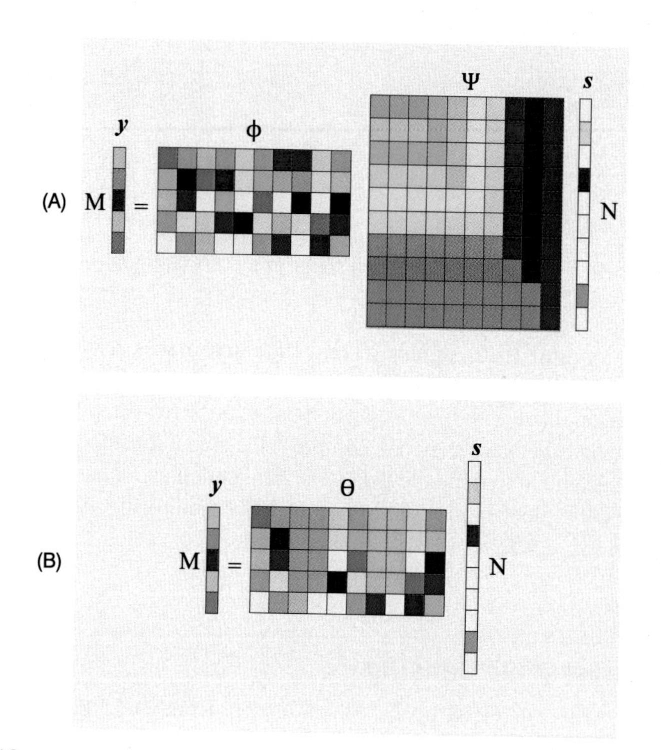

FIGURE 13.10

Visual example of CS theory.

the other representation (**s**) is in **Ψ** domain. Moreover, substituting Eq. (13.15) into (13.14), we get

$$\mathbf{y} = \mathbf{\Phi x} = \mathbf{\Phi \Psi s}; \tag{13.16}$$

Φ and **Ψ** can be combined in a single matrix **Θ** as shown in Eq. (13.17):

$$\mathbf{y} = \mathbf{\Phi x} = \mathbf{\Phi \Psi s} = \mathbf{\Theta s}. \tag{13.17}$$

The acquisition is not adaptive. Therefore, the measurement matrix is not dependent on the signal. The acquisition process shown in Eq. (13.16) is represented in Fig. 13.10A, and 13.10B shows the acquisition according to Eq. (13.17).

In order to reconstruct the original signal from the compressed measurements, an optimization problem, shown in Eq. (13.18), must be solved.

$$\hat{\mathbf{s}} = \min_{\mathbf{s}} \|\mathbf{s}\|_0 \qquad \text{subject to } \mathbf{y} = \Phi\Psi\mathbf{s}. \tag{13.18}$$

However, minimization of norm-0 is an NP-hard problem which requires nonlinear programming and meta-heuristic evolutionary optimization [62–72]. Because of that, the replacement by norm-1 has been suggested in the literature.

The Restricted Isometry Property (RIP) property is used to ensure the signal recovery. This property is represented by

$$1-\delta \leq \frac{\|\Phi\Psi u\|_2}{\|u\|_2} \leq 1+\delta \tag{13.19}$$

where $\delta > 0$ is defined as restricted isometrics constant and u is a vector with the same k nonzero entries as $\mathbf{x}$.

However, the RIP property is challenging to be verified. In this situation, another alternative is to use the inconsistency property. This property is shown in Eq. (13.20):

$$\mu(\Phi, \Psi) = \sqrt{n} \max_{1 \leq i,j \leq N} |\langle \Phi_i, \Psi_j \rangle|. \tag{13.20}$$

13.4.1 Compressive sensing metrics

In the CS technique, in order to measure the quality of signal reconstruction and the amount of compression that has been obtained, four import metrics are usually employed. The Compression Ratio (CR) [8], Eq. (13.21), the percentage root-mean-squared difference (PRD) [8], Eq. (13.22), the signal to noise and distortion ratio (SNDR) [73], Eq. (13.23), and the Percentage Sparsity (PS), Eq. (13.24). We have

$$CR = \frac{n}{m}, \tag{13.21}$$

$$PRD(\%) = \frac{\|\mathbf{x}-\hat{\mathbf{x}}\|_2^2}{\|\mathbf{x}\|_2^2} \times 100, \tag{13.22}$$

$$SNDR = 10\log_{10}\left(\frac{\|\mathbf{x}\|_2^2}{\|\mathbf{x}-\hat{\mathbf{x}}\|_2^2}\right), \tag{13.23}$$

$$PS(\%) = \left(\frac{N-K}{N}\right) \times 100. \tag{13.24}$$

13.5 Applications

CS has been applied to different sorts of biomedical signals [74,75]. In particular, there are a lot of applications of CS in EEG signals. In this section, an overview of the current state of the art applications of CS for EEG signals will be presented.

13.5.1 Block sparse Bayesian learning

The CS approach can only be successfully applied to signals that are sparse in some domain. However, the EEG signal does not comply with this requirement.

In order to overcome this issue, the authors of [76] applied Block Sparse Bayesian Learning (BSBL) to recover EEG signals acquired using the CS paradigm. The results showed that the BSBL approach presents better performance than other CS reconstruction algorithms, e.g., l_1 and CoSaMP.

In [77], the authors have also applied BSBL for EEG signals. Moreover, they applied a faster version of the BSBL algorithm and the results showed significant speed increase. Besides, the paper implements the algorithm in Field Programmable Gate Array (FPGA).

13.5.2 Bayesian compressive sensing

In 2010, an interesting paper was published [78]. The purpose was to use CS to compress the EEG signal. Before applying the compressive sensing measurement matrix, the authors applied a wavelet transform to the input signal in order to make it sparse. For reconstruction, they applied *Bayesian compressive sensing*. This paper shows that the CS approach is also suitable for denoising effects.

13.5.3 Slepian basis

Different sorts of basis have been used in order to map the EEG signal to a domain that makes it sparse. In [58], the authors proposed the use of slepian basis. The results show that slepians are a good choice to sparsify EEG signals.

13.5.4 Monitoring

Due to the increase in chronic diseases, there is an increasing demand for new continuous monitoring devices. In this context, CS plays a critical role in order to decrease battery usage and, therefore, increase continuous monitoring devices lifetime.

In [79], the authors proposed a noninvasive method for monitoring of Alzheimer's disease. The authors compare the performance between CS and Permutation Entropy (PE) for this monitoring. However, PE needs signal window overlap to ensure a smooth profile. Nevertheless, for a correct comparison with the CS approach, they use a signal window with no-overlap. The authors also use four electrodes for this comparison. The result shows that the CS approach has a smaller MSE than PE. Furthermore, the application was effective in monitoring of Alzheimer's disease.

In [80], the authors proposed a detection method for absence seizures. The results show that changes in compressibility using CS are good features to distinguish seizure-free, per-seizure, and seizure state. The approach proposed by the authors shows better accuracy (76.7%) than other approaches, e.g., Hurst index (55.3%), sample entropy (71%) and permutation entropy (73%).

13.5.5 Wireless body area network

Many applications of CS are composed of miniaturized hardware with low energy expenditure, complexity reduction, and other requirements. This sort of implementation is essential for wearable systems of health monitoring [8].

In the same context, in [81], the authors focused in WBAN. The monitoring networks must be compact. The approach of low power is fundamental in this case and the paper shows another example of adequate CS use in a BAN.

In [82], the authors have focused on applying CS for multi-channel EEG signals in order to reduce energy consumption for WBAN. This has also proposed the use of the Blind Compressed Sensing (BCS) framework in order to reconstruct the original signal. The BCS approach eliminates the need to know the sparse basis beforehand. In [83], the authors have also attempted to use BCS for EEG signals for the WBAN paradigm.

In [83], the authors use the BCS approach. The proposed method is a sort of WBAN implementation. Again, the application exploit the signals sparsity and thus the authors conclude that the energy expenditure is reduced.

13.5.6 Comparing reconstruction algorithms

There are different reconstruction algorithms in order to reconstruct signals acquired using the CS approach. In [84], the authors have compared different reconstruction algorithms, e.g., basis pursuit (BP), basis pursuit denoising (BPDN), orthogonal matching pursuit (OMP), and compressive sampling matching pursuit (CoSaMP). The research shows an interesting result of performance. For the same data, the BPND had a fewer iterations required for convergence and OMP algorithm had the smallest convergence time.

13.5.7 Biometric identification

In [85], the authors developed a wearable EEG for biometric identification. This device uses CS to compress the data for wireless transmission. The data is transmitted to a backend data center that reconstructs the signal and extracts features for user identification.

13.5.8 Dictionary learning

In [86], the authors discuss that to build an ideal dictionary is necessary to obtain excellent reconstruction accuracy and low computational complexity. Most methodologies has focused on static dictionaries such as Gabor, Fourier and wavelet, motivated by the dynamic of the EEG signals. In this paper, is presented how to create dictionaries learned with K-SVD and how to adopt them in EEG compression with CS. The results compared to the dictionaries cited above demonstrate that a learned K-SVD dictionary provides high reconstruction accuracy with a short computational time.

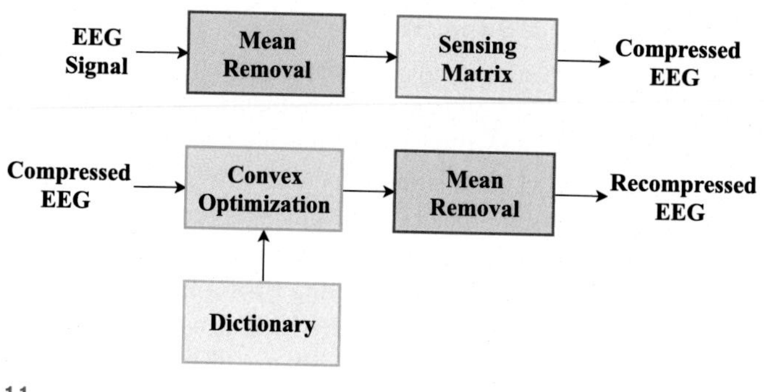

FIGURE 13.11

EEG CS flowchart proposed by [87].

In [87], a quantitative comparison of EEG CS performance is made using two different types of dictionaries, the Gabor dictionary versus the learned K-SVD dictionaries. In Fig. 13.11 is presented the diagram for signal compression used in this work. The EEG signal is first segmented into without over-lap sections. The average will be removed before processing to improve signal dispersion. This average result must be stored for restoration in the recovery phase. In the compression step, the Gaussian white noise random matrix [88] was chosen as our detection matrix to minimize the computational cost. For recovery step, the Basis Pursuit De-Noise algorithm (BPDN) [89] was used.

13.5.9 General applications

In 2007, one of the first papers on CS applications for EEG signals was published [61]. In particular, this paper focused on collection and storage of EEG signals. A compacted detection framework has been introduced by the author to efficiently represent multi-channel and multi-evaluation EEG data. The proposed structure is based on the sparsity of EEG signals in a Gabor framework. A simultaneous orthogonal matching pursuit algorithm was shown to be effective in retrieving the original multi-track EEG signals from a small number of projections.

Another application is shown in [90]. In this paper, CS is applied to Event-Related Potential (ERP) signals. ERP signals are unique EEG-hidden waveforms corresponding to specific physical, mental, or mentor events. In this paper, random and deterministic sensory matrices were applied to the input signal. The results showed that the proposed methodology can compress the ERP signals up to 75%. For signal recovery, a preprocessing approach called the Kronecker-based technique [91] was also used. It was possible to reconstruct the original signal with high accuracy up to 30 dB using the Kronecker technique.

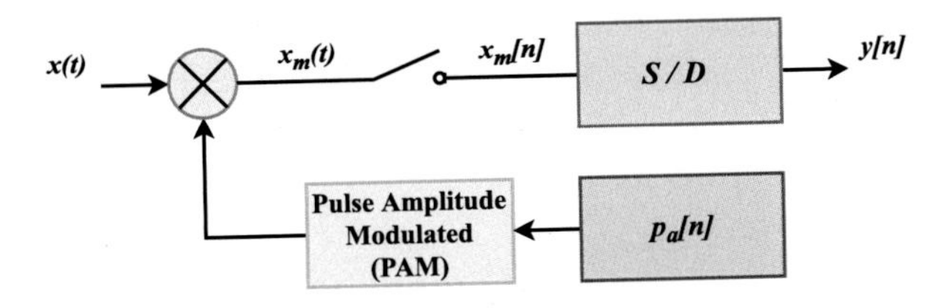

FIGURE 13.12

Proposed architecture for one bit CS.

13.5.10 Hardware implementation

In [73], is presented a study to solve computational complexity problems of the Wireless Body Area Network (WBAN) application through the development of CS for multi-channel EEG signals. The paper presents the optimization of the model from a system to compress and collect EEG signals in real time. Based on the CS theory, the EEG signal sparsity was analyzed, the measurement matrix was selected, the reconstruction algorithm was designed and the optimization analysis was performed. In the Field Programmable Gate Array (FPGA), the collection, storage, compression and transmission were performed, configuring a compression and acquisition system. Each system function module is inspected and the performance of the compressed multi channel EEG system is evaluated from the perspective of computational complexity and reconstruction accuracy. The evaluated results show that the real-time performance improvement is contributed by the application of the binary permutation block diagonal matrix (BPBD), which converts the CS multiplications into single circuit additions and drastically reduces the computational time. The average signal reconstruction rate (SNDR) reaches 21.74 dB below compression ratio 2, which also meets WBAN requirements. The authors state that the proposed method has faster compression, higher accuracy and simpler coding, and can be used in many multi-channel EEG-related applications, especially in situations where system power consumption and real-time performance are critical.

In [92] is presented a converter architecture that produces one bit compression measurements. Architectural performance can be improved if the acquired signal has a classic sparse and some kind of power locator. The effectiveness of the architecture and its enhancement is demonstrated by measuring the EEG, which has a non-uniform spectral profile. The proposed architecture is described by Fig. 13.12. The input signal $x(t)$ is first multiplied by the antipodal signal $p_a(t)$. After Pulse Amplitude Modulated (PAM), it is sampled to produce the discrete sequence $x_m[n]$ that feeds a modulator Σ/Δ [93]. At system output, we get a single sequence of bits $y[n]$.

13.5.11 Feasibility of compressive sensing for EEG signals

In [94], the authors analyze the advantages and disadvantages for using the CS implemented in a mobile EEG system. This system is presented in Fig. 13.13. Various

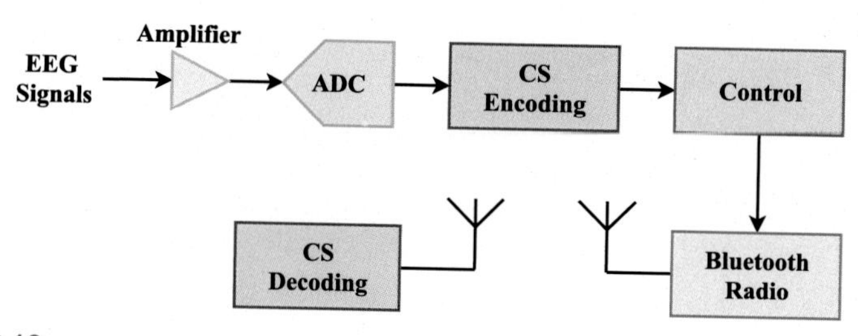

FIGURE 13.13

CS System architecture of mobile EEG sensor.

spatiotemporal coding methods for CS were proposed and the performance of each encoder was evaluated using a spontaneous EEG data set recorded during moderate movement. In the work, it was possible to conclude that the reconstruction performance depends strongly on the compression rate and weakly on the spatiotemporal coding method. This suggests a small spatial correlation between different EEG data channels.

In [95], the appropriateness of CS to provide online data reduction is investigated to reduce the amount of power required for system operation. In the results presented, the CS system offers no benefit when using a system with only a few channels. However, it can lead to significant energy savings in situations where more than approximately 20 channels are used. The result shows that further investigation and optimization of CS algorithms for EEG data is warranted. The simplified model of the EEG system proposed by this paper is presented by Fig. 13.14A. This model incorporates an input instrumentation amplifier to amplify small EEG signals, an Analog to Digital Converter (ADC) to convert EEG signals and a transmitter. For comparison, Fig. 13.14B illustrates the modifications required to incorporate compressive detection in the EEG system. For the Compressive detection is necessary a block to generate the measurement matrix that would be used to select a random set of samples. The elements in the measurement matrix form a pseudorandom sequence after a specific probability distribution.

13.6 Conclusion

Continuous monitoring of the EEG signal enables the creation of different devices to both diagnosis disease and monitor disease development. However, some challenges must be addressed. For example, reducing noise that interferes with the EEG signal, creating capture devices that enable greater user comfort for continuous use, and especially reducing battery consumption. The CS technique, introduced in this chapter, is up-and-coming when applied to devices that have battery restriction. It enables the acquisition of a much smaller amount of data that is required by the Nyquist–Shannon

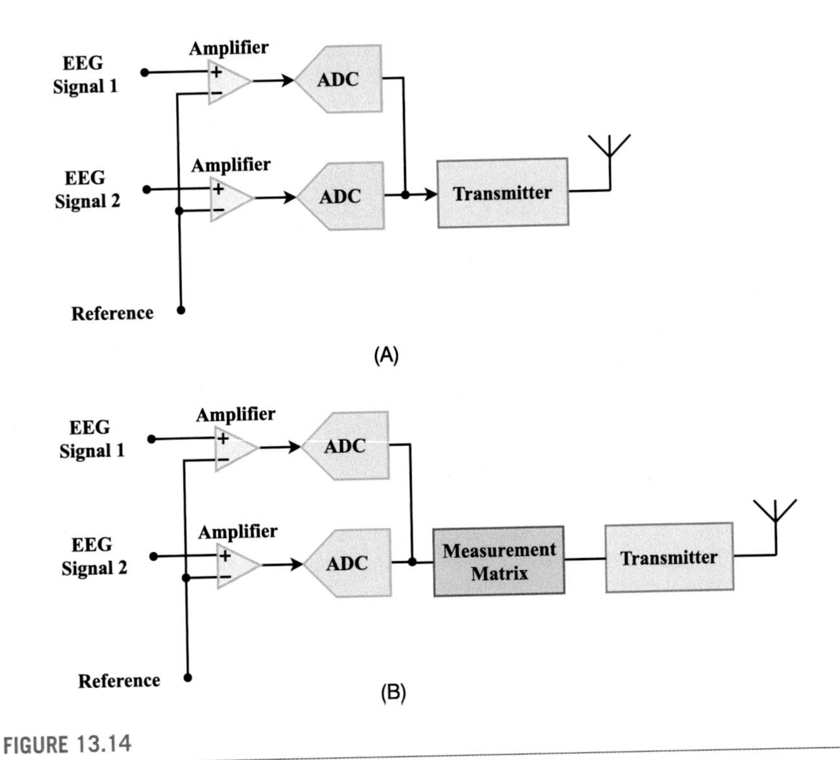

FIGURE 13.14

EEG system diagram to enable power modeling.

content. This way, it is possible to use fewer data storage space and less battery for sending data.

The biomedical technology for analyzing signals is an important research area. One such technology is the electroencephalogram, which is to measure the brain potential in order to help the disabled people and obtain an accurate diagnosis of diseases. EEG records brain waves with respect to a specific frequency by placing electrodes on the scalp. This chapter presented the importance of EEG signals and their characteristics.

This book chapter shows a quick introduction to CS theory. It reviews the main applications already made in the literature on CS applications to EEG signals. After reading this review, the reader should be able to understand the intricacies related to the theme and perceive the vastness of applications.

References

[1] D.P. Subha, P.K. Joseph, R. Acharya, C.M. Lim, EEG signal analysis: a survey, Journal of Medical Systems 34 (2) (2010) 195–212.

[2] A.G. Correa, E. Laciar, H. Patiño, M. Valentinuzzi, Artifact removal from EEG signals using adaptive filters in cascade, Journal of Physics Conference Series 90 (2007) 012081.
[3] E. Candes, J. Romberg, T. Tao, Robust uncertainty principles: exact signal reconstruction from highly incomplete frequency information, preprint, arXiv:math/0409186, 2004.
[4] D.L. Donoho, et al., Compressed sensing, IEEE Transactions on Information Theory 52 (4) (2006) 1289–1306.
[5] E. Candes, T. Tao, Near optimal signal recovery from random projections: universal encoding strategies?, preprint, arXiv:math/0410542, 2004.
[6] H. Mamaghanian, N. Khaled, D. Atienza, P. Vandergheynst, Compressed sensing for real-time energy-efficient ECG compression on wireless body sensor nodes, IEEE Transactions on Biomedical Engineering 58 (9) (2011) 2456–2466.
[7] K. Melo, M. Khosravy, C. Duque, N. Dey, Chirp code deterministic compressive sensing: analysis on power signal, in: 4th International Conference on Information Technology and Intelligent Transportation Systems (ITITS 2019), 2019.
[8] T.W. Cabral, M. Khosravy, F.M. Dias, H.L.M. Monteiro, M.A.A. Lima, L.R.M. Silva, R. Naji, C.A. Duque, Compressive sensing in medical signal processing and imaging systems, in: Sensors for Health Monitoring, Elsevier, 2019, pp. 69–92.
[9] S. Gupta, M. Khosravy, N. Gupta, H. Darbari, N. Patel, Hydraulic system onboard monitoring and fault diagnostic in agricultural machine, Brazilian Archives of Biology and Technology 62 (2019).
[10] S. Gupta, M. Khosravy, N. Gupta, H. Darbari, In-field failure assessment of tractor hydraulic system operation via pseudospectrum of acoustic measurements, Turkish Journal of Electrical Engineering & Computer Sciences 27 (4) (2019) 2718–2729.
[11] M.H. Sedaaghi, R. Daj, M. Khosravi, Mediated morphological filters, in: 2001 International Conference on Image Processing, Proceedings, vol. 3, IEEE, 2001, pp. 692–695.
[12] M. Khosravy, N. Gupta, N. Marina, I.K. Sethi, M.R. Asharif, Morphological filters: an inspiration from natural geometrical erosion and dilation, in: Nature-Inspired Computing and Optimization, Springer, Cham, 2017, pp. 349–379.
[13] M. Khosravy, M.R. Asharif, K. Yamashita, A PDF-matched short-term linear predictability approach to blind source separation, International Journal of Innovative Computing, Information & Control (IJICIC) 5 (11) (2009) 3677–3690.
[14] M. Khosravy, M.R. Asharif, K. Yamashita, A theoretical discussion on the foundation of Stone's blind source separation, Signal, Image and Video Processing 5 (3) (2011) 379–388.
[15] M. Khosravy, M.R. Asharif, K. Yamashita, A probabilistic short-length linear predictability approach to blind source separation, in: 23rd International Technical Conference on Circuits/Systems, Computers and Communications (ITC-CSCC 2008), Yamaguchi, Japan, ITC-CSCC, 2008, pp. 381–384.
[16] M. Khosravy, M.R. Alsharif, K. Yamashita, A PDF-matched modification to Stone's measure of predictability for blind source separation, in: International Symposium on Neural Networks, Springer, Berlin, Heidelberg, 2009, pp. 219–222.
[17] M. Khosravy, Blind source separation and its application to speech, image and MIMO-OFDM communication systems, PhD thesis, University of the Ryukyus, Japan, 2010.
[18] M. Khosravy, M. Gupta, M. Marina, M.R. Asharif, F. Asharif, I. Sethi, Blind components processing a novel approach to array signal processing: a research orientation, in: 2015 International Conference on Intelligent Informatics and Biomedical Sciences, ICIIBMS, 2015, pp. 20–26.
[19] M. Khosravy, N. Punkoska, F. Asharif, M.R. Asharif, Acoustic OFDM data embedding by reversible Walsh–Hadamard transform, AIP Conference Proceedings 1618 (2014) 720–723.
[20] M.H. Sedaaghi, M. Khosravi, Morphological ECG signal preprocessing with more efficient baseline drift removal, in: 7th IASTED International Conference, ASC, 2003, pp. 205–209.
[21] N. Dey, S. Samanta, X.-S. Yang, A. Das, S.S. Chaudhuri, Optimisation of scaling factors in electrocardiogram signal watermarking using cuckoo search, International Journal of Bio-Inspired Computation 5 (5) (2013) 315–326.
[22] M. Khosravy, M.R. Asharif, M.H. Sedaaghi, Morphological adult and fetal ECG preprocessing: employing mediated morphology, IEICE Technical Report, IEICE 107 (2008) 363–369.

[23] N. Dey, A.S. Ashour, F. Shi, S.J. Fong, R.S. Sherratt, Developing residential wireless sensor networks for ECG healthcare monitoring, IEEE Transactions on Consumer Electronics 63 (4) (2017) 442–449.

[24] M. Khosravi, M.H. Sedaaghi, Impulsive noise suppression of electrocardiogram signals with mediated morphological filters, in: 11th Iranian Conference on Biomedical Engineering, ICBME, 2004, pp. 207–212.

[25] N. Dey, S. Mukhopadhyay, A. Das, S.S. Chaudhuri, Analysis of P-QRS-T components modified by blind watermarking technique within the electrocardiogram signal for authentication in wireless telecardiology using DWT, International Journal of Image, Graphics and Signal Processing 4 (7) (2012) 33.

[26] M. Khosravy, N. Gupta, N. Patel, T. Senjyu, C.A. Duque, Particle swarm optimization of morphological filters for electrocardiogram baseline drift estimation, in: Applied Nature-Inspired Computing: Algorithms and Case Studies, Springer, 2020, pp. 1–21.

[27] E. Santos, M. Khosravy, M.A. Lima, A.S. Cerqueira, C.A. Duque, A. Yona, High accuracy power quality evaluation under a colored noisy condition by filter bank ESPRIT, Electronics 8 (11) (2019) 1259.

[28] E. Santos, M. Khosravy, M.A. Lima, A.S. Cerqueira, C.A. Duque, Esprit associated with filter bank for power-line harmonics, sub-harmonics and inter-harmonics parameters estimation, International Journal of Electrical Power & Energy Systems 118 (105) (2020) 731.

[29] A.A. Picorone, T.R. de Oliveira, R. Sampaio-Neto, M. Khosravy, M.V. Ribeiro, Channel characterization of low voltage electric power distribution networks for PLC applications based on measurement campaign, International Journal of Electrical Power & Energy Systems 116 (105) (2020) 554.

[30] M. Foth, R. Schroeter, J. Ti, Opportunities of public transport experience enhancements with mobile services and urban screens, International Journal of Ambient Computing and Intelligence (IJACI) 5 (1) (2013) 1–18.

[31] N. Gupta, M. Khosravy, N. Patel, T. Senjyu, A bi-level evolutionary optimization for coordinated transmission expansion planning, IEEE Access 6 (2018) 48455–48477.

[32] N. Gupta, M. Khosravy, K. Saurav, I.K. Sethi, N. Marina, Value assessment method for expansion planning of generators and transmission networks: a non-iterative approach, Electrical Engineering 100 (3) (2018) 1405–1420.

[33] M. Yamin, A.A.A. Sen, Improving privacy and security of user data in location based services, International Journal of Ambient Computing and Intelligence (IJACI) 9 (1) (2018) 19–42.

[34] C.E. Gutierrez, P.M.R. Alsharif, M. Khosravy, P.K. Yamashita, P.H. Miyagi, R. Villa, Main large data set features detection by a linear predictor model, AIP Conference Proceedings 1618 (2014) 733–737.

[35] C. Castelfranchi, G. Pezzulo, L. Tummolini, Behavioral implicit communication (BIC): communicating with smart environments, International Journal of Ambient Computing and Intelligence (IJACI) 2 (1) (2010) 1–12.

[36] G.V. Kale, V.H. Patil, A study of vision based human motion recognition and analysis, International Journal of Ambient Computing and Intelligence (IJACI) 7 (2) (2016) 75–92.

[37] S. Hemalatha, S.M. Anouncia, Unsupervised segmentation of remote sensing images using FD based texture analysis model and isodata, International Journal of Ambient Computing and Intelligence (IJACI) 8 (3) (2017) 58–75.

[38] M. Khosravy, N. Gupta, N. Marina, I. Sethi, M. Asharifa, Perceptual adaptation of image based on Chevreul–Mach bands visual phenomenon, IEEE Signal Processing Letters 24 (5) (2017) 594–598.

[39] A.S. Ashour, S. Samanta, N. Dey, N. Kausar, W.B. Abdessalemkaraa, A.E. Hassanien, Computed tomography image enhancement using cuckoo search: a log transform based approach, Journal of Signal and Information Processing 6 (03) (2015) 244.

[40] M. Khosravy, N. Gupta, N. Marina, I. Sethi, M. Asharif, Brain action inspired morphological image enhancement, in: Nature-Inspired Computing and Optimization, Springer, Cham, 2017, pp. 381–407.

[41] M. Khosravy, M.R. Asharif, M.H. Sedaaghi, Medical image noise suppression: using mediated morphology, IEICE, Technical Report 107 (461) (2008) 265–270.

[42] N. Dey, A.S. Ashour, A.S. Ashour, A. Singh, Digital analysis of microscopic images in medicine, Journal of Advanced Microscopy Research 10 (1) (2015) 1–13.

[43] M. Khosravy, M.R. Asharif, M.H. Sedaaghi, Medical image noise suppression using mediated morphology, in: IEICE Tech. Rep., IEICE, 2008, pp. 265–270.
[44] B. Alenljung, J. Lindblom, R. Andreasson, T. Ziemke, User experience in social human–robot interaction, in: Rapid Automation: Concepts, Methodologies, Tools, and Applications, IGI Global, 2019, pp. 1468–1490.
[45] M. Khosravy, M.R. Alsharif, M. Khosravi, K. Yamashita, An optimum pre-filter for ICA based multi-input multi-output OFDM system, in: 2010 2nd International Conference on Education Technology and Computer, vol. 5, IEEE, 2010, pp. V5–129.
[46] F. Asharif, S. Tamaki, M.R. Alsharif, H. Ryu, Performance improvement of constant modulus algorithm blind equalizer for 16 QAM modulation, International Journal on Innovative Computing, Information and Control 7 (4) (2013) 1377–1384.
[47] M. Khosravy, M.R. Alsharif, K. Yamashita, An efficient ICA based approach to multiuser detection in MIMO OFDM systems, in: Multi-Carrier Systems & Solutions 2009, Springer, 2009, pp. 47–56.
[48] M. Khosravy, M.R. Alsharif, B. Guo, H. Lin, K. Yamashita, A robust and precise solution to permutation indeterminacy and complex scaling ambiguity in BSS-based blind MIMO-OFDM receiver, in: International Conference on Independent Component Analysis and Signal Separation, Springer, 2009, pp. 670–677.
[49] M. Khosravy, A blind ICA based receiver with efficient multiuser detection for multi-input multi-output OFDM systems, in: The 8th International Conference on Applications and Principles of Information Science (APIS), Okinawa, Japan, 2009, 2009, pp. 311–314.
[50] M. Khosravy, S. Kakazu, M.R. Alsharif, K. Yamashita, Multiuser data separation for short message service using ICA, SIP, IEICE Technical Report 109 (435) (2010) 113–117.
[51] M. Baumgarten, M.D. Mulvenna, N. Rooney, J. Reid, Keyword-based sentiment mining using Twitter, International Journal of Ambient Computing and Intelligence (IJACI) 5 (2) (2013) 56–69.
[52] C.E. Gutierrez, M.R. Alsharif, K. Yamashita, M. Khosravy, A tweets mining approach to detection of critical events characteristics using random forest, International Journal of Next-Generation Computing 5 (2) (2014) 167–176.
[53] N. Kausar, S. Palaniappan, B.B. Samir, A. Abdullah, N. Dey, Systematic analysis of applied data mining based optimization algorithms in clinical attribute extraction and classification for diagnosis of cardiac patients, in: Applications of Intelligent Optimization in Biology and Medicine, Springer, 2016, pp. 217–231.
[54] P. Sosnin, Precedent-oriented approach to conceptually experimental activity in designing the software intensive systems, International Journal of Ambient Computing and Intelligence (IJACI) 7 (1) (2016) 69–93.
[55] A.V. Oppenheim, Discrete-Time Signal Processing, Pearson Education India, 1999.
[56] P.T. Dao, A. Griffin, X.J. Li, Compressed sensing of EEG with Gabor dictionary: effect of time and frequency resolution, in: 2018 40th Annual International Conference of the IEEE Engineering in Medicine and Biology Society (EMBC), IEEE, 2018, pp. 3108–3111.
[57] M. Mohsina, A. Majumdar, Gabor based analysis prior formulation for EEG signal reconstruction, Biomedical Signal Processing and Control 8 (6) (2013) 951–955.
[58] S. Şenay, L.F. Chaparro, M. Sun, R.J. Sclabassi, Compressive sensing and random filtering of EEG signals using slepian basis, in: 2008 16th European Signal Processing Conference, IEEE, 2008, pp. 1–5.
[59] J.S. Ebersole, T.A. Pedley, Current Practice of Clinical Electroencephalography, Lippincott Williams & Wilkins, 2003.
[60] T. Alotaiby, F.E.A. El-Samie, S.A. Alshebeili, I. Ahmad, A review of channel selection algorithms for EEG signal processing, EURASIP Journal on Advances in Signal Processing 2015 (1) (2015) 66.
[61] S. Aviyente, Compressed sensing framework for EEG compression, in: Proc. IEEE/SP 14th Workshop Stat. Signal Process, 2007, pp. 181–184.
[62] M. Khosravy, N. Gupta, N. Patel, T. Senjyu, Frontier Applications of Nature Inspired Computation, Springer, 2020.
[63] N. Gupta, M. Khosravy, O.P. Mahela, N. Patel, Plant biology-inspired genetic algorithm: superior efficiency to firefly optimizer, in: Applications of Firefly Algorithm and Its Variants, Springer, 2020, pp. 193–219.

[64] C. Moraes, E. De Oliveira, M. Khosravy, L. Oliveira, L. Honório, M. Pinto, A hybrid bat-inspired algorithm for power transmission expansion planning on a practical Brazilian network, in: Applied Nature-Inspired Computing: Algorithms and Case Studies, Springer, 2020, pp. 71–95.
[65] G. Singh, N. Gupta, M. Khosravy, New crossover operators for real coded genetic algorithm (RCGA), in: 2015 International Conference on Intelligent Informatics and Biomedical Sciences (ICIIBMS), IEEE, 2015, pp. 135–140.
[66] N. Gupta, N. Patel, B.N. Tiwari, M. Khosravy, Genetic algorithm based on enhanced selection and log-scaled mutation technique, in: Proceedings of the Future Technologies Conference, Springer, 2018, pp. 730–748.
[67] N. Gupta, M. Khosravy, N. Patel, I. Sethi, Evolutionary optimization based on biological evolution in plants, Procedia Computer Science 126 (2018) 146–155.
[68] J. Kaliannan, A. Baskaran, N. Dey, A.S. Ashour, M. Khosravy, R. Kumar, ACO based control strategy in interconnected thermal power system for regulation of frequency with HAE and UPFC unit, in: International Conference on Data Science and Application (ICDSA-2019), in: LNNS, Springer, 2019.
[69] N. Gupta, M. Khosravy, N. Patel, O. Mahela, G. Varshney, Plants genetics inspired evolutionary optimization: a descriptive tutorial, in: Frontier Applications of Nature Inspired Computation, Springer, 2020.
[70] M. Khosravy, N. Gupta, N. Patel, O. Mahela, G. Varshney, Tracing the points in search space in plants biology genetics algorithm optimization, in: Frontier Applications of Nature Inspired Computation, Springer, 2020.
[71] N. Gupta, M. Khosravy, N. Patel, S. Gupta, G. Varshney, Evolutionary artificial neural networks: comparative study on state of the art optimizers, in: Frontier Applications of Nature Inspired Computation, Springer, 2020.
[72] N. Gupta, M. Khosravy, N. Patel, S. Gupta, G. Varshney, Artificial neural network trained by plant genetics inspired optimizer, in: Frontier Applications of Nature Inspired Computation, Springer, 2020.
[73] D. Liu, Q. Wang, Y. Zhang, X. Liu, J. Lu, J. Sun, FPGA-based real-time compressed sensing of multichannel EEG signals for wireless body area networks, Biomedical Signal Processing and Control 49 (2019) 221–230.
[74] G. Da Poian, R. Bernardini, R. Rinaldo, Separation and analysis of fetal-ECG signals from compressed sensed abdominal ECG recordings, IEEE Transactions on Biomedical Engineering 63 (6) (2015) 1269–1279.
[75] G. Da Poian, N.A. Letizia, R. Rinaldo, G.D. Clifford, A low complexity photoplethysmographic systolic peak detector for compressed sensed data, Physiological Measurement (2019).
[76] Z. Zhang, T.-P. Jung, S. Makeig, B. Rao, Compressed sensing of EEG for wireless telemonitoring with low energy consumption and inexpensive hardware, IEEE Transactions on Biomedical Engineering 60 (Jan. 2013) 221–224.
[77] B. Liu, Z. Zhang, G. Xu, H. Fan, Q. Fu, Energy efficient telemonitoring of physiological signals via compressed sensing: a fast algorithm and power consumption evaluation, Biomedical Signal Processing and Control 11 (2014) 80–88.
[78] H.-x. Zhang, H.-q. Wang, X.-m. Li, Y.-h. Lu, L.-k. Zhang, Implementation of compressive sensing in ECG and EEG signal processing, The Journal of China Universities of Posts and Telecommunications 17 (6) (2010) 122–126.
[79] F. Morabito, D. Labate, G. Morabito, I. Palamara, H. Szu, Monitoring and diagnosis of Alzheimer's disease using noninvasive compressive sensing EEG, in: Independent Component Analyses, Compressive Sampling, Wavelets, Neural Net, Biosystems, and Nanoengineering XI, vol. 8750, International Society for Optics and Photonics, 2013, 87500Y.
[80] K. Zeng, J. Yan, Y. Wang, A. Sik, G. Ouyang, X. Li, Automatic detection of absence seizures with compressive sensing EEG, Neurocomputing 171 (2016) 497–502.
[81] R. Hussein, A. Mohamed, M. Alghoniemy, Scalable real-time energy-efficient EEG compression scheme for wireless body area sensor network, Biomedical Signal Processing and Control 19 (2015) 122–129.

[82] A. Majumdar, R.K. Ward, Energy efficient EEG sensing and transmission for wireless body area networks: a blind compressed sensing approach, Biomedical Signal Processing and Control 20 (2015) 1–9.

[83] A. Shukla, A. Majumdar, Row-sparse blind compressed sensing for reconstructing multi-channel EEG signals, Biomedical Signal Processing and Control 18 (2015) 174–178.

[84] M. Rani, S. Dhok, R. Deshmukh, EEG monitoring: performance comparison of compressive sensing reconstruction algorithms, in: Information Systems Design and Intelligent Applications, Springer, 2019, pp. 9–17.

[85] Y. Dai, X. Wang, X. Li, Y. Tan, Sparse EEG compressive sensing for web-enabled person identification, Measurement 74 (2015) 11–20.

[86] P.T. Dao, X.J. Li, A. Griffin, H.N. Do, K-SVD dictionary learning applied in clinical EEG compressed sensing, in: 2018 International Conference on Advanced Technologies for Communications (ATC), IEEE, 2018, pp. 179–183.

[87] P.T. Dao, X.J. Li, A. Griffin, Quantitative comparison of EEG compressed sensing using Gabor and K-SVD dictionaries, in: 2018 IEEE 23rd International Conference on Digital Signal Processing (DSP), IEEE, 2018, pp. 1–5.

[88] D. Craven, B. McGinley, L. Kilmartin, M. Glavin, E. Jones, Compressed sensing for bioelectric signals: a review, IEEE Journal of Biomedical and Health Informatics 19 (2) (2014) 529–540.

[89] Y.C. Pati, R. Rezaiifar, P.S. Krishnaprasad, Orthogonal matching pursuit: recursive function approximation with applications to wavelet decomposition, in: Proceedings of 27th Asilomar Conference on Signals, Systems and Computers, IEEE, 1993, pp. 40–44.

[90] S.A. Khoshnevis, S. Ghorshi, Recovery of event related potential signals using compressive sensing and Kronecker technique, in: 7th IEEE Global Conference on Signal and Information Processing (GlobalSIP), Ottawa, Canada, 2019.

[91] H. Zanddizari, S. Rajan, H. Zarrabi, Increasing the quality of reconstructed signal in compressive sensing utilizing Kronecker technique, Biomedical Engineering Letters 8 (2) (2018) 239–247.

[92] J. Haboba, M. Mangia, R. Rovatti, G. Setti, An architecture for 1-bit localized compressive sensing with applications to EEG, in: 2011 IEEE Biomedical Circuits and Systems Conference (BioCAS), IEEE, 2011, pp. 137–140.

[93] R. Schreier, G.C. Temes, et al., Understanding Delta-Sigma Data Converters, vol. 74, IEEE Press, Piscataway, NJ, 2005.

[94] B. Senevirathna, P. Abshire, Spatio-temporal compressed sensing for real-time wireless EEG monitoring, in: 2018 IEEE International Symposium on Circuits and Systems (ISCAS), IEEE, 2018, pp. 1–5.

[95] A.M. Abdulghani, A.J. Casson, E. Rodriguez-Villegas, Quantifying the feasibility of compressive sensing in portable electroencephalography systems, in: International Conference on Foundations of Augmented Cognition, Springer, 2009, pp. 319–328.

CHAPTER

14 Calibrationless parallel compressed sensing reconstruction for rapid magnetic resonance imaging

Sumit Datta, Bhabesh Deka

Department of Electronics and Communication Engineering, Tezpur University, Tezpur, Assam, India

14.1 Introduction

Magnetic resonance imaging (MRI) is one of the most widely used medical imaging modalities in clinical practice. Unlike other well-known medical imaging modalities, like X-ray, ultrasound, and computed tomography (CT), it has several advantages; it gives good contrast for soft-tissue imaging, non-invasive, and produces no-ionizing radiations. For these reasons, radiologists commonly prefer to use MRI over other imaging modalities. However, it has two major limitations/drawbacks, one is the relatively longer scan time compared to other techniques and other is the acoustic noise due to rapid variation of magnetic gradients. Although, in parallel MRI (pMRI) data acquisition time is reduced significantly, it still requires 15 to 40 minutes for the body imaging depending on the associated field-of-view (FoV) [1]. In parallel MRI (pMRI), theoretically the acceleration factor is equal to the number of receiver coils. However, there are some practical limitations of the number of parallel receiver coils as the arrangement of coils should provide varying sensitivity information in the phase encode direction in which data are to be undersampled. The acceleration factor cannot be greater than the number of elements in that direction, subject to the geometrical arrangement of the coil array such that sensitivities from different elements are not highly correlated [2].

All consumer electronics and medical equipment follow the Shannon–Nyquist sampling theorem for data acquisition, which results in a significant amount of data for the handling of all practical signals. In some cases, signal acquisition itself is a very expensive process, for example, in MRI. Candes [3] and Donoho [4] introduced the concept of Compressed sensing (CS) in signal/image processing. According to the CS theory, one can accurately reconstruct a signal/image from just a few linear measurements in contrast to the traditional Shannon–Nyquist sampling theorem provided

Compressive Sensing in Healthcare. https://doi.org/10.1016/B978-0-12-821247-9.00019-6

the signals follow a few basic conditions. First, the signal should be sparse/compressible in some transform domain, second, the acquisition should be incoherent with respect to the transform domain [5].

MRI naturally fits into the fundamental requirements posed by the CS theory, i.e. MR signal is sparse/compressible in a transform domain and their physical measurements are done in the Fourier domain, which is fairly incoherent with most of the image transformation bases [6]. For example, MR image is compressible in the wavelet domain, and the k-space in which raw image samples are naturally acquired is sufficiently incoherent with the wavelets. Lustig et al. [7] demonstrate that application of CS in MRI reduces the scan time 4–5 times compared to the traditional approach. They refer to it as sparse MRI or Compressed Sensing MRI (CS-MRI). Application of CS in pMRI has been successfully demonstrated in clinical settings, which can further reduce the scan time of pMRI [8,9]. However, unlike conventional MRI, linear reconstruction is not possible in the case of CS-MRI. As the measured data are randomly undersampled, so aliasing occurs; one needs to incorporate non-linear optimization techniques as post-processing methods in order to avoid aliasing for rapid MR image reconstruction. Reconstruction of diagnostically relevant MR images with minimum number of measurements and time is obviously the major research issue in CS-MRI [10,11], which will be addressed in this chapter.

In clinical practice, volume MRI is very common for the investigation of human anatomical structures. Generally, a number of thin 2D slices are obtained in the slice-select direction with 2–3 mm gaps in between every two slices followed by volume rendering for 3D imaging of the target area for clinical study and analysis. Due to very small inter-slice gaps, images of adjacent slices in multi-slice MRI are structurally highly correlated. On the other hand, spatial sensitivity encoded aliased images of multiple receiver coils of the target field-of-view (FoV) are similar. To exploit spatial data correlation both in the slice-select as well as in coil directions of multi-slice multi-coil MRI at a time, we have exploited the group-sparsity as prior knowledge in both the wavelet as well as gradient domains [12–14].

In this chapter, we propose a calibrationless 3D CS-pMRI reconstruction model using transform as well as spatial domain group-sparsity based regularization constraints. Since computational time is a major concern for the practical implementation of CS-pMRI reconstruction, we have used a hybrid CPU-GPU platform for acceleration in the proposed CS reconstruction. The rest of the chapter is organized as follows: Section 14.2 briefly presents the background and related work. Then the proposed group-sparsity based CS pMRI reconstruction technique is detailed in Sect. 14.3. Next, experimental results and performance evaluations are presented in Sect. 14.4, followed by conclusions in Sect. 14.5.

14.2 Background

MR angiography images are commonly sparse in spatial domain itself and other images, like, T1, T2 and proton density are highly compressible, when they are projected on to a suitable basis set. Further, MR images are mostly piecewise smooth, so their spatial domain gradients are also sparse. For the reconstruction of high-quality images from undersampled data, one needs to solve a highly nonlinear and ill-posed inverse problem. To solve this problem, conventional CS-MRI reconstruction model includes transform and gradient domain sparsities as regularization constraints [6] into it. The same model has been also extended to implement CS in pMRI. Let us define the CS-based pMRI reconstruction problem as follows—suppose $\mathbf{y}_q \in C^m$ is the measured k-space data corresponding to the cth coil image, $\mathbf{x}_q \in C^n$ i.e. $\mathbf{y}_q = \mathbf{F}_u \mathbf{x}_q$, here $m << n$, $q = 1, 2, ..., Q$; Q is the number of coils, and $\mathbf{F}_u \in C^{m \times n}$ represents partial Fourier operator [15–17]. Aliased image corresponding to the qth receiver coil, $\mathbf{x}_q$ can be obtained from $\mathbf{y}_q$, by minimizing the following optimization problem:

$$\hat{\mathbf{x}}_q = \underset{\mathbf{x}_q}{\operatorname{argmin}} \frac{1}{2} \|\mathbf{F}_u \mathbf{x}_q - \mathbf{y}_q\|_2^2 + \lambda_1 \| \Psi \mathbf{x}_q \|_1 + \lambda_2 \|\mathbf{x}_q\|_{\mathrm{TV}}, \tag{14.1}$$

where λ_1 and λ_2 are the two regularization parameters to maintain the trade-off between regularization terms and the data fidelity, Ψ denotes the wavelet transform operator, and $\|\mathbf{x}_q\|_{\mathrm{TV}}$ denotes the TV-norm of $\mathbf{x}_q$, i.e.,

$$\sum_{i,j} \sqrt{\left\{(\nabla_h \mathbf{x}_q)_{i,j}\right\}^2 + \left\{(\nabla_v \mathbf{x}_q)_{i,j}\right\}^2},$$

where ∇_h and ∇_v denotes the first-order gradient operators, respectively, in the horizontal and vertical directions [10].

There are a number of algorithms which can solve the above optimization problem. Lustig et al. [6] used the *Nonlinear Conjugate Gradient* (NCG) algorithm to solve the optimization problem. However, it is computationally inefficient due to the high computational complexity. Ma et al. [18] and Yang et al. [19] later on introduced two relatively faster algorithms, namely, the *Total Variation ℓ_1 Compressed MR Imaging* (TVCMRI) and the *Reconstruction from Partial Fourier data* (RecPF) based on operator and variable splitting techniques, respectively. Huang et al. [20] combined the concepts of both operator and variable splitting techniques in their technique and named it as the *Fast Composite Splitting Algorithm* (FCSA). In this method, they split the composite regularization problem into two relatively smaller subproblems by introducing new variables, which is followed by operator splitting technique on each individual subproblem. In each iteration, solutions of individual subproblems are linearly combined to estimate the solution of the composite problem.

According to Chen and Huang [21], wavelet coefficients of MR images have the typical quadtree structure. In the wavelet decomposition, if any wavelet coefficient in the coarser scale contain a significant value then there is a high probability that its four counterparts in the adjacent finer scale are also significant. This particular structured sparsity is also known as the wavelet tree sparsity. They call their reconstruction algorithm as the *Wavelet Tree Sparsity MRI* (WaTMRI). In pMRI, aliased images corresponding to multiple receiver coils of a particular FoV are also highly correlated.

they represent the same FoV with different spatial sensitivity profiles. Since they belong to the same FoV, edges and structural details in them would also appear in the same position. For visualization of inter-coil correlation, a set of clinical pMRI are shown in Fig. 14.1. So, it can be assumed that wavelet coefficients corresponding to different coil images of the same location are highly correlated. Chen et al. [22] exploit this inter-channel correlation during the CS-based pMRI reconstruction. The idea of tree sparsity in 2D image can be extended to multi-coil MRI, where multiple wavelet trees of identical positions corresponding to multi-coil images may be grouped together to demonstrate joint tree sparsity. They term it as the forest sparsity and corresponding reconstruction method as the FCSA Forest. As mentioned earlier, in 3D MRI a number of thin slices are acquired with very negligible inter-slice gaps. Since inter-slice gap is less and slices are thin, adjacent slices are highly correlated. For the visualization of the inter-slice correlation, a set of clinical multi-slice MRI are shown in Fig. 14.1. Recently, the authors in [22–25], exploited this inter-slice correlation to improve the reconstruction performance of multi-slice MRI.

As mentioned earlier, in pMRI reconstruction, images corresponding to multiple receiver coils represent the same FoV with different spatial coil sensitivity functions depending on their physical locations. Parallel MRI reconstruction methods are mainly classified into two broad categories depending on how the coil sensitivity profiles are used for the reconstruction of images—(a) Methods requiring the coil sensitivity information explicitly, for example, the SENSE [26] and the SMASH [27]. Here, it is assumed that coil sensitivity profiles are already estimated. The SENSE is able to produce optimal results if accurate coil sensitivity profiles are given. However, practically it is very difficult to accurately estimate the coil sensitivity information. (b) Methods requiring the coil sensitivity information implicitly during reconstruction, for example, the AUTO-SMASH [28] and the GRAPPA [29]. They estimate the coil sensitivity profiles from the measured raw data itself and known popularly as the auto-calibrating methods. Some of the well-known methods in this category are – the SPIRiT [30], the CS-SENSE [15], and the ESPIRiT [31]. The main drawback associated with this category is that a little error in estimation of the sensitivity information leads to non-removable artifacts in the target MR image.

There is yet another approach of pMRI reconstruction, which is relatively new; it does not require the coil sensitivity information either explicitly or implicitly during reconstruction and known as the calibrationless methods, for example, the *Calibration-Less Multi-coil MRI* (CalM MRI) [32], the *Joint Total Variation MRI* (JTVMRI) [33].

14.3 Proposed method

In this chapter, we have proposed a calibrationless pMRI reconstruction technique using group-sparsity. Intra- and inter-channel similarities in the wavelet domain are exploited using the concept of joint forest sparsity i.e. wavelet trees of identical positions from multiple channels as well as slices are grouped together. This grouping

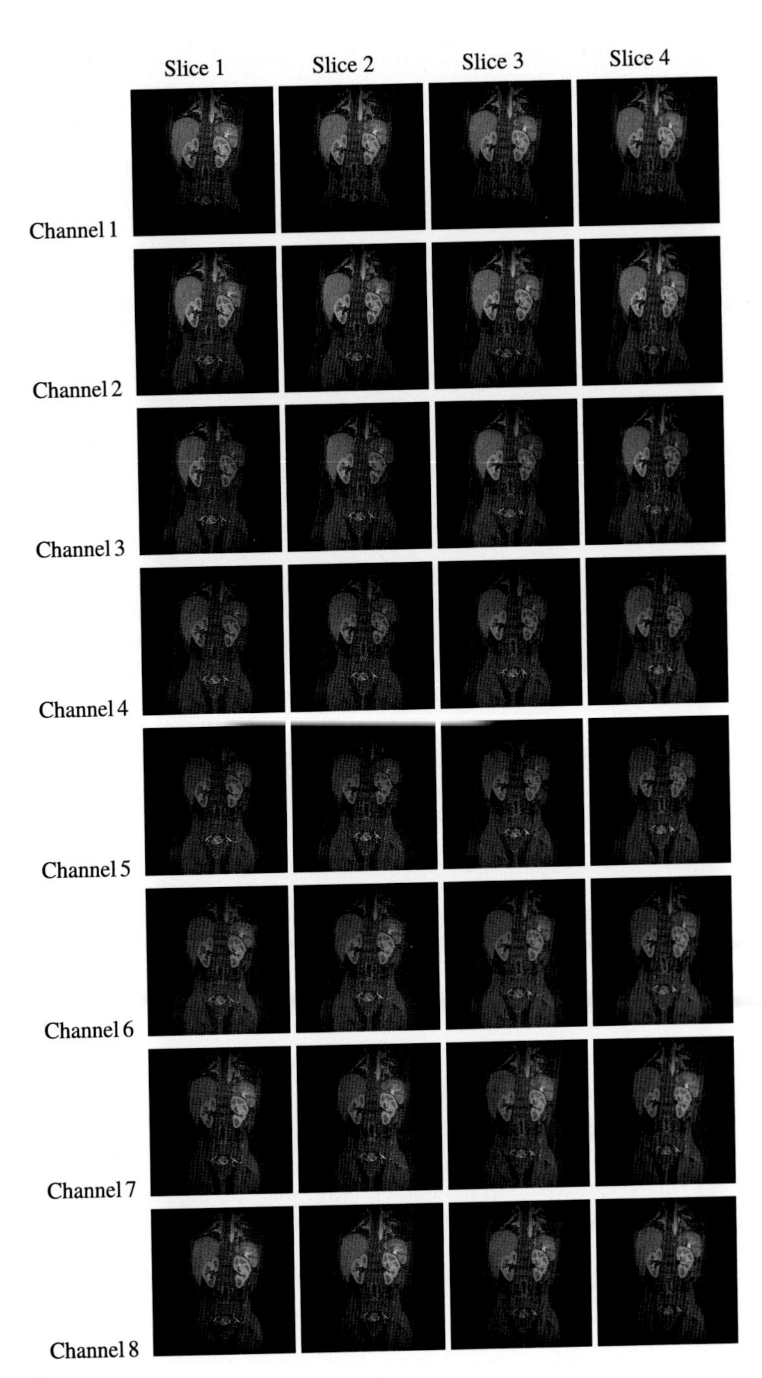

FIGURE 14.1

Multi-channel multi-slice magnitude MRI data.

arrangement induces similarity among the group members of a particular forest group (i.e. a group contains multi-channel multi-slice MRI data having similar wavelet tree structures). Additionally to exploit data redundancy in the spatial domain, the joint total variation (JTV)-norm is applied directly on the multi-channel multi-slice data.

Consider $\mathbf{Y} = \{(\mathbf{y}_{1,1}, \mathbf{y}_{1,2}, \cdots, \mathbf{y}_{1,Q}); (\mathbf{y}_{2,1}, \mathbf{y}_{2,2}, \cdots, \mathbf{y}_{2,Q}); \cdots; (\mathbf{y}_{P,1}, \mathbf{y}_{P,2}, \cdots, \mathbf{y}_{P,Q})\}$, the undersampled k-space measurements corresponding to the spatial domain multi-channel multi-slice MRI data, $\mathbf{X} = \{(\mathbf{x}_{1,1}, \mathbf{x}_{1,2}, \cdots, \mathbf{x}_{1,Q}); (\mathbf{x}_{2,1}, \mathbf{x}_{2,2}, \cdots, \mathbf{x}_{2,Q}); \cdots; (\mathbf{x}_{P,1}, \mathbf{x}_{P,2}, \cdots, \mathbf{x}_{P,Q})\}$ generated by the sampling model $\mathbf{Y} = \mathbf{F}_u\mathbf{X}$, where $\mathbf{F}_u$ is the undersampled Fourier operator, P and Q are the total number of slices and channels, respectively. Now, the joint CS-MRI reconstruction of multi-channel multi-slice data $\mathbf{X}$ from the given undersampled measurements $\mathbf{Y}$ may be done by solving the following composite minimization problem:

$$\widehat{\mathbf{X}} = \arg\min_{\mathbf{X}} \tfrac{1}{2} \sum_{p=1}^{P} \sum_{q=1}^{Q} \|\mathbf{F}_u\mathbf{x}_{p,q} - \mathbf{y}_{p,q}\|_2^2 + \lambda_1 \sum_{g_i \in G} \|(\Psi\mathbf{X})_{g_i}\|_2 + \lambda_2 \|\mathbf{X}\|_{\mathrm{JTV}}. \tag{14.2}$$

Wavelet forest sparsity is mathematically denoted by the $\ell_{1,2}$-norm, where g_i denotes the indices of wavelet coefficients of ith group and G represents the set of indices belonging to different forest groups, i.e. $G = [g_1, \ldots, g_i, \ldots, g_p]$. The joint TV-norm is denoted by

$$\|\mathbf{X}\|_{\mathrm{JTV}} = \sum_{i=1,j=1}^{\sqrt{n},\sqrt{n}} \sqrt{\sum_{p=1,q=1}^{P,Q} \left\{(\nabla_h \mathbf{x}_{p,q})_{i,j}^2 + (\nabla_v \mathbf{x}_{p,q})_{i,j}^2\right\}}.$$

The optimization problem defined in Eq. (14.2) is a composite regularization problem. To solve it efficiently, we use the concept of variable splitting technique and introduce an auxiliary variable $\mathbf{Z}$. Rewriting the above problem after introducing a new variable, we get

$$\hat{\mathbf{X}} = \arg\min_{\mathbf{X},\mathbf{Z}} \tfrac{1}{2} \sum_{p=1}^{P} \sum_{q=1}^{Q} \|\mathbf{F}_u\mathbf{x}_{p,q} - \mathbf{y}_{p,q}\|_2^2 + \lambda_1 \|\mathbf{X}\|_{\mathrm{JTV}} + \lambda_2 \sum_{g_i \in G} \|\mathbf{Z}_{g_i}\|_2 + \tfrac{\beta}{2} \|G\Psi\mathbf{X} - \mathbf{Z}\|_2^2 \tag{14.3}$$

where β is a small regularization parameter. Now, one can decompose this problem into two relatively simpler subproblems with respect to $\mathbf{Z}$ and $\mathbf{X}$ and defined as follows:

$$\mathbf{Z} = \arg\min_{\mathbf{Z}} \lambda_1 \sum_{g_i \in G} \|\mathbf{Z}_{g_i}\|_2 + \tfrac{\beta}{2} \|\mathbf{Z} - G\Psi\mathbf{X}\|_2^2,$$

$$\widehat{\mathbf{X}} = \arg\min_{\mathbf{X}} \tfrac{1}{2} \sum_{p=1}^{P} \sum_{q=1}^{Q} \|\mathbf{F}_u\mathbf{x}_{p,q} - \mathbf{y}_{p,q}\|_2^2 + \tfrac{\beta}{2} \|\mathbf{Z} - G\Psi\mathbf{X}\|_2^2 + \lambda_2 \|\mathbf{X}\|_{\mathrm{JTV}}.$$

The above two subproblems are solved efficiently using existing techniques, namely, the FCSA [20], the FCSA Forest [22] and the WaTMRI [21]. The subproblem associated with the forest sparsity has a closed form solution and is defined as

$$\hat{\mathbf{Z}}_{g_i} = \max\left(\left\| (\Psi \mathbf{X})_{g_i} \right\|_2 - \frac{\lambda_2}{\beta}, 0 \right) \frac{(\Psi \mathbf{X})_{g_i}}{\left\| (\Psi \mathbf{X})_{g_i} \right\|_2}$$

which is also popularly known as the *shrinkgroup*(.) operator [22] and stated as

$$\hat{\mathbf{Z}} = shrinkgroup\left(G\Psi \mathbf{X}, \frac{\lambda_2}{\beta} \right).$$

With $\mathbf{Z}$ as obtained above, the second subproblem can be rewritten as follows:

$$\hat{\mathbf{X}} = \arg\min_{\mathbf{X}} \; h(\mathbf{X}) + \gamma(\mathbf{X})$$

where

$$h(\mathbf{X}) = \tfrac{1}{2} \sum_{p=1}^{P} \sum_{q=1}^{Q} \| \mathbf{F}_u \mathbf{x}_{p,q} - \mathbf{y}_{p,q} \|_2^2 + \tfrac{\beta}{2} \| G\Psi \mathbf{X} - \mathbf{Z} \|_2^2$$

and $\gamma(\mathbf{X}) = \lambda_1 \|\mathbf{X}\|_{\mathrm{JTV}}$. The dual form of this subproblem can be solved using the Fast Joint-Gradient Projection (FJGP) method [34, Algorithm 2]. We summarize the proposed method in Algorithm 14.1.

Algorithm 14.1 Joint multi-slice multi-channel MRI reconstruction.

Input: $\Psi, \mathbf{F}_u, \mathbf{Y}, \lambda_1, \lambda_2, \beta, L_f$

Initialization: $\mathbf{X}^0 \leftarrow \mathbf{F}_u^T \mathbf{Y},\ \mathbf{r}^1 \leftarrow \mathbf{X}^0,\ t^1 \leftarrow 1,\ k \leftarrow 0$

1: **while** not converge **do**

2: $k \leftarrow k + 1$

3: $\mathbf{Z}^k \leftarrow shrinkgroup\left(G\Psi \mathbf{X}^{k-1}, \frac{\lambda_2}{\beta} \right)$

4: $\mathbf{R}^k \leftarrow \mathbf{r}^k - \frac{1}{L_f} \left[\sum_{p=1,p=1}^{P,Q} \mathbf{F}_u^T \left(\mathbf{F}_u \mathbf{r}_{p,q}^k - \mathbf{Y}_{p,q} \right) + \beta \Psi^T G^T \left(G\Psi \mathbf{r} - \mathbf{Z}^k \right) \right]$

5: $\mathbf{X}^k \leftarrow \arg\min_{\mathbf{X}} \frac{L}{2} \|\mathbf{X}^{k-1} - \mathbf{R}^k\|_2^2 + \lambda_1 \|\mathbf{X}^{k-1}\|_{\mathrm{JTV}}$

6: $t^{k+1} \leftarrow \left(\frac{1 + \sqrt{1 + 4(t^k)^2}}{2} \right)$

7: $\mathbf{r}^{k+1} \leftarrow \mathbf{X}^k + \frac{t^k - 1}{t^{k+1}} \left(\mathbf{X}^k - \mathbf{X}^{k-1} \right)$

8: **end while**

Output: $\hat{\mathbf{X}} \leftarrow \mathbf{X}^k$

Table 14.1 Comparison of computational time for different methods (min).

Dataset	SPIRiT	CaLM MRI	JTV MRI	FCSA Forest	ESPIRiT	CaL JS CS SENSE	Proposed	Proposed GPU
Dataset I	236	404	145	244	439	832	184	3.35
Dataset II	186	350	110	195	383	540	135	2.93

14.4 Results and discussion

Experiments are performed in the MATLAB® environment on a DELL workstation having Intel Xeon processor E5-2650, 128 GB RAM, and NVIDIA Quadro P5000 GPU card having 16 GB GDDR5X memory, up to 288 GB/s bandwidth and 2560 CUDA Cores. Two raw complex MRI datasets from publicly available sources are considered as experimental data. Dataset I: Stanford Fully sampled 3D FSE Knees[1] ($320 \times 320 \times 8 \times 172$). Dataset II: NYU machine learning data[2] ($768 \times 770 \times 15 \times 31$). We compare our results with seven existing CS pMRI reconstruction techniques. Among them three are well-known auto-calibrating techniques, namely, the CS-SENSE [15], the SPIRiT [30], and the ESPIRiT [31], and four are calibrationless techniques, namely, the CaLM MRI [32], the FCSA-Forest [22], the JTV MRI [33], and the CaL JS CS SENSE [35].

For objective evaluation, we consider two well-known image quality assessment (IQA) metrics, namely, the Mean Structural Similarity Index (MSSIM) and the Signal to Noise ratio (SNR) besides the computational time. In our experiments, we have used 25% undersampling ratio for both datasets. We use "db2" wavelet and four decomposition levels for sparse representation.

A hybrid CPU-GPU environment is used for parallel implementation of the proposed method. To solve the 4D CS MRI reconstruction efficiently, we split the 4D reconstruction problem of the target dataset into multiple smaller 4D subproblems. Each such subproblem contains three adjacent slices and their corresponding coil images. These subproblems are solved in parallel.

Computational time for reconstruction using different methods are shown in Table 14.1. In the case of sequential implementation, the proposed method takes the least computational time except for JTV MRI. JTV MRI takes slightly less time because it does not use the forest sparsity in the wavelet domain. Parallel implementation of the proposed technique requires only 2 to 3 minutes for reconstruction of clinical datasets, which is highly significant from the clinical point of view.

For visual comparison, reconstructed images using different techniques on slice #10 of Dataset II and their corresponding error images are shown in Figs. 14.2 and 14.3, respectively. We can observe that the reconstructed image using the proposed method gives the least visible artifacts. Reconstruction performances in terms IQA metrics are shown in Figs. 14.4 and 14.5. Figures show that the proposed technique

[1] http://mridata.org/list?project=Stanford%20Fullysampled%203D%20FSE %20Knees.
[2] http://mridata.org/list?project=NYU%20machine%20learning%20data.

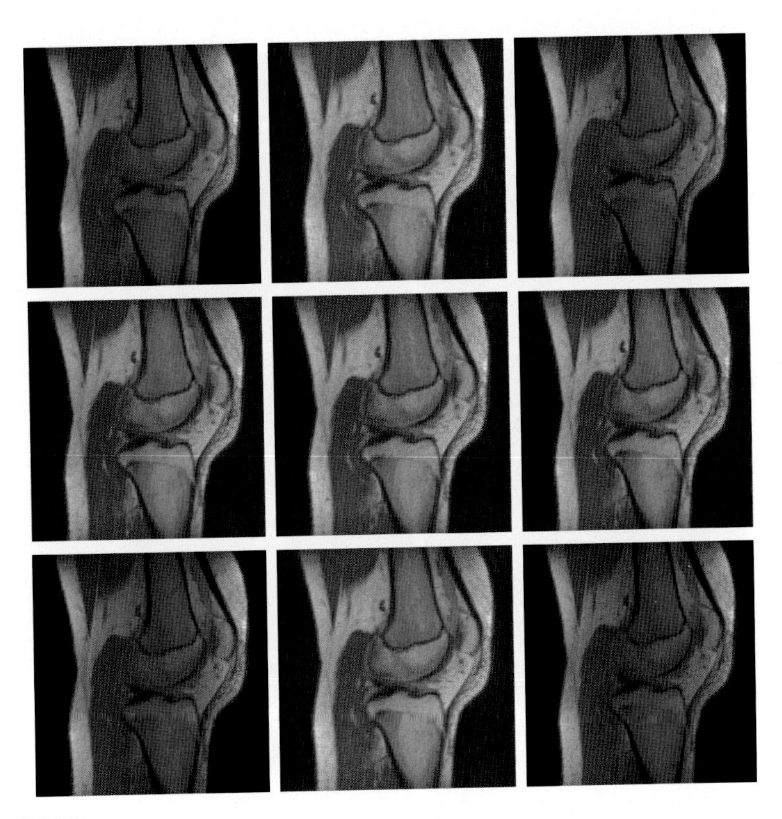

FIGURE 14.2

Comparison of a reconstructed images (slice #10) from dataset II using different techniques at 25% undersampling ratio. Left to right and top to bottom: original image (fully-sampled) and reconstructed images using the CS SENSE, the SPIRiT, the CaLM MRI, the JTV-MRI, the FCSA Forest, the ESPIRiT, the CaL JS CS SENSE, and the proposed technique, respectively.

gives a better performance in terms of both SNR and MSSIM values among all compared methods, irrespective of datasets.

14.5 Conclusions

In this chapter, we have proposed a joint CS based reconstruction technique for multi-slice calibrationless pMRI. Performance evaluations are carried out with two raw clinical pMRI datasets. Experimental results show that the proposed technique outperforms others in terms of quality of reconstruction. To check the clinical feasibility in terms of computational time, we have also implemented the proposed technique in a hybrid CPU-GPU environment and observe that it is quite efficient to reconstruct multi-dimensional clinical MRI data within few minutes.

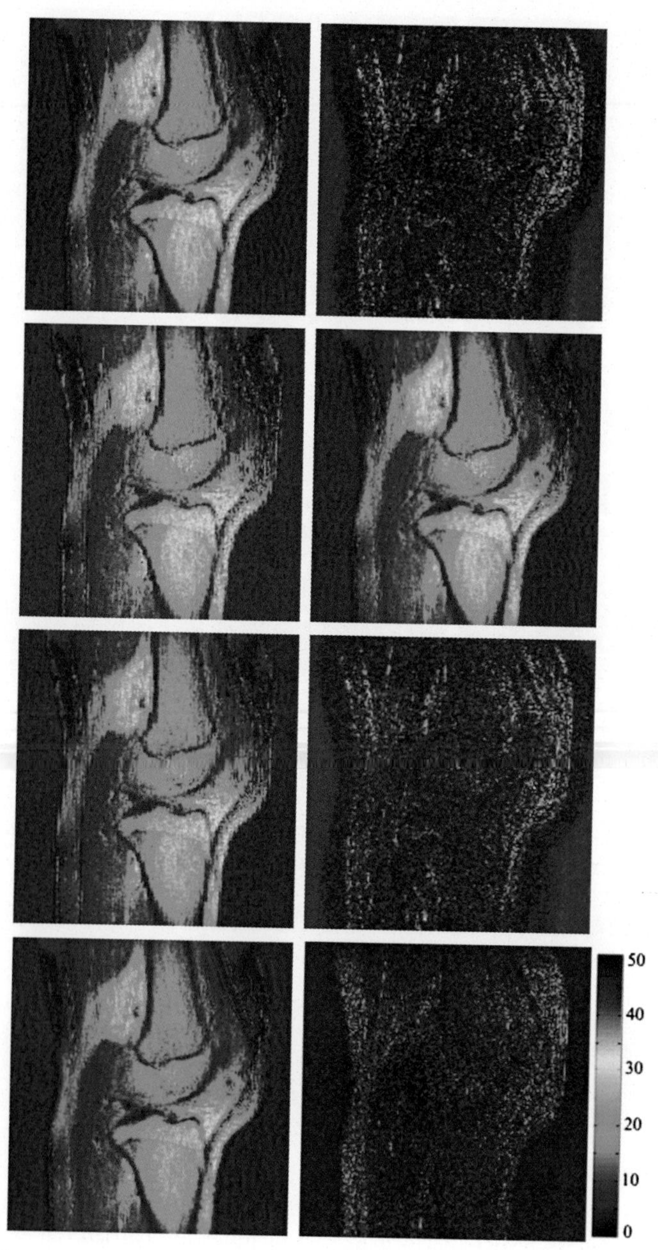

FIGURE 14.3

Comparison of error images obtained by finding difference between the original image and each reconstructed image. Left to right and top to bottom: error images corresponding to the CS SENSE, the SPIRiT, the CaLM MRI, the JTV-MRI, the FCSA Forest, the ESPIRiT, the CaL JS CS SENSE, and the proposed technique, respectively.

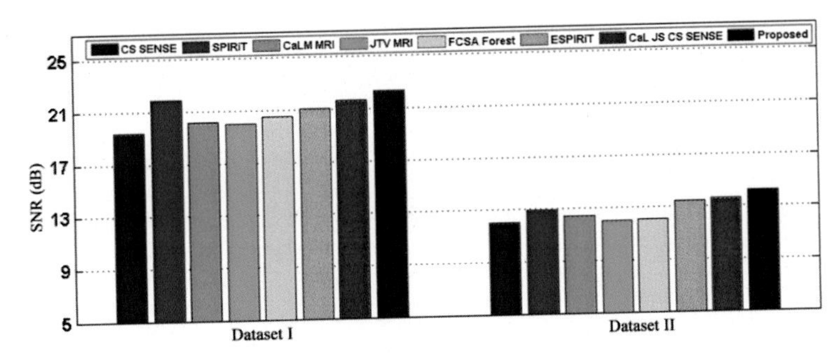

FIGURE 14.4

Reconstruction performance in terms of SNR for both datasets.

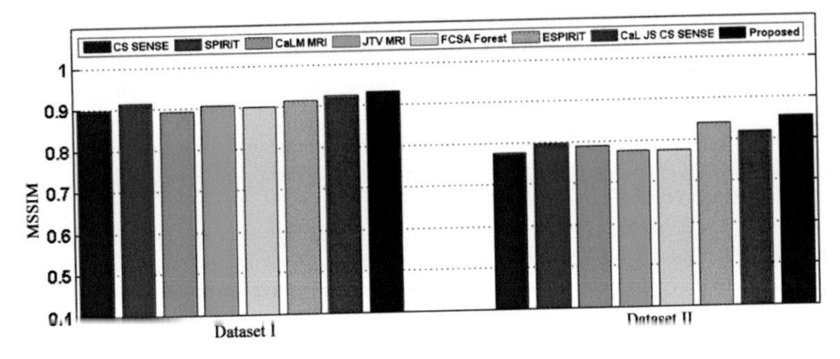

FIGURE 14.5

Reconstruction performances in terms of MSSIM for both datasets.

Acknowledgment

The authors would like to thank Ministry of Electronics and Information Technology (MeiTY), GoI for providing financial support under the Visvesvaraya Ph.D. Scheme for Electronics & IT (Ph.D./MLA/ 04(41)/2015-16/01) to carry out the above research work.

References

[1] D.W. McRobbie, E.A. Moore, M.J. Graves, M.R. Prince, MRI From Picture to Proton, 2nd edition, Cambridge University Press, Cambridge, UK, 2006.

[2] K.G. Hollingsworth, Reducing acquisition time in clinical MRI by data undersampling and compressed sensing reconstruction, Physics in Medicine and Biology 60 (21) (2015) R297.

[3] E.J. Candes, J.K. Romberg, T. Tao, Robust uncertainty principles: exact signal reconstruction from highly incomplete frequency information, IEEE Transactions on Information Theory 52 (2) (2006) 489–509.

[4] D. Donoho, Compressed sensing, IEEE Transactions on Information Theory 52 (4) (2006) 1289–1306.

[5] B. Deka, S. Datta, S. Handique, Wavelet tree support detection for compressed sensing MRI reconstruction, IEEE Signal Processing Letters 25 (5) (2018) 730–734.

[6] M. Lustig, D. Donoho, J.M. Pauly, Sparse MRI: the application of compressed sensing for rapid MR imaging, Magnetic Resonance in Medicine 58 (2007) 1182–1195.
[7] M. Lustig, D. Donoho, J. Santos, J. Pauly, Compressed sensing MRI, IEEE Signal Processing Magazine 25 (2) (2008) 72–82.
[8] S.S. Vasanawala, M. Lustig, Advances in pediatric body MRI, Pediatric Radiology 41 (2) (2011) 549–554.
[9] S. Vasanawala, M. Alley, B. Hargreaves, R. Barth, J. Pauly, M. Lustig, Improved pediatric MR imaging with compressed sensing, Radiology 256 (2) (2010) 607–616.
[10] B. Deka, S. Datta, Compressed Sensing Magnetic Resonance Image Reconstruction Algorithms: A Convex Optimization Approach, Springer Series on Bio- and Neurosystems, Springer, Singapore, 2019.
[11] T.W. Cabral, M. Khosravy, F.M. Dias, H. Luis, M. Monteiro, M. Antônio, A. Lima, L. Rodrigues, M. Silva, R. Naji, C.A. Duque, Compressive Sensing in Medical Signal Processing and Imaging Systems, Advances in Ubiquitous Sensing Applications for Healthcare, Sensors for Health Monitoring, vol. 5, Academic Press, 2019, pp. 69–92.
[12] S. Datta, B. Deka, Efficient interpolated compressed sensing reconstruction scheme for 3D MRI, IET Image Processing 12 (11) (2018) 2119–2127.
[13] S. Datta, B. Deka, Magnetic resonance image reconstruction using fast interpolated compressed sensing, Journal of Optics 47 (2) (2017) 154–165.
[14] S. Datta, B. Deka, H.U. Mullah, S. Kumar, An efficient interpolated compressed sensing method for highly correlated 2D multi-slice MRI, in: International Conference on Accessibility to Digital World (ICADW-2016), Guwahati, India, 2016, pp. 187–192.
[15] D. Liang, B. Liu, J. Wang, L. Ying, Accelerating SENSE using compressed sensing, Magnetic Resonance in Medicine 62 (6) (2009) 1574–1584.
[16] B. Deka, S. Datta, Weighted wavelet tree sparsity regularization for compressed sensing magnetic resonance image reconstruction, in: Engineering, A. Kalam, S. Das, K. Sharma (Eds.), Advances in Electronics, Communication and Computing, in: Lecture Notes in Electrical Engineering, vol. 443, Springer, Singapore, 2017, pp. 449–457.
[17] S. Datta, B. Deka, Efficient adaptive weighted minimization for compressed sensing magnetic resonance image reconstruction, in: Tenth Indian Conference on Computer Vision, Graphics and Image Processing (ICVGIP -2016), Guwahati, Assam, India, 2016, pp. 1–8.
[18] S. Ma, W. Yin, Y. Zhang, A. Chakraborty, An efficient algorithm for compressed MR imaging using total variation and wavelets, in: IEEE Conference on Computer Vision and Pattern Recognition (CVPR-2008), Anchorage, AK, 2008, pp. 1–8.
[19] J. Yang, Y. Zhang, W. Yin, A fast alternating direction method for TVL1-L2 signal reconstruction from partial Fourier data, IEEE Journal of Selected Topics in Signal Processing 4 (2) (2010) 288–297.
[20] J. Huang, S. Zhang, D.N. Metaxas, Efficient MR image reconstruction for compressed MR imaging, Medical Image Analysis 15 (5) (2011) 670–679.
[21] C. Chen, J. Huang, Exploiting the wavelet structure in compressed sensing MRI, Magnetic Resonance Imaging 32 (2014) 1377–1389.
[22] C. Chen, Y. Li, J. Huang, Forest sparsity for multi-channel compressive sensing, IEEE Transactions on Signal Processing 62 (11) (2014) 2803–2813.
[23] S. Datta, B. Deka, Multi-channel, multi-slice, and multi-contrast compressed sensing MRI using weighted forest sparsity and joint TV regularization priors, in: J.C. Bansal, K.N. Das, A. Nagar, K. Deep, A.K. Ojha (Eds.), Soft Computing for Problem Solving (SocProS) 2017, in: Advances in Intelligent Systems and Computing, vol. 1, Springer, 2017, pp. 821–832.
[24] S. Datta, B. Deka, Interpolated compressed sensing for calibrationless parallel MRI reconstruction, in: National Conference on Communications (NCC) 2019, 2019, pp. 1–6.
[25] S. Datta, B. Deka, Parallel MRI reconstruction using group sparsity based CS-MRI reconstruction, in: Pattern Recognition and Machine Intelligence (PReMI) 2019, in: Lecture Notes in Computer Science, vol. 11942, Springer, 2019, pp. 70–77.
[26] K.P. Pruessmann, M. Weiger, M.B. Scheidegger, P. Boesiger, SENSE: sensitivity encoding for fast MRI, Magnetic Resonance in Medicine 42 (5) (1999) 952–962.

[27] D.K. Sodickson, W.J. Manning, Simultaneous acquisition of spatial harmonics (SMASH): fast imaging with radiofrequency coil arrays, Magnetic Resonance in Medicine 38 (4) (1997) 591–603.
[28] P.M. Jakob, M.A. Grisowld, R.R. Edelman, D.K. Sodickson, AUTO-SMASH: a self-calibrating technique for SMASH imaging, Magnetic Resonance Materials in Physics, Biology and Medicine 7 (1) (1998) 42–54.
[29] M.A. Griswold, P.M. Jakob, R.M. Heidemann, M. Nittka, V. Jellús, J. Wang, B. Kiefer, A. Haase, Generalized autocalibrating partially parallel acquisitions (GRAPPA), Magnetic Resonance in Medicine 47 (6) (2002) 1202–1210.
[30] M. Lustig, J. Pauly, SPIRiT: iterative self-consistent parallel imaging reconstruction from arbitrary k-space, Magnetic Resonance in Medicine 64 (2) (2010) 457–471.
[31] M. Uecker, P. Lai, M.J. Murphy, P. Virtue, M. Elad, J.M. Pauly, S.S. Vasanawala, M. Lustig, ESPIRiT – an eigenvalue approach to autocalibrating parallel MRI: where SENSE meets GRAPPA, Magnetic Resonance in Medicine 71 (3) (2014) 990–1001.
[32] A. Majumdar, R.K. Ward, Calibration-less multi-coil MR image reconstruction, Magnetic Resonance Imaging 30 (7) (2012) 1032–1045.
[33] C. Chen, Y. Li, J. Huang, Calibrationless parallel MRI with joint total variation regularization, in: K. Mori, I. Sakuma, Y. Sato, C. Barillot, N. Navab (Eds.), Medical Image Computing and Computer Assisted Intervention (MICCAI- 2013), in: Lecture Notes in Computer Science, vol. 8151, Springer, Berlin, Heidelberg, 2013, pp. 106–114.
[34] J. Huang, C. Chen, L. Axel, Fast multi-contrast MRI reconstruction, Magnetic Resonance Imaging 32 (10) (2014) 1344–1352.
[35] I.Y. Chun, B. Adcock, T.M. Talavage, Efficient compressed sensing SENSE pMRI reconstruction with joint sparsity promotion, IEEE Transactions on Medical Imaging 35 (1) (2016) 354–368.

Index

Index

D

Index

Index

S

T

U

V

Index

W

ıe United States
ısters